AAPG Treatise of Petroleum Geology
Reprint Series

The American Association of Petroleum Geologists
gratefully acknowledges and appreciates the leadership and support
of the AAPG Foundation in the development of the
Treatise of Petroleum Geology.

Treatise of Petroleum Geology Reprint Series

Compiled by Edward A. Beaumont and Norman H. Foster

1. Geologic Basins I: Classification, Modeling, and Predictive Stratigraphy
2. Geologic Basins II: Evaluation, Resource Appraisal, and World Occurrence of Oil and Gas
3. Reservoirs I: Properties
4. Reservoirs II: Sandstones
5. Reservoirs III: Carbonates
6. Traps and Seals I: Structural/Fault-Seal and Hydrodynamic Traps
7. Traps and Seals II: Stratigraphic/Capillary Traps
8. Geochemistry
9. Structural Concepts and Techniques I: Basic Concepts, Folding, and Structural Techniques
10. Structural Concepts and Techniques II: Basement-Involved Deformation
11. Structural Concepts and Techniques III: Detached Deformation
12. Geophysics I: Seismic Methods
13. Geophysics II: Tools for Seismic Interpretation
14. Geophysics III: Geologic Interpretation of Seismic Data
15. Geophysic IV: Gravity, Magnetic, and Magnetotelluric Methods
16. Formation Evaluation I: Log Evaluation
17. Formation Evaluation II: Log Interpretation
18. Photogeology and Photogeomorphology
19. Remote Sensing
20. Oil is First Found in the Mind: The Philosophy of Exploration

OIL IS FIRST FOUND IN THE MIND
THE PHILOSOPHY OF EXPLORATION

COMPILED BY
NORMAN H. FOSTER
AND
EDWARD A. BEAUMONT

TREATISE OF PETROLEUM GEOLOGY
REPRINT SERIES, NO. 20

PUBLISHED BY
THE AMERICAN ASSOCIATION OF PETROLEUM GEOLOGISTS
TULSA, OKLAHOMA 74101-0979, U.S.A.

ISBN: 0-89181-419-1
ISSN: 1043-6103

Available from:
The AAPG Bookstore
P. O. Box 979
Tulsa, OK 74101-0979
U.S.A.

Phone: 918-584-2555
Telex: 49-9432
Fax: 918-584-0469

Association Editor: Susan Longacre
Science Director: Gary D. Howell
Publications Manager: Cathleen P. Williams
Special Projects Editor: Anne H. Thomas
Project Editor: Michael H. Blechner

Library of Congress Cataloging-in-Publication Data

Oil is first found in the mind : the philosophy of exploration /
    compiled by Norman H. Foster and Edward A. Beaumont.
        p. cm. — (Treatise of petroleum geology reprint series, ISSN
    1043-6103 ; no. 20)
    Includes bibliographical references.
    ISBN 0-89181-419-1
    1. Petroleum—Prospecting—Philosophy. 2. Natural gas—
Prospecting—Philosophy. 3. Petroleum geologists. [1. Science—
Methodology.] I. Foster, Norman H.   II. Beaumont, E.A. (Edward A.)
III. Series.
TN271.P4044 1992   92-15418
622'.1828'01—dc20   CIP

AMERICAN ASSOCIATION OF PETROLEUM GEOLOGISTS FOUNDATION
TREATISE OF PETROLEUM GEOLOGY FUND*

Major Corporate Contributors
($25,000 or more)

Amoco Production Company
BP Exploration Company Limited
Chevron Corporation
Exxon Company, U.S.A.
Mobil Oil Corporation
Oryx Energy Company
Pennzoil Exploration and Production Company
Shell Oil Company
Texaco Foundation
Union Pacific Foundation
Unocal Corporation

Other Corporate Contributors
($5,000 to $25,000)

ARCO Oil & Gas Company
Ashland Oil, Inc.
Cabot Oil & Gas Corporation
Canadian Hunter Exploration Ltd.
Conoco Inc.
Marathon Oil Company
The McGee Foundation, Inc.
Phillips Petroleum Company
Transco Energy Company
Union Texas Petroleum Corporation

Major Individual Contributors
($1,000 or more)

John J. Amoruso
Thornton E. Anderson
C. Hayden Atchison
Richard A. Baile
Richard R. Bloomer
A. S. Bonner, Jr.
David G. Campbell
Herbert G. Davis
George A. Donnelly, Jr.
Paul H. Dudley, Jr.
Lewis G. Fearing
Lawrence W. Funkhouser
James A. Gibbs
George R. Gibson
William E. Gipson
Mrs. Vito A. (Mary Jane) Gotautas
Robert D. Gunn
Merrill W. Haas
Cecil V. Hagen
Frank W. Harrison
William A. Heck

Roy M. Huffington
J. R. Jackson, Jr.
Harrison C. Jamison
Thomas N. Jordan, Jr.
Hugh M. Looney
Jack P. Martin
John W. Mason
George B. McBride
Dean A. McGee
John R. McMillan
Lee Wayne Moore
Grover E. Murray
Rudolf B. Siegert
Robert M. Sneider
Estate of Mrs. John (Elizabeth) Teagle
Jack C. Threet
Charles Weiner
Harry Westmoreland
James E. Wilson, Jr.
P. W. J. Wood

The Foundation also gratefully acknowledges the many who have supported this endeavor with additional contributions.

*Based on contributions received as of March 31, 1992.

INTRODUCTION

The *Treatise of Petroleum Geology* is AAPG's Diamond Jubilee project, commemorating the 75th anniversary of the organization of the Association in 1916. *The Treatise of Petroleum Geology* is made up of three series: the Handbook of Petroleum Geology, the Atlas of Oil and Gas Fields, and the Reprint Series. With input from more than 260 geologists, geophysicists, geochemists, and engineers from around the world, we designed the *Treatise* to represent the cutting edge in petroleum exploration knowledge and application. The Handbook of Petroleum Geology series is a professional explorationist's guide to the methodology and technology used to find oil and gas. The Atlas of Oil and Gas Fields series collects detailed field studies to illustrate the myriad ways oil and gas are trapped. The third part of the *Treatise*, the Reprint Series, provides landmark papers from diverse worldwide geological, geophysical, geochemical, and engineering publications. In some cases, papers are reprinted from very obscure sources. In this volume, we also include several important papers that have not been published before.

The papers in the various volumes of the Reprint Series complement the subjects covered in the Handbook series. Papers were selected on the basis of their current usefulness in petroleum exploration and development. Many "classic papers" that led to our present state of knowledge have not been included because of space limitation. In some cases, a paper covers several topics, thus making it difficult for us to decide in which Reprint Series volume that paper should be published. We suggest, therefore, that interested readers become familiar with all the Reprint Series volumes if they are looking for a particular paper.

The title of this volume, *Oil is First Found in the Mind: The Philosophy of Exploration*, paraphrases the famous 1952 quotation from Wallace Pratt, which concludes the first paper in this volume: "Where oil is first found, in the final analysis, is in the minds of men. The undiscovered oil field exists only as an idea in the mind of some oil-finder. When no man any longer believes more oil is left to be found, no more oil fields will be discovered, but so long as a single oil-finder remains with a mental vision of a new oil field to cherish, along with freedom and incentive to explore, just so long new oil fields may continue to be discovered."

This volume collects papers on the philosophy of exploration. The authors give their overall visions and attitudes toward oil finding, discussing fundamental and motivating beliefs, concepts, and principles. The great oil finders (and, of course, we include natural gas finders and mineral finders as well) have characteristics in common that have enabled them to succeed. As you read these papers, you will be struck by many shared threads of thought.

Oil finders are positive thinkers (negative-thinking people do not find oil), they develop creativity through visual thinking, they have vivid imaginations controlled by facts, they have a great desire to find oil, they are self-motivating and self-starting, they are optimistic, they are persistent, and, above all, they love the thrill of discovery and the deep satisfaction of being able to use science and art to find a valuable deposit for the benefit of all mankind.

Some authors also discuss the importance of a working environment in which creativity can flourish, and the critical factors of favorable economic and political settings.

It is well known that only a small percentage of people exploring for oil and other natural resources actually find them. We fervently believe, however, that one learns to be an oil finder, and that one of the best ways to learn is to associate with other oil finders. In reading these papers, you will be able to experience the collective wisdom of some of the greatest oil finders in history. We believe that this volume is the most important one in the Reprint Series, and that it should be required reading for anyone exploring for natural resources.

We hope this book will be of great benefit to you in your search for the elusive trap.

Norman H. Foster
Denver, Colorado

Edward A. Beaumont
Tulsa, Oklahoma

OIL IS FIRST FOUND IN THE MIND

THE PHILOSOPHY OF EXPLORATION

PHILOSOPHY OF OIL AND GAS DISCOVERY

PETROLEUM EXPLORATION AND THE ROLE OF THE GEOLOGIST

THE SCIENTIFIC METHOD

PHILOSOPHY OF OIL AND GAS DISCOVERY

Volume 36   Number 12

BULLETIN
of the
AMERICAN ASSOCIATION OF PETROLEUM GEOLOGISTS

DECEMBER, 1952

TOWARD A PHILOSOPHY OF OIL-FINDING[1]

WALLACE E. PRATT[2]
Carlsbad, New Mexico

ABSTRACT

The United States—a country whose total petroleum resources constitute but a minor fraction of the total for the earth as a whole—has supplied from oil fields within its own boundaries two-thirds of all the oil the world has consumed in the past. This remarkable record might plausibly be attributed to the proficiency of Americans in the art of oil-finding. But petroleum geologists, on whose science the art of oil-finding is based, make no such claim. Geologists generally have failed to anticipate the magnitude of the petroleum resources now proved to have existed in this country. On the contrary, our best informed authorities have consistently and grossly underestimated the size of these resources in the past. No one has been more surprised than they at our achievement. As a matter of fact, a finished art of prospecting is not the only factor—not even the most important factor—in making possible the discovery of the oil fields which make up the petroleum resources of a nation. There exist more formidable barriers to success in oil-finding than the lack of perfected methods and techniques of exploration: the ultra-conservatism of the trained scientist and engineer, the tendency of the human mind to discount or to ignore the significance of what remains unknown to it, the restriction of free enterprise, these have been greater handicaps to success in the search for oil fields over the world. Where oil fields are really found, in the final analysis, is in the minds of men and we have found an unparalleled number of oil fields in the United States, not because our petroleum resources were exceptionally abundant, but because in our economic and political climate men have enjoyed unparalleled freedom to devote themselves to the search for oil fields.

As a geologist long engaged in the search for oil, I have witnessed the development over the last 40 years of an amazingly effective art of prospecting based on the science of geology, vastly aided since the middle 1920's by geophysics. I share the satisfaction of the geological profession in this splendid achievement. We have gone far beyond the utmost that seemed possible to me at the beginning of my work in petroleum geology.

Nevertheless, my experience has forced me to the conclusion that even the most finished art of prospecting, by itself, is not adequate to the task of finding

---

[1] This and the following three papers were presented as part of the program of the regional meeting of the Association and the 18th annual meeting of the South Texas Geological Society, at Austin, October 18–20, 1951. Presiding officers were Frank A. Morgan, president of the A.A.P.G., and John R. Sandidge, president of the South Texas Geological Society. Details of the meeting were reported, in the A.A.P.G. *Bulletin*, December, 1951.

[2] Geologist.

the earth's oil. There are other factors which constitute barriers to success in oil-finding of such a nature that no perfection of methods and techniques of search can remove or surmount them. These factors are fundamental; they are innate in our very habits of thought. They are greatly re-enforced by the existing social order in many countries. Their nature and the way in which they operate to defeat the oil-finder are revealed in two incidents—familiar to all of us—which are typical of the activities of the oil-producing industry. These incidents may be briefly recited.

In May, 1920, when the total past production of this country amounted to only 5 billion barrels, David White, the competent and highly respected chief geologist of the United States Geological Survey, predicted that "the production of natural petroleum in the United States must pass its peak at an early date, probably within five years and possibly within three years, due to the exhaustion of our reserves." "Our domestic production," he continued, "is not likely ever to exceed 450 million barrels" annually. If we did produce so much we would "exhaust the estimated 7 billion barrels of natural petroleum remaining available in the ground in the United States in 18 years."

To-day, 30 years after this prediction was made, we are producing at roughly five times the rate David White set as the maximum we could ever attain, and over the intervening years we have produced five times as much as the total he estimated as remaining available in the ground. Yet we still have a proved reserve four times larger than David White's estimate of total remaining resource in 1920, to say nothing of an additional undiscovered resource of unknown volume.

It is clear in retrospect, therefore, that David White's prediction was absurdly pessimistic. He was firmly convinced that there was very little oil in the United States. His opinion was based on his knowledge of the occurrence of oil and he was unusually well informed on this subject. Most of us agreed with him at the time. In 1921, this society, the American Association of Petroleum Geologists, joined with him and his colleagues in the United States Geological Survey in the preparation of a similarly pessimistic estimate of the remaining petroleum resources of the United States.

There are many instances where our knowledge, supported in some cases by elaborate and detailed studies, directed specifically to the question at issue, has convinced us that no petroleum resources were present in areas which subsequently became sites of important oil fields. The second incident here to be recited, is a case in point.

Over a period of 15 years preceding its discovery in 1937, the largest known oil field in the world—Kuwait, in the Middle East—was offered for a nominal consideration to a number of the larger units in the petroleum industry, including the three leading oil companies—British, Dutch, and American—in the world; and all these companies, in turn, declined the offer. Not one of them thought it worth while even to undertake to explore an area which, once it was explored, proved to be the site of an oil field larger—far larger—than any previously known.

In other words, at the end of 90 years of vigorous search for oil over the earth and of intensive study of the occurrence of oil in the crust of the earth, the best minds in the oil-finding industry failed to recognize the earth's greatest oil field (to date) before it was proved by drilling: worse than that, they were convinced that it was no oil field at all!

Why do we so completely misjudge the potentialities of oil-bearing rocks? It is not that this incredible error arose out of unfamiliarity with the occurrence of oil in the region in question. The Middle East has been famous for its seeps of oil and gas for hundreds of years. Kuwait itself boasts prolific oil seeps. In adjacent Iran, great oil fields had already been producing for more than 20 years. Moreover, these adjacent Iranian oil fields had been developed and were currently administered by one of the very companies which insisted that Kuwait was not worth exploring. Two of the other leading companies had for years been participants in large oil-producing operations in Irak, on the other side of Kuwait.

These companies knew more than anybody else about oil in the Middle East. They commanded the services of the best geologists in the world. They had carried out special investigations—long and painstaking geologic research on the occurrence of oil in the Middle East. They were not deterred from exploiting Kuwait by any feeling that they knew too little about the occurrence of oil there to justify the risk of loss the venture would involve. On the contrary, they were deterred by their conviction, based on long experience and extensive surveys in the Middle East, that there was no chance of success in Kuwait. They knew, in short, that "there is no oil in Arabia."

Because of this knowledge, because of what they thought they knew, the three best informed oil companies in the world declined the opportunity to explore for oil in Kuwait, where our greatest known oil field was to be discovered only a few years later. But another, smaller company—a company much less familiar with the occurrence of oil in the Middle East—took the initiative in acquiring the Kuwait concession and exploring it. This smaller company was willing to drill for oil in Kuwait because it did not know that "there is no oil in Arabia."

Before I leave this incident, I should complete the record by recalling to you that the company which finally had the courage to take up the concession on Kuwait, ten years after it had first been offered to the leading oil producers of the world, was Gulf Oil Corporation. At the same time Gulf Oil Corporation acquired also a concession on the neighboring island of Bahrein, in the Persian Gulf, which it later sold to Standard Oil of California. The California company promptly initiated drilling operations on Bahrein Island and, in 1932, discovered a major oil field. This discovery made it abundantly clear to everyone that there is oil in Arabia, after all, and one of the three great companies which had spurned the Kuwait concession a few years earlier, now re-entered the field and finally (1934) secured a half interest with Gulf Oil Corporation in the Kuwait concession.

It would be consoling to us as geologists and oil-finders if we could convince ourselves that our experience at Kuwait constitutes merely an isolated mistake in judgment, out of keeping with our general and normal performance. But this solace is denied us. If we examine the record, we find that Kuwait is only one more instance in a long series of similar misjudgments. And these errors are frequently the errors of our best informed authorities.

Each of the incidents just recited illustrates a mental attitude which becomes a formidable barrier to oil-finding. The first is an example of the natural conservatism of the trained scientific mind—a trait which, it has been said, has grievously impeded the quest for truth, everywhere. It has impeded the search for oil. David White was conservative. He dealt with facts—so far as they were known; and he would not venture beyond the known. Because so much was unknown to him, and therefore ignored by him, David White erred. We all erred with him. Consistently, from the birth of the industry to the inception of the Second World War, the scientist has grossly underestimated the petroleum resources of the United States. This conviction of our best minds that little or no oil remained to to be found has continuously handicapped the search for oil. Unless men can believe that there is more oil to be discovered, they will not drill for oil.

The second incident illustrates an even more pernicious habit of mind. Joseph Pogue has emphasized this singular trait in a recent unpublished discussion of estimates of petroleum reserves. That which becomes known to us in a particular field, he says, however small our knowledge may be, tends often, not only to color, but actually to obscure what remains unknown to us in the same field. A "little learning" about our proved reserves, for example, makes it impossible for us to measure our total resources by any other yardstick. If the part we know is small the unknown extension of it must also be small. What we learn, instead of illuminating what we have still to learn, sometimes casts a mental shadow over it, rendering it less discernible and impelling us to ignore it.

Those who are trained and experienced in the art of oil-finding appear to be particularly susceptible to this blindness which comes with the acquisition of a little knowledge. Oil-finders, of necessity, are guided by their observation of the facts—their knowledge of the occurrence of oil in the area they have to explore. But the oil-finder can almost never know all the factors with which he has to deal —the actual conditions at depth beneath the surface of an unexplored, or incompletely explored, segment of the earth's crust. For him, the disparity between the known and the unknown is unusually large. For him, the need to be always alert to the potentialities of what he does not know is paramount. He must maintain a constant awareness that he does not know everything that may enter into his problem. To assume that our knowledge of an area is complete when it is not, may be to conclude that there is no oil where there is oil. So it is that we conclude from an inadequate knowledge of the facts that "there is no oil in Arabia."

The record demonstrates that, as oil-finders, we have persistently underestimated the amount of oil and gas that have been stored up in the earth's crust.

We have been too conservative always. Despite all our mistakes, however, we have already found in the United States far more oil than our knowledge of the occurrence of oil permitted us to anticipate. Our discoveries since 1920 are already almost 9 times greater than David White's estimate of the total remaining resource; and we know there are still new oil fields to be discovered.

We have produced in the United States nearly two-thirds of all the oil the world has consumed in the past. This means a past production in this country nearly 10 times as great per unit of area of earth's crust favorable for the occurrence of oil, as the rest of the world.

It was long held that our ability to find and produce oil more rapidly and more abundantly than the rest of the world arose from an exceptionally rich endowment of natural resources in oil. But it is clear to-day that oil resources far richer and larger than ours exist elsewhere over the earth.

How, then, does it come about that we have been able to produce so much more oil in the past than the rest of the world? And how have we surmounted or circumvented the mental barriers which, as we have noted, stand in the way of mankind's search for oil?

Since the very inception of the industry, the finding and producing of oil in the United States have been carried on by literally thousands of independent enterprises; thousands of individuals, each an oil-finder in his own right; each free to put to the practical test of the drill his own ideas and theories of where oil might be found; and each spurred on to the drilling of exploratory wells by the assurance that if he made a discovery, he would reap a reward commensurate with its value to society.

Under these circumstances, hundreds of thousands—more than a million— wells have been drilled in the United States in search for oil. And many thousands of them have been drilled at places which were believed by everyone except the driller, to be unfavorable for the occurrence of oil. Yet, time after time, these wells at unfavorable locations discovered major oil fields. This is the way we have found oil in the United States—the drilling of many exploratory wells all over the land. By this procedure more oil fields have been discovered than anybody had dreamed were possible.

Under these circumstances, also, the mental barriers that normally stand in the way of the oil-finder were dissipated. The conservatism of the trained scientist was nullified by the action of the untrained oil-finder, unhampered by "a little learning." If one man said there is no oil in Arabia (or West Texas) another man promptly drilled a well there to see—and often he found oil.

The freedom of every citizen to explore by drilling wells for oil and the assurance to every citizen of a generous reward for success in finding oil have accrued to us out of our social and political environment. These are advantages which have contributed more to our achievement in oil-finding than the perfection of the art of prospecting. The citizens of few other countries possess these advantages. To their lack is due, in large part, the relatively poor showing in the search for oil over much of the earth. Oil-finding is at once the task of the individual and of the community at large. If the action of either is inappropriate, the search languishes.

The qualities which mark the individual oil-finder are faith, persistence, the venture spirit, and vision. If he is informed and trained in the art of oil-finding, so much the better; unless his knowledge makes him over-conservative, or obscures for him the fact that much remains always unknown to him. If his knowledge blinds him to the unknown, his discoveries will be fewer.

One indispensable attribute of the successful oil-finder is vision. Levorsen has said "until a discovery well has been drilled the undiscovered oil or gas field exists at best only as an idea in the mind of the geologist." If it is in the mind of geologist, or the oil-finder, that new fields first take form, then discovery must wait on our mental visualization—our imagination.

Where oil is first found, in the final analysis, is in the minds of men. The undiscovered oil field exists only as an idea in the mind of some oil-finder. When no man any longer believes more oil is left to be found, no more oil fields will be discovered, but so long as a single oil-finder remains with a mental vision of a new oil field to cherish, along with freedom and incentive to explore, just so long new oil fields may continue to be discovered.

AAPG Explorer, v. 10, no. 5 (May 1989), p. 1, 8.

President's Column

Creativity Vital for Successful Exploration

The Creative Process

1. **First Insight**
2. **Saturation**
3. **Incubation**
4. **Illumination**
5. **Verification**

Figure 1.

By NORMAN H. FOSTER

One of the best things we can do in these times of depressed prices is develop our creative thinking skills. Creativity is the most important ingredient in successful petroleum and mineral exploration. That is why Ted Beaumont, Dick Vincelette and I have organized an AAPG seminar called "Creative Exploration." The next offering will be a two-day presentation at the Rocky Mountain Section meeting in Albuquerque, N.M., on Sept. 30 - Oct. 1.

In 1981, Dr. Roger Sperry received the Nobel Prize in medicine for his brain hemisphere research, which was first published in 1968. Sperry found that the right and left hemisphere of the cerebrum process data differently. The conscious left side processes information like a computer, in a linear step-by-step fashion, so that verbal, logical and analytical thinking occur there. Left functions include speech, reading, writing, math and a sense of time.

The subconscious right side is nonverbal and processes large amounts of data simultaneously. The right side attempts to interpret the whole from fragments of information, so that creative thinking in most individuals occurs there. For left-handed people the thought locations may be switched.

It really does not matter to us where thought is occurring. What is important is to realize that there are two modes of thought, as Dr. Betty Edwards has shown in her excellent book *Drawing on the Artist Within* (Simon and Schuster, 1986). These are a sequential, verbal, analytical mode, generally located on the left side and a global, visual, perceptual mode, generally located on the right.

The process of creative thought occurs in stages that have been well established in the literature and have been summarized by Edwards (Fig. 1). We can develop and use the creative process to help us find petroleum and other natural resources.

With "first insight," which is primarily a right brain activity, one becomes aware of an area's potential, perhaps because of good hydrocarbon shows and reservoirs, or the presence of an accumulation which may have analogs nearby, or because some new technique might change the economics of a play.

"Saturation" follows first insight and involves the complete study of all available information pertaining to the problem. This is mainly a left brain activity. When the mind becomes fully saturated with all the available data, such as well control, surface geology and seismic data, then it is time to incubate, which involves switching back to the subconscious right side and allowing the data to be analyzed.

One of the main blocks to creativity comes at the end of the saturation stage. Our educational system trains mainly the left side with subjects such as reading, writing and arithmetic. We become conditioned to believe that once the data have been gathered and studied, that we should be able to plug these into a formula and come up with a quick answer. That is not the way creative thought occurs. The information must be processed on the right side to find patterns and solutions to the problem. After saturation, it is best to relax and allow the subconscious mind to work on an answer.

At some point in time, usually at a quiet moment, maybe in the middle of the night or on a walk, or when you have your feet on the desk and are gazing out the window with your eyes glazed over, the answer will come as a flash of insight (Fig. 2). Suddenly, the left side has become aware of the solution to the problem that the right side has developed. This relatively short time period is called the "illumination stage." The answer usually is in almost complete form.

The new insight to the problem may or may not be correct. Therefore, it is imperative to switch back to the left side and rigorously test the idea against the data. In petroleum exploration this would include, among other things, all well and surface control. If after thorough verification it is still possible that the idea is correct, which it may not be, then we have added a final stage — application. If the idea can be right, then we must find a way to drill a well or, at an earlier stage, to do field work or perhaps shoot seismic data.

This final stage is another major block to creativity in exploration because so many outstanding prospects go untested. Managers and individuals must find the funds to get the good ideas drilled, because, of course, no petroleum will be found without drilling the creative plays.

Both Betty Edwards and Robert H. McKim, in his book *Experiences in Visual Thinking*, published by PWS Engineering and available through the Stanford University Bookstore, have stressed the importance of learning to draw and diagram to aid visual-perceptual thinking. Through drawing and other visual exercises one can learn to bring the right side to a conscious level and thus greatly improve our creative abilities.

The great oilfinders have long stressed the development of creativity through visualization. Wallace Pratt, in a 1952 paper, said, "Where oil is first found, in the final analysis, is in the minds of men. The undiscovered oil field exists only as an idea in the mind of some oilfinder. When no one any longer believes that more oil is left to be found, no more oil fields will be discovered, but so long as a single oilfinder remains with a mental vision of a new oil field to cherish, along with freedom and incentive to explore, just so long new oil fields may continue to be discovered."

Pratt recognized the necessity of visualization saying, "One indispensable attribute of the oilfinder is vision. If it is in the mind of the geologist or oilfinder that new fields first take form, then discovery must wait on our mental visualization — our imagination.''

In 1943, A.I. Levorsen said, ''There are many geological tools rusting away in our kit for want of use. The need is for 'creative' geology as compared with what may be called 'routine' geology.''

And in 1938 Parke Dickey reminded us that ''We usually find oil in an old place with a new idea, but we seldom find much oil in an old place with an old idea. Several times in the past we have thought we were running out of oil, whereas actually we were only running out of ideas.''

More recently (1983) Jack Parker stated, ''Creativity comes from our brains. It involves making connections — connections between concepts and facts we have learned and remembered. Creativity occurs because we overcome obstacles — because we dare.''

And finally Ted Bear, who, along with the geologists above, has been an inspiration to us, said in 1985, ''It is my firm belief that given all these advances in science and technology, the greatest single factor left in the exploration equation is creative thinking. The best science in the world cannot develop a prospect. Only the petroleum geologist, through creative assembly of all the disparate bits of scientific information available to him, can accomplish this. New ideas in older provinces are not generated by data manipulators. They are generated by creative thinking.''

Most geologists are creative visual thinkers because of the way we are trained and have to think. However, we believe that creativity can be enhanced by visualization exercises that help teach use and control of imagination. May we suggest that if you are interested in developing your creative skills and finding more petroleum or other natural resources, you obtain copies of the above mentioned books by Edwards and McKim.

During my term as president of AAPG I have had the opportunity to visit many of the productive areas of the United States and other parts of the world.

Figure 2 — ''Illumination''

''One indispensable attribute of the oil finder is vision ... discovery must wait on our mental visualization — our imagination.''
— Wallace E. Pratt

During these travels I have been made aware of the great opportunities in exploration and production which still exist, even in today's economic climate. I have seen firsthand how creative thinking has led, over the years, to significant discoveries. In the future it will be the creative geologist that continues to find the elusive field.

It is incumbent upon all of us to develop our creative thinking skills to their fullest extent, so that we may continue to contribute to the wealth of all nations for the benefit of mankind.

Norm Foster

Adapted from an address given by L. W. Funkhouser during his term as President of the American Association of Petroleum Geologists, 1987–1988. Printed by permission of the author.

Petroleum Exploration— A Science, A Business, An Art

L. W. Funkhouser[1]

In terms of scientific applications leading to economic progress, the practice of petroleum exploration is a relative newcomer. Many of the current members of the American Association of Petroleum Geologists began their careers before or near the midpoint in chronological development of the science as we know it today.

From the earliest use of field geology to outline areas favorable for the occurrence of oil and gas to the present-day array of sophisticated technology, the science of petroleum exploration has been one of the great success stories of all time. Tracking down the vast and elusive stores of oil and gas that have fueled the industrial revolution of the past century has been a major contribution to the well-being and quality of life of much of the world's population.

Through the years, the structural and stratigraphic parameters pursued through field geology studies have been supplemented and extended by continuous advances in surface and subsurface data collecting, primarily as a result of new concepts in geology and its important scientific partners: geophysics and geochemistry.

In less than 60 years, reflection seismic surveys have progressed from the crude collecting of 4-way dip data to 3-dimensional studies which provide structural, stratigraphic, and hydrocarbon information in incredible detail and accuracy. During the same period, the science of borehole logging has advanced from a one-curve resistivity reading to the complex technical systems now available for the practice of formation evaluation.

Geochemical investigations related to the source, maturation, and migration of hydrocarbons has become an important exploration tool only within the past 30 years. Today, organic analyses are critical to our understanding of source-reservoir relationships, basin evaluation, and the geology of fluids.

Geology as a science has been dramatically advanced in the past 25 years by the geological and geophysical observations that have led to our current understanding of plate tectonics. Application of plate tectonic principles has given us the ability to understand aspects of the earth's geology that were incomprehensible only a few decades ago.

The purpose of relating this chronology has been to point out how new and dynamic the science of petroleum exploration has remained through its relatively short history. What passed for the leading edge of exploration technology in the mid-1950s is even more obsolete and outmoded than the Edsel of that era. The new exploration models that will be unveiled in the next 50 years will no doubt make today's version equally antiquated.

One requirement for realizing the promise of continued advancement in petroleum exploration methods is to have the will to continue funding research that looks ahead to eventual resurgence in exploration programs. The research leading to the technical breakthroughs of the future must be nourished during the lean years. Much of the world's undiscovered oil exists in stratigraphic traps and in masked areas, such as those under complicated thrust systems, under volcanics or salt, or under thick surface carbonates. Research leading to resolution of these unsolved problem areas will be a key factor in tomorrow's discovery of new hydrocarbon supplies.

Better use of data stored in the files is an area of applied research that deserves much more attention. Without question, old well records, logs, and core repositories are full of unrecognized prospects. A new look at old data, particularly with the advantage of increased insight gained from subsequent control, should be a most profitable assignment for some of the excess geologic talent now available. Database management tools need strengthening in order to recover in real time the massive amounts of invaluable information that crowd the files.

Recently, major fields have been found through a team approach of compiling and assimilating old data (well records, drill-stem test results, electric log analyses, and the like) and careful petrographic studies of old cores and cutting samples. Some of these new fields were bypassed by several generations of explorers, who did not pursue the thorough analyses that finally led to their recognition. This is an example of applied research of the highest quality—and of the most immediate economic impact.

---

[1] I thank Chevron Corporation for permission to use many of the illustrations and examples included in this paper.

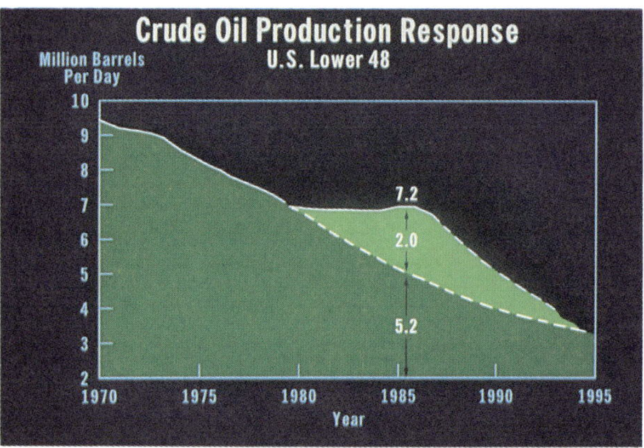

Figure 1—Daily oil production response, lower 48 states.

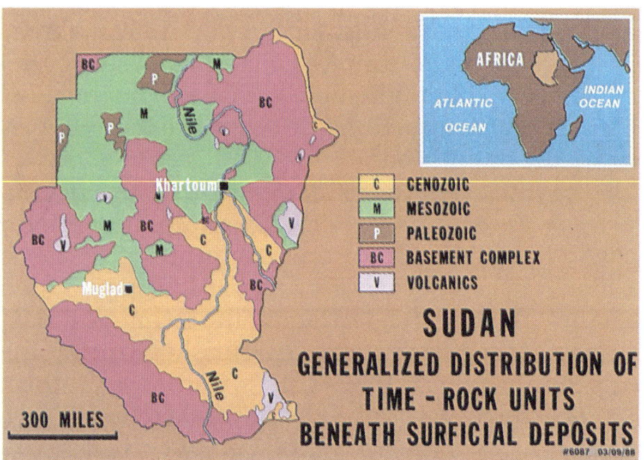

Figure 2—Pre-discovery regional map of interior Sudan.

Full and open communication between research scientists and line earth-science professionals must be stressed, so that real-life problems are solved by the researchers and new technology is quickly applied by earth scientists at operational levels.

One unique characteristic of petroleum exploration is that its practitioners are paid to be wrong more often than right. But that is a reflection of the fact that a dry wildcat can be an immensely successful research project. Every dry hole—much as we all hate them— adds significant data to understanding the true geologic picture in its vicinity. I think we all tend to forget that the real sampling area of an exploratory well is little more than 9 inches across. The eventual discovery of major oil and gas provinces in places such as the North Sea, Canada's Scotian Shelf, Wyoming's overthrust belt, and Western Australia came about as the result of data gathered by many individual research projects called dry holes.

Throughout its history, petroleum exploration has been constrained by the economics of supply and demand. Except for the aberration in oil and gas prices that resulted from the 1979 oil embargo, the price of oil has been relatively constant in real terms since the early part of the twentieth century.

In the 1950s and 1960s, exploration was successful in its mission to find major new hydrocarbon reserves, even under the limitations imposed by $2–3 per barrel oil prices. The keys to that success included: (1) a careful selection and ranking of plays and prospects; (2) a thorough analysis of all geological and geophysical data upon which the play or prospect was based; (3) the evolutionary improvement of the technology available to solve exploration puzzles; and (4) a long-term commitment to funding exploration programs that would lead to major discoveries in the future.

That same set of criteria should result in successful exploration programs under conditions of today's oil prices—which are actually somewhat better than those of the 1950s in terms of dollars of the day. The weak link in the key attributes appears to be condition four, the long-term commitment needed to pursue aggressive exploration of major long-range projects.

Adequate exploration funding is a must if the benefits to be derived from using the full talents of the exploration community are to be realized. Budget restrictions in the major and middle-sized companies and lack of outside financing for the independents have resulted in record lows in the use of seismic crews and exploratory drilling rigs. Overhead costs have been chopped too, but the limited amount of discretionary money severely impedes the geologists who remain. More funding is needed to drill the wells that provide the data leading to prospect recognition. More money should be earmarked to explore new trends, new ideas, and unconventional new concepts.

In 1970, Kenneth Crandall made an observation in his AAPG Presidential Address that is remarkably relevant today: "When the oil economy sneezes, the exploration segment usually catches pneumonia. Because it is, or should be, a long range activity with lead times of from 3 to 10 years or more from initiation to conclusion of a project, a sustained, balanced exploratory program is desirable and has been demonstrated to be most successful. However, because of the practice of expensing exploration expenditures on a fairly current basis, a sharp reduction in them will cause a quick rise in current profits, but at the risk of reducing long range earnings and stability."

One of the arguments that has led to the current malaise in U.S. exploration and to the ominously repetitive "restructurings" of the past five years has been that the onshore United States is an over-mature, exhausted area that has, for all practical purposes, been drilled out. The Bureau of Economic Geology at the University of Texas has recently completed a resource appraisal study that contradicts that view.

An elegantly simple chart from that study (Figure 1) tells a dramatic story. As a result of price incentives and accelerated exploration efforts, increased U.S. drilling activity from 1980 to 1985 led to oil well completions that were double the number completed in the late 1970s. Domestic reserves were added at an average of nearly 2.5 billion bbl/year, more than double those added during the prior period of reduced drilling. Crude oil production in the lower 48 states at the end of 1985 was 2 million bbl/day—or 40%—higher than it would have been had earlier decline rates persisted. Declining success in replacing reserves had not been so much a lack of drilling opportunities as the lack of a sustained drilling effort.

There is a real lesson here. The United States can choose a course of stabilizing production at the current 7.5 million bbl/day rate by acting now to encourage new exploration and production drilling—or it can wait several years and then take steps to stabilize at 7 or 6 or 5 million bbl/day, depending on how long it takes for the issue of domestic production rates to reach crisis proportions. Once production levels have slipped, it is unlikely that even a crash program will be able to do more than match the normal decline of the depleting fields.

The Bureau of Economic Geology study also identified the potential resource base that is available in the United States for sustaining future production. The study concluded that about 300 billion barrels of oil have been found that do not yet fit in the category of proved reserves. Adding in an estimated 85 billion barrels yet to be discovered in large to modest-size fields adds up to a total U.S. resource target (excluding oil shales) of more than 400 billion barrels. As much as 280 billion barrels of this resource are considered recoverable, given two conditions: (1) prices must reach $40 per barrel in 1988 dollars, in order to proceed with the segment of enhanced oil recovery projects in the higher cost range; and (2) improved efficiency will be required in terms of expanding technology and increased skills in reservoir description and management.

Two hundred eighty billion barrels is 100 years supply for the United States at today's daily production rate. This untapped wealth of oil resources promises an active oil industry for several more generations of geologists. We need to communicate the facts about this national treasure to all levels of government and society. We must encourage the adoption of a sane national energy policy that removes the shackles from exploration professionals and lets them go back to work.

In response to the Sidney Powers Medal award in 1948, A. I. Levorsen put it this way: "One thing is certain—we are not going to find oil if we give up with-

out trying. Oil is like fishing—one does not catch fish with his hook out of the water, and oil is not found if the drill is not kept going. A favorable economic climate is one of the essentials of keeping the drills going. Another is a solid geological conviction that oil is to be found if the drills are kept going. This requires a lot more geological vigor than does the conclusion we are about to run out of oil discoveries. The industry has always explored at the edge of its knowledge and the fact that some in the industry see little favorable ground for future prospecting is nothing new." Nothing new in 1920, nothing new in 1948, and certainly nothing new in 1992.

Petroleum exploration can be regarded as a science or as a business, but it is my conviction that it is most of all an art. While all three aspects of the profession are critical to its practice, the one that provides the ultimate reward is the genius, the ingenuity, the creativity of the individual explorationist. Whereas teamwork is an essential part of using all the modern tools applicable to an exploration problem, it is the spark of the creative geologist or geophysicist that ignites the process.

During my career, I had the pleasure and privilege of managing and participating in many imaginative geologic programs that eventually led to major hydrocarbon discoveries. Many of these projects began as a gleam in the eye of a single explorationist. These individual oil finders had the ability to view geologic data with controlled optimism. This attitude is in striking contrast to the viewpoint of the less creative explorationist who considers a dry hole—or even a number of dry holes—as a big negative rather than as something to study for the positive leads they may have to offer.

From the catalogue of successful plays I have witnessed, I have selected three to discuss in some detail because they are examples of major discoveries that can be specifically traced back to their inception in the creative thought process of one geologist. While many explorationists contributed to the eventual culmination of the idea as a discovery, the program would not have been begun without the exploration art displayed by the geologist who came up with the initial concept.

The first example involves the genesis of the exploration program that preceded the discovery of a new oil province in the interior of Sudan in 1979. Figure 2 shows the surface and subsurface control that existed in that very large area prior to the decision to take a major land position. The surface outcrops in the area of interest are all Precambrian metamorphic rocks and granites or Pleistocene sands and gravels. The deepest of the few existing dry holes was only 1117 ft, and since all of the wells had been drilled in search of ground water, subsurface control was completely lacking in the area.

The area was conventionally considered to be a part of the African craton. However, John Miller, an experienced photogeologist engaged in a reconnais-

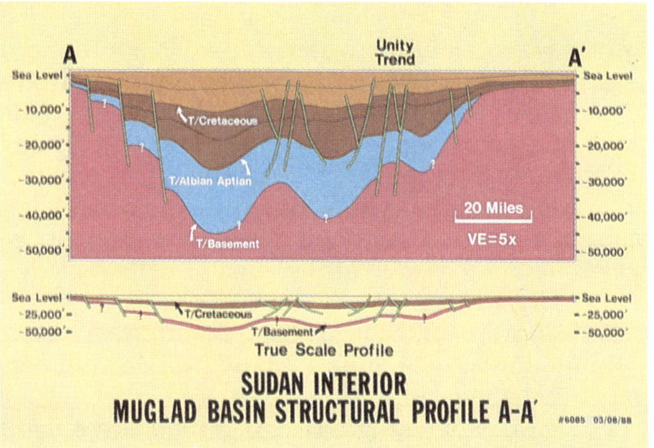

Figure 3—Regional structural cross section of interior Sudan.

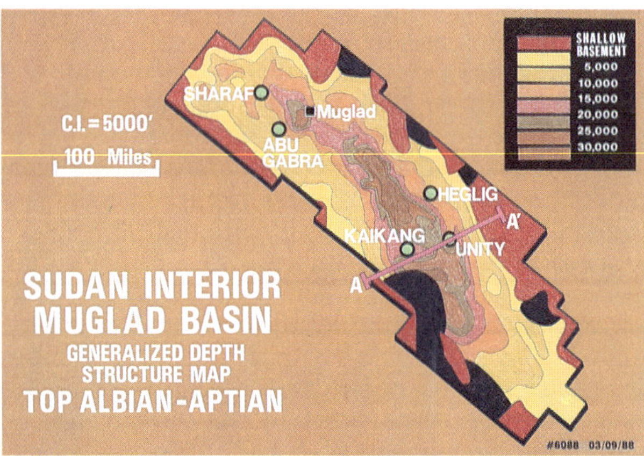

Figure 4—Regional structural map of interior Sudan.

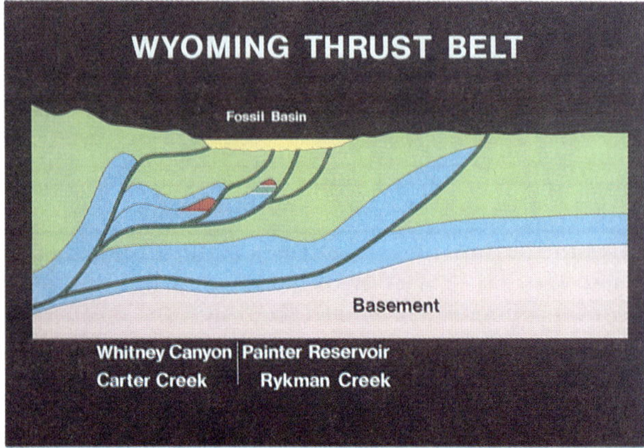

Figure 5—Pre-discovery concept cross section of the Wyoming thrust belt.

sance review of that part of the world, noted that the southwest Sudan is a major surface basin, traversed by the White Nile River. This deduction, together with the projection of regional faults known to exist in Kenya, led him to speculate that the surface basin could be an expression of a buried basin formed as the result of a failed arm of a triple junction. A search for geophysical confirmation turned up a United Nations seismic refraction and gravity profile (run for the purpose of evaluating water resources) across the western part of the area of interest. A study of seismic velocity data from that profile indicated that a previously unknown sedimentary section more than 12,000 ft thick was present on the traverse. That profile happened to be the only geophysical program ever run in this vast area—an area generally assumed to be nonprospective.

Following land acquisition, a comprehensive program of regional magnetic, gravity, and reconnaissance seismic surveys was conducted to outline the configuration of the postulated rifted basin. Figures 3 and 4 are based on that data and subsequent subsurface well control. They show the general structure of this lacustrine basin which had remained hidden through so many waves of worldwide exploration. Five to ten miles from the Precambrian outcrops, the surface gravels had concealed a rift-related sedimentary section as much as 45,000 ft thick.

Proved reserves in the Interior Sudan basin exceed 300 million bbl, with much of the area remaining to be explored and tested. In assessing Sudan's potential, many pessimistic geologists had seen a craton. One creative and optimistic geologist saw an unexplored area that could contain major oil reserves.

Another area where imaginative geologic thought and advanced seismic technology combined to turn a nonprospective area into a significant new productive complex is the thrust belt trend in western Wyoming. Through the years, many expensive dry holes had been drilled along this trend, and resource estimates of the area, made both by the U. S. Geological Survey and many oil companies, were highly unfavorable.

Despite that gloomy history and outlook, Frank Royse, a geologist who had spent more than 25 years in unravelling the complex geology of the thrust belt, refused to give up. He finally convinced his company to take a major lease position along the trend and to conduct a survey using modern seismic technology to check his creative interpretation.

The cross section shown in Figure 5 illustrates the subsurface model that he had conceived to justify the initiation of a major exploration program. He postulated that Paleozoic and Mesozoic reservoir beds in the hanging wall of several thrust faults known from surface control had been superimposed on Cretaceous source rocks in the footwall, setting up a perfect source-reservoir relationship. The overburden of the thrusted section, according to his concept, would bury the Cre-

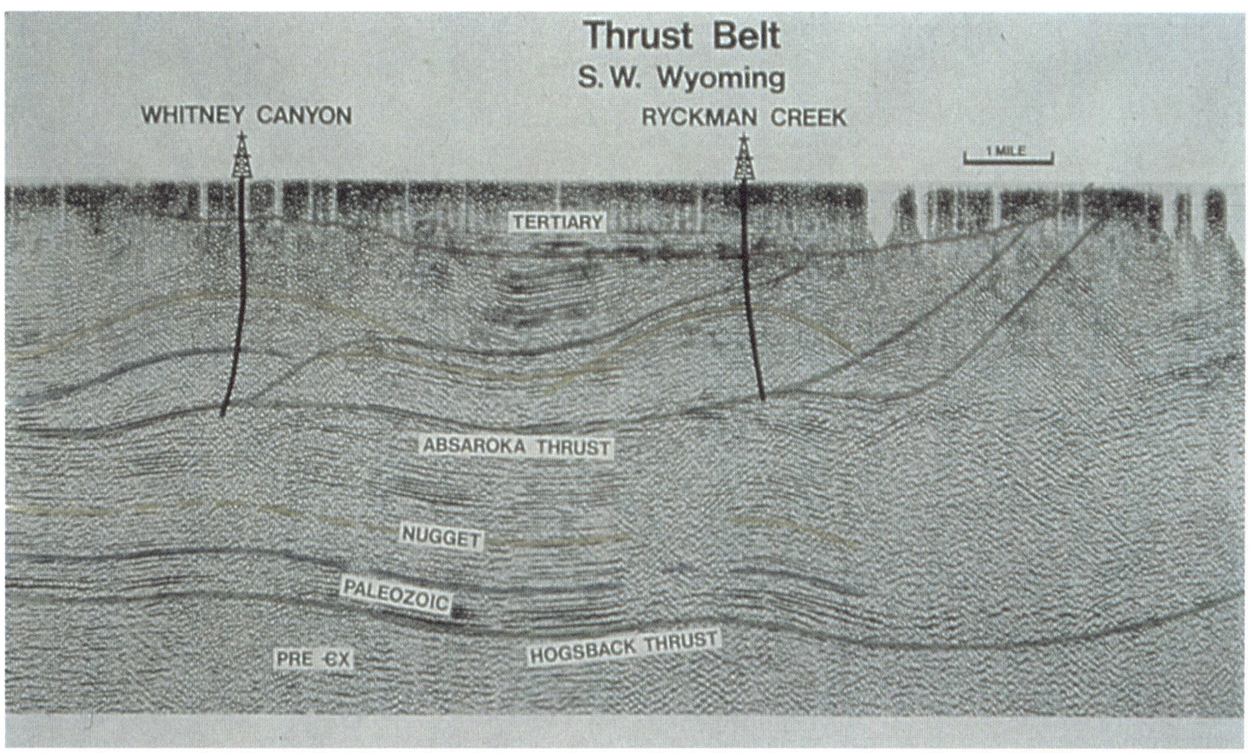

Figure 6—Seismic cross section through Wyoming thrust belt discoveries.

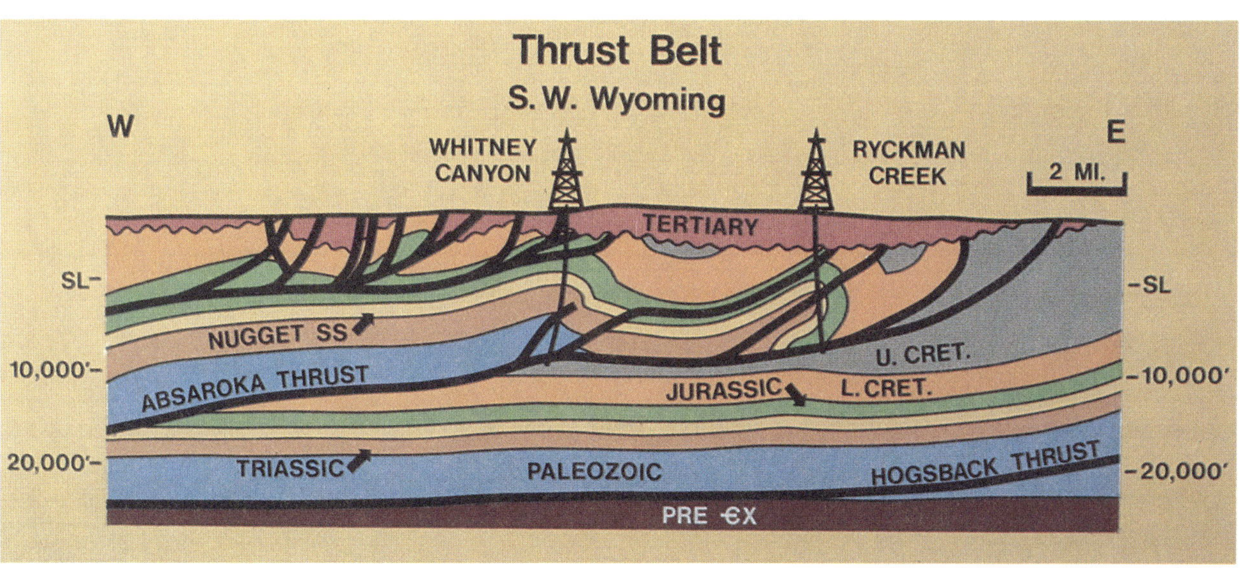

Figure 7—Post-discovery structural cross section of the Wyoming thrust belt.

taceous shales to an optimum generating depth.

Available seismic data were of insufficient quality to verify the structural model. Fortunately, however, state-of-the-art seismic data gathering and process- ing technology in the mid 1970s had evolved suffi- ciently to map the complicated structures associated with the objective thrust faults. Figures 6 shows a seismic profile and Figure 7 a structural cross section

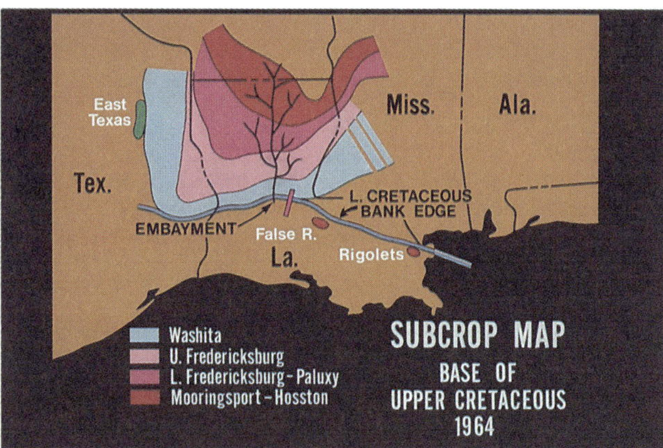

Figure 8—Pre-discovery subcrop map on the base of the Upper Cretaceous, north Louisiana.

the initial concept, which had been based almost entirely on surface control.

Fifteen oil and gas fields were found along this trend which is 60 miles long and 15 miles wide. Reserves estimated to total more than 8 tcf of gas and 750 million bbl of oil were added, thanks in large part to the perseverance and deductive ability of a geologist who understood the art of exploration.

The deep Tuscaloosa gas trend in south Louisiana is another producing province whose discovery can be traced back to the imagination of one geologist, Bill Crain. His curiosity about the sedimentary processes in the Late Cretaceous in that part of the world led directly to the acquisition of modern seismic data—and a strong land position—that resulted in discovery of hydrocarbon resources that had been totally unsuspected prior to his work.

Figure 8 is a subcrop map at the base of the Upper Cretaceous, drawn from subsurface control available in 1964. The observant geologist who studied this pattern noted that enormous quantities of sediment had been eroded in north Louisiana during the Early Cretaceous. He wondered where it had all been deposited—and decided that it had probably been moved into deep water in front of the Lower Creta-

across the area of interest following completion of the seismic studies and exploratory drilling. In comparing these to Figure 5, it is obvious that although the actual geologic configuration is more complex than the original model, it is in substantial agreement with

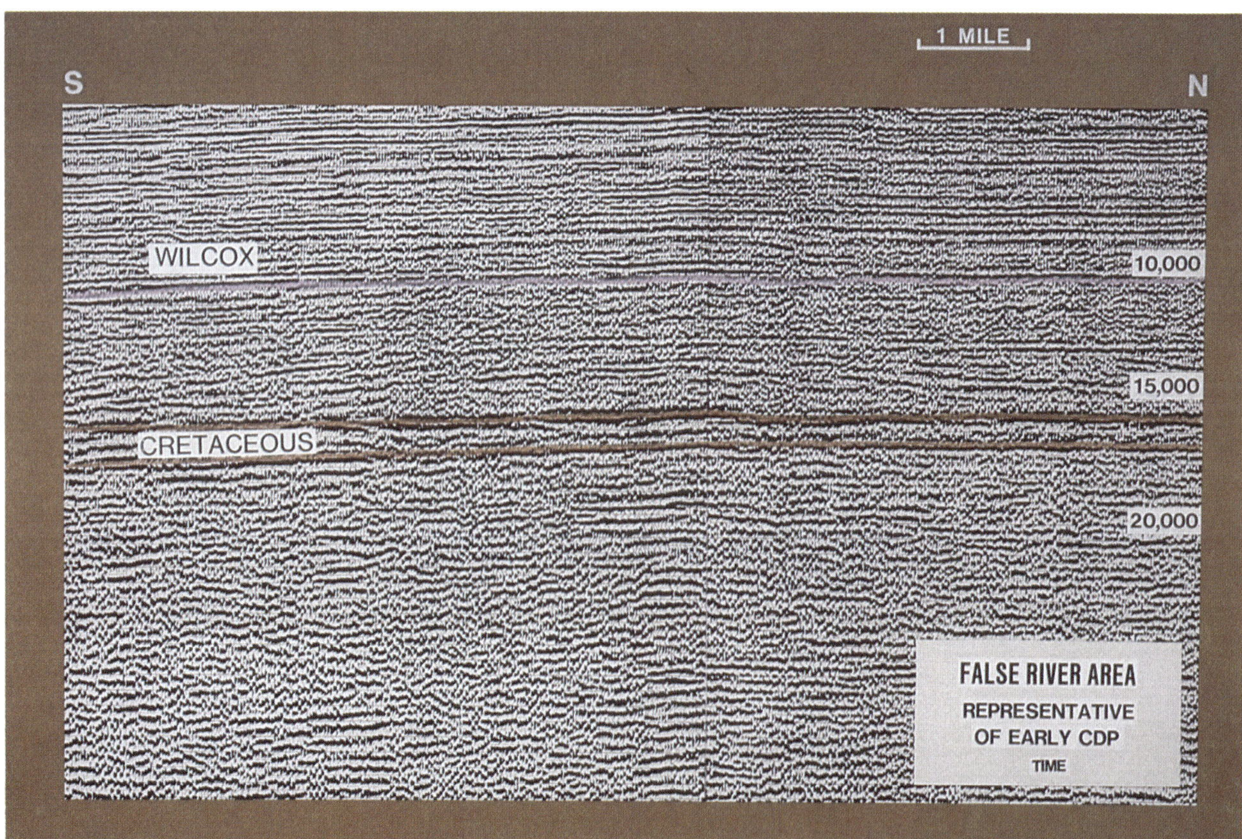

Figure 9—Pre-discovery seismic cross section through the discovery well of the False River deep Tuscaloosa gas trend.

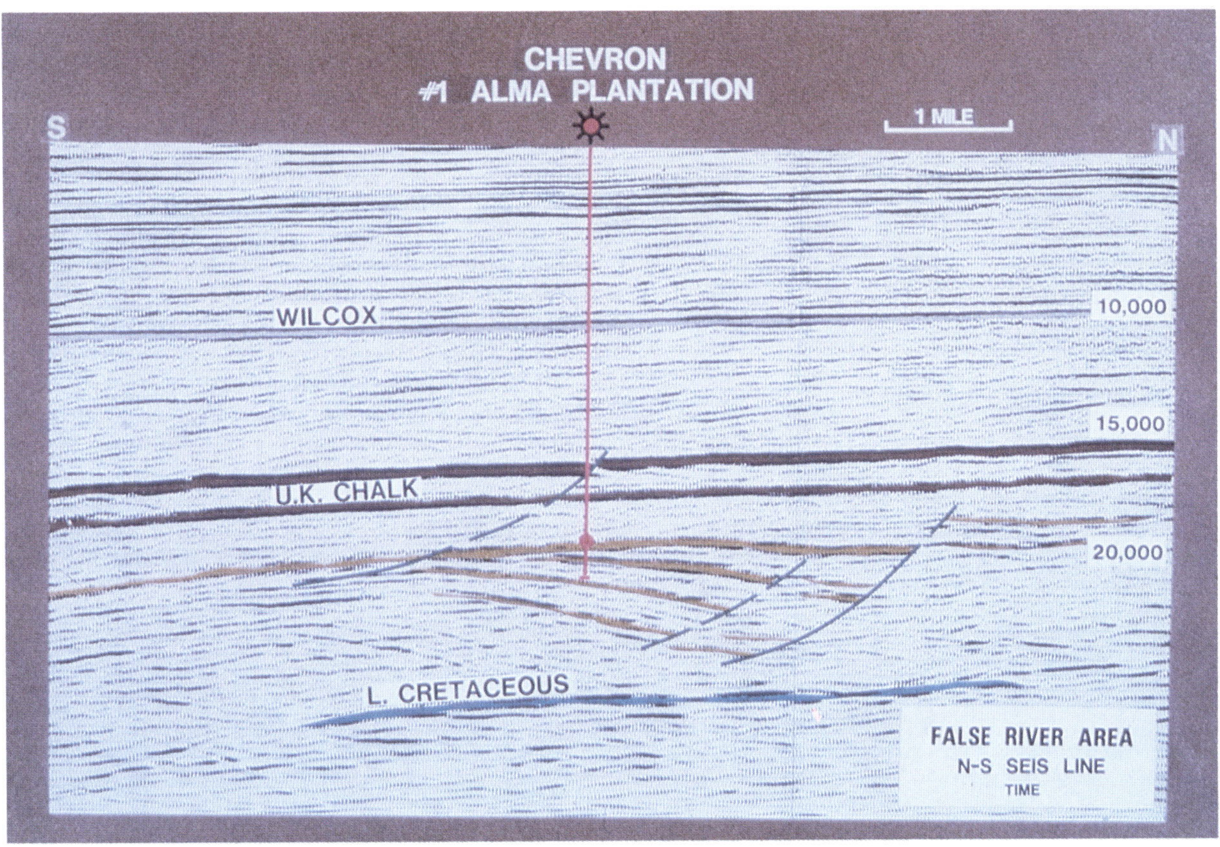

Figure 10—Seismic cross section through the discovery well of the False River deep Tuscaloosa gas trend.

ceous bank edge through breaks in the reef.

This was an intriguing concept which obviously could lead to major structural or stratigraphic accumulations in unknown deep water Cretaceous sediments. Unfortunately, in this case, it was an idea which came too early for verification by the then-current seismic technology. Figure 9 is a seismic cross section shot in the mid 1960s across the area of interest. No evidence for the postulated deep Tuscaloosa trend could be seen because no data existed below the regionally dipping Upper Cretaceous chalk horizon.

The area was kept under surveillance for several years, and the shooting of a modern seismic reconnaissance line was recommended in the mid 1970s. Figure 10 shows the seismic breakthrough that finally confirmed the concept originated a decade earlier. This line, through the subsequent first discovery along this new deep trend, gave good evidence for growth faulting and structural culmination in the part of the geologic section previously hidden beneath the regional dip at chalk depth.

After the first discovery of deep Tuscaloosa gas in False River field in 1975, continued exploration found 20 new oil and gas fields along a trend 50 miles long. Figures 11 is a cross section through False River field and Figure 12 is a map at the top of the Tuscaloosa

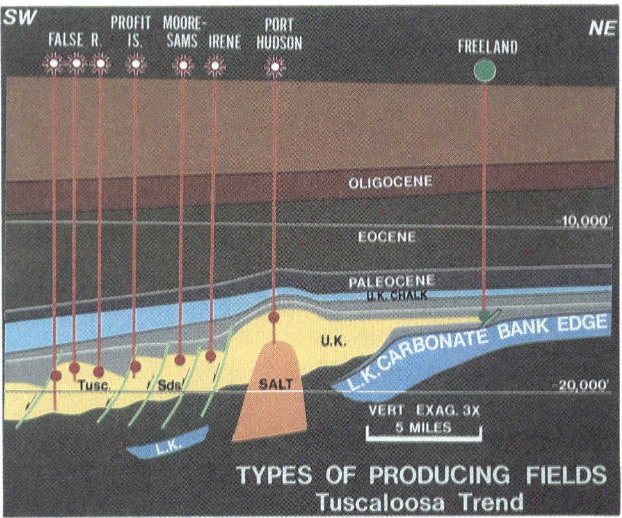

Figure 11—Post-discovery structural cross section of the deep Tuscaloosa gas trend.

sandstone. The complex structure and multiple traps concealed at depth validated the original thought pro-

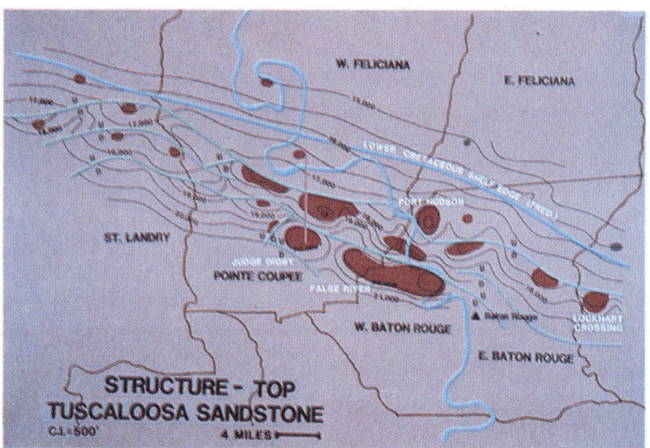

Figure 12—Regional structure map of the deep Tuscaloosa gas trend.

cess that had stimulated one geologist to speculate about their presence. Estimated reserves from the trend total approximately 2 tcf of gas and 100 million bbl of oil and condensate.

One common link in these three examples is that, in each case, an unconventional explorer chose to believe the unbelievable. Once the concept hurdle was crossed, he used his professional skill and exploration art to find the unfindable. Art is the essential talent that a true explorationist melds with science and business to attain success in one of the most fascinating disciplines available as a profession.

Creativity, leading-edge technology, and communication are all vital elements in working the geologic puzzles that remain unsolved. Critical to the task is maintenance of the academic system that generates creative professionals—and that system needs careful nurturing during the current exploration downturn. Continued in-depth research in earth science processes and further evolution of exploration technology is required for the discovery of the significant hydrocarbon resources that have not yet been found.

Most important of all, today's explorationist should be encouraged to pursue his art and to speculate about the original and unconventional ideas that may be the key to tomorrow's discoveries.

AAPG Explorer, v. 4, no. 10 (June 1983), p. 12–13.

Parker Speaks On Concepts And Creativity

by John M. Parker (AC '44)
AAPG President

(Editor's Note: The following is the presidential address made by Parker April 18, during AAPG's 68th annual convention in Dallas.)

Concepts and creativity. Thirty thousand years ago, magnificent drawings were made on the walls of caves in Spain, France and Africa. At that time, Homo sapiens had already been wandering the earth for more than 2 million years. Why did this new subspecies we call Cro-Magnon suddenly create art? What was the impetus to create?

Creativity comes from our brains. It involves making connections — connections between concepts and facts we have learned and remembered. Creativity occurs because we overcome obstacles — because we dare.

Compare a complex protein molecule in a microscopic sea animal — and the same hydrogen and carbon atoms in a complex molecule in a crude oil. To bridge that gap in time and space between the two molecules we need concepts and creativity. That is how we find oil.

We develop concepts about birth, life and death of the sea animal; we consider its burial, preservation, transformation, migration — and its final end in a trap. From those concepts we create a mental picture of what we want to find. We use tools to map horizons and characteristics in rock strata. We drill a hole — and usually do not find oil.

Our concepts were wrong; we did not map correctly. We did not achieve the right combination of facts and speculation.

Mapping New Ground

John Wesley Powell lived from 1834 to 1902. After losing his right arm below the elbow from a minie ball early in the Civil War, he returned to active duty in 1862 and was discharged in 1865 as a lieutenant colonel. A largely self-taught geologist, Powell was a professor at Illinois-Wesleyan University after the war. In 1866 he promoted money and surveying instruments from the University of Illinois, the Chicago Academy of Science, the Museum of Natural History and the Smithsonian Institution. Accompanied by his wife,

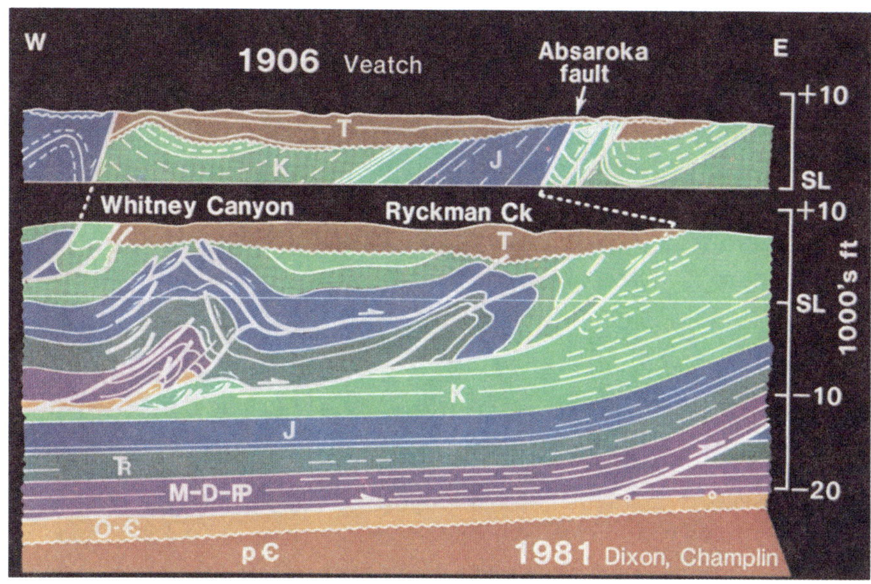

The upper section illustrates Arthur Veatch's 1907 findings. The section below is the same area as identified by surface seismic and well data.

students and neighbors, he headed west to map and collect specimens. In 1867, the expedition climbed Pikes Peak. In 1868, they climbed Long's Peak; and in the winter of 1868-1869, Powell mapped in northwestern Colorado.

In Green River, Wyo., May 24, 1869, Powell started his first wild trip down the Colorado River and through the Grand Canyon. Two years later, May 22, 1871, he made a second trip down the Colorado. He continued to work in the Colorado Plateau region and the West in subsequent years. In 1878, Powell lobbied to change the public land system and to establish the United States Geological Survey. It was established in 1879. Powell became the second director of the USGS two years later. He gathered able men around him and gave freely of his masterful intellect and ideas. Based on his field work, Powell wrote about concepts of erosion and uplift, land forms, stream drainage development and a "base level of erosion."

Powell was extremely active in politics. He had strong views on the role of government — particularly the geological survey. He was continually working to influence Congress. In 1889, he addressed the constitutional conventions of the Territories of North Dakota and Montana as they met to draw their constitutions to become

states. He told them how geology and geography and the arid climate would affect them. He urged them to drop the old system of section, township and range. He felt land subdivision should conform to drainage divides and basins. Some water rights should be controlled by the states, and individual water rights should go with title to the land. He wanted the basic facts about water and man's dependence on it understood. He made many speeches, wrote many articles but lost the campaign.

A Daring Concept

Arthur C. Veatch lived from 1878 to 1939. He started his career in 1899 doing field work for Louisiana State University, working on the petroleum geology of the Gulf Coast and mapping salt domes. From 1901-1902, he was a consulting geologist for a predecessor of Kirby Petroleum Co. and may have been responsible for leasing some tracts on Barber's Hill, Hockley and Pierce Junction. He worked for the USGS from 1905 to 1910 in northern Louisiana, southern Arkansas and southwest Wyoming.

From 1910 to 1912, he worked for General Asphalt in Trinidad and Venezuela. For the next six years, Veatch headed Lord Cowdray's Aguila Co. exploration and development operations from their London office.

Geologist Arthur Veatch stands by the mold of a tree in a Tertiary sandstone outcrop at Point of Rocks, Wyo., in 1905. Veatch, who was leading a USGS mapping party in the area, named the Absaroka Fault and mapped it as a thrust fault, a "daring new concept" at the time. (Photo courtesy of Steve Oriel (AC '48), USGS, Denver.)

Their primary activity in Mexico and other foreign countries was very successful. Closer to the office, Veatch helped to locate a wildcat in Derbyshire, northwest of London, that flowed 12 barrels of oil per day. Aguila had a large exploration program in Portuguese Angola, and Veatch personally did some of the field work along the Congo River.

From 1919 to 1928, Veatch was director of exploration for Sinclair-Consolidated in their New York City office, primarily involving foreign exploration. William B. Heroy had much of the responsibility in that office. Veatch operated as a consultant out of New York City until his death. Along with some partners, he obtained concessions in the lower Saxony Basin of Germany in 1929. These blocks were sold to Mobil, Shell and Standard Oil Co. of New Jersey. Production was found in the area after World War II.

Veatch was perhaps the first to hypothesize huge submarine canyons on the continental shelves and slopes.

In 1900, in the southwest corner of Wyoming, the Union Pacific Railroad began drilling a water well in the tiny railroad settlement of Spring Valley. By the spring of 1901, they had struck small amounts of oil at 49 feet, again at 573 feet and again at 1,148 feet.

Oil fever gripped the area. From 1901 to 1904 and later, oil stocks were sold in Salt Lake City, Denver, Pittsburgh and New York City. The Union Pacific land grant odd sections were not patented yet, and most of the rest of the area was unoccupied government land. Soon the whole countryside was staked with petroleum claims under the placer mining laws.

High on the desolate, windswept, sage-covered ridges there was chaos. Claim jumping was rampant. Companies hired armed men to guard claims. The U.S. General Land Office first sent a special agent, followed by a special investigator, to make inquiries. Congress was in an uproar. To add to the mess, some of the General Land Office land surveys were fraudulent.

In 1905, the USGS sent a field party to the area to map the geography, to check land surveys and to map the geology with special reference to coal and oil. The party chief was Arthur C. Veatch.

He and his assistants mapped an extremely complicated area 32 miles wide and 70 miles long between June 30 and Oct. 10, 1905. While mapping on horseback Aug. 3, Veatch was hit with appendicitis. When he did not return to camp, his assistants began a search. They found him and brought him back to camp. But by Sept. 3, he was back in the field mapping again.

Veatch named the large fault in the central part of this southwest Wyoming area the Absaroka, and he correctly mapped it as a thrust fault. At that time, a thrust fault was a daring new concept. In much of the area mapped, flat-lying late Cretaceous and Tertiary beds covered the faulted and folded Paleozoic and Mesozoic beds. In 1879, A.C. Peale, a USGS geologist, had mapped this fault as a normal fault. But by piecing together structural trends and concepts from the scattered outcrops below the unconformity, Veatch was able to identify the thrust fault. It was an astounding job.

Ignoring The Obvious

Today's seismic data shows the picture below the unconformity. Above the Absaroka thrust is a large fold with Paleozoic rocks just above the thrust — the Whitney Canyon gas-condensate field with more than 3 trillion cubic feet of gas and 100 million barrels of liquids. The recumbent fold axis to the east with Mesozoic rocks above the Absaroka thrust is the Ryckman Creek-Clear Creek field complex with more than half a trillion cubic feet of gas and more than 100 million barrels of oil.

In the Whitney Canyon fold axis, the folding and faulting are extremely complex with thrust faults going in opposite directions and small displacement thrusts located above other thrusts.

One mile west of what is now the south end of the Whitney Canyon field, the Michigan-Wyoming Oil Co. had a camp back in 1904. The Standard Reserve Oil Co. had their camp 11 miles east of the location of the nearby Painter Reservoir field, which was not discovered until 1977. Jager Oil Co., American Consolidated and the Pittsburg-Salt Lake Oil Co. all had camps within three miles of Standard Reserve's camp in 1904.

The Absaroka fault continues 16 miles south of the cross sections to a railroad cut on Sulfur Creek. A well-exposed section of the lower Cretaceous Bear River formation is visible on the upper side of an overturned syncline. The beds are fresh-water deposits and contain typical fresh-water clam fossils. Four well-known geologists mapped this outcrop — Henry Engelmann and F. B. Meek in 1859 and 1860, and C. A. White and John Wesley Powell in 1876. They all made the same mistake. Since the Bear River beds were at the top, they assumed they were the youngest rocks in the section. Since they were fresh-water beds, they thought they were Eocene in age. Thus they misplaced these lower Cretaceous Bear River formation beds 20,000 feet too high in the stratigraphic column and assigned an age 70 million years too young.

T.W. Stanton mapped the age of this outcrop correctly in 1891 and Arthur Veatch mapped the details of the stratigraphy and structure correctly in 1905.

At the top of the cut, there is downhill creep. The overturned beds dip east, while the hill slopes west. At the top of the hill is west dip (slope creep) in the same beds that dip east down the face of the cut. Engelmann, Meek, White and Powell all mapped this creep phenomenon as an anticline.

In 1948, I mapped the Sulfur Creek railroad cut with Veatch's map in my hand. I did it correctly and added some detailed data in the vicinity. Stanolind Oil and Gas Co. moved a seismic crew into the area on my recommendation. But my interpretive structure cross sections — based on surface and seismic work — were wrong. We did not have all of the concepts we needed and we did not have good enough seismic reflections.

The Common Thread

There are other examples of creative concepts. The front of the Mackenzie Mountains are at 65 degrees north latitude in the Northwest Territories of Canada. Just north of Powell Creek — a tempestuous stream I named in honor of John Wesley Powell — is a low shale hogback.

The hogback turns into a huge ridge capped with 400 feet of coral reef. The real rock data in the field destroyed the old concept that coral reefs could not develop in high latitudes and could not be found in Western Canada, particularly in the upper Devonian.

Offshore marine bars — even 120 miles offshore — were a concept of mine in the 1950s. Until then the geology of the bottom of the ocean was almost unknown in the far offshore continental shelf areas of present oceans and ancient seas. Geologists were afraid to hypothesize something in ancient seas that they had not seen in modern seas.

Solution and removal of bedded salts in the central parts of huge basins, at depths of 5,000 feet to 12,000 feet, was a daring concept in the 1950s. Those of us who pioneered this concept still do not know all of the mechanics of the solution and removal.

It takes facts to be able to derive a concept just as it takes concepts to be creative. New concepts can sweep through our profession like a prairie fire. They become fads. In the 1950s, all sorts of sedimentary environments were called turbidites. Much poor geology was published by people who picked out part of the facts to support their theses. Turbidites went out of fashion in the 1960s, although they came back in the 1970s.

What are the common threads in these stories of exploration, hardship, failure and success? Concepts and creativity do not spring up in a vacuum. They occur to geologists with a purpose, a mission, to those who are looking for oil, not playing games — to geologists who take chances. As Pasteur said, "Chance comes to the prepared mind."

Creativity comes from minds full of concepts, full of carefully observed facts. It is the outpouring of people who are energetic and curious, who suffer adversity and hardships, who have known failure — from people who share ideas as they travel about. It is the spark of those who dare.

AAPG President Addresses Opening Session

Creativity Indispensable, Bear Says

The United States is entering a period in which geologists need to be motivated, optimistic and creative to meet the needs of growing energy demands, Ted Bear (AC '46), president of the American Association of Petroleum Geologists, told the Annual Convention Monday in New Orleans.

"The oil is out there to be found," Bear, an independent California geologist, said. "The need is there, established by an energy-hungry United States, the prospects are there to be created, and the oil and gas companies of the United States will find a way to establish the budgets," he said.

Bear said geologists need to be well educated by both the universities and the AAPG to step forward and be ready to answer the nation's energy challenge.

Bear pointed out that with the exception of the Prudhoe Bay discovery, the U.S. reserves discovery rate has dropped gradually since 1965.

"The price of oil and the number of wells drilled have an effect on the decline, but since 1965 there has been no reversal of the discovery drop except in 1979 where the minor increase was undoubtedly due to the extreme increase in active drilling rigs," Bear said.

"It is time that we (petroleum geologists) turn the declining discovery rate around," Bear said. "We can do it with motivation, optimism, creativity and overcoming prejudices and mental barriers."

Bear said that a fundamental axiom of petroleum exploration expressed in 1942 by Wallace E. Pratt is all-important to the petroleum geologist of the past, present and future — "Oil is first sought in our minds." Bear said that quote was important then and remains an important goal today, because oil "can only be found in our minds by creativity and not in any book or paper or talk."

"This hypothesis is forever guiding all exploratory efforts. As the exploration progresses, we accumulate geology, geophysics, engineering data, well data and the prospects developed in our minds." he said.

Bear said "creativity will show us prospects favorable or unfavorable for entrapment of oil and gas.

"Creativity in the exploration for oil and gas is indispensable," Bear said. "It is not the only ingredient for a successful prospect but it is necessary for a discovery, unless the geologist is blessed with super luck — and we all know a lot about luck and no luck," Bear told the convention.

The AAPG president pointed out four requirements for a successful exploration petroleum geologist:

☐ He must have all the knowledge available and certainly as much as his competitor.

☐ He must be without prejudices and mental barriers or at least have the ability to overcome the prejudices that inhibit the search for petroleum.

☐ He must be optimistic and remain optimistic though the drilling of prospects has a tendency to change this attitude.

☐ Above all, he must be creative.

"We should be very concerned about the mistaken facts given the news media and politicians that our nation has more energy than we know what to do with and that there is little need to replace the oil and gas that we produce from day to day throughout the U.S.," Bear said.

The AAPG president said geologists need to get better control of these issues because "if we let a decade of well-educated geologists leave the industry, our domestic energy industry will be left permanently weaker."

"I will not venture to predict what might happen to our petroleum industry if it becomes too weak to meet the demands placed upon it by an energy-hungry United States," Bear said.

"I believe that at the present time there is very good evidence that the oil companies should be out looking for the best geologists available and getting their exploration departments ready for the future," Bear said.

"It is important for all of us to accept the challenge and do everything to increase our oil finding abilities," he said.

"First, we must recognize that the oil is there to be found. We must also recognize that the remaining oil is difficult to find," Bear said.

"The prospects are complex; we must unravel the geology to create prospects that will find the oil. We must all remain optimistic and motivated, and above all, we must face the challenge with a positive creative attitude," Bear said.

"This will take more well-trained geologists," he said. "We can improve our discovery records by becoming better geologists. AAPG is continually looking for new ideas to teach and will continue to search and improve.

"It is now time for all of us to recognize the importance of creativity and believe that we can enhance our own creative ability," Bear said.

Reprinted by permission of Chevron Corporation from
Chevron World, v. 56, no. 2 (Spring 1979), p. 14-17.

CREATIVITY... THE QUANTUM LEAP

Consider creativity. Under what conditions and in what manner is it accomplished? It is indeed a *human* resource—no other living thing has such an ability to create new things, to construct an image in their mind of something they have never seen, though often in a form similar to something they *have* seen.

The human brain is considered to be divided into two sides—the left-hand side which controls movements of the right side of the body and the right-hand side of the brain controlling the left side of the body. The left side of the brain is supposed to be more adept in thinking in precise terms, in numbers, in facts. The right side of the brain is inclined to use some form of image and attempt to solve a problem by a process of visualization.

Be that as it may, I suppose that all of us have abilities to use either or both approaches from time to time, but I would also believe that in *some* persons there is a stronger tendency to *conceptualize* a problem while other persons tend to approach it in a more *exact, mathematical* way. If this is so, it is fairly

Few persons in the oil industry are better qualified to examine the unique talent of man to think creatively than Kenneth Crandall. He has been doing just that for nearly six decades in the study and practice of his profession as a geologist.

His still active career has included 44 years with Standard of California, from which he retired in 1969 as Director and Vice-President, Exploration. During this time he was instrumental in developing the then-new field of geophysics and the related area of seismic technology.

We believe the accompanying thoughts of Geologist Crandall—expressed recently at a meeting of explorationists—are most appropriate to the creativity theme of this, our Centennial year.

obvious that the latter person would make a better engineer or banker, and the former a better artist, novelist, detective or—you guessed it—geologist.

Creativity usually comes about from considering many facts and ideas and using them to construct an idea, concept, proposition or explanation that is new or heretofore unknown. Geological science began with the search for concepts or explanations for surface features. But soon, of course, it was forced to project the consequences of these well-observed features into the subsurface, which could not be observed. Geologists in those early days were therefore forced to create mental images in order to make rational explanations for the observed features.

New Theories

Having been engaged in the study and practice of geology for almost 60 years now, I have seen the acquisition of new data explode theories, hypotheses or dogmatic statements often enough to gain a feeling that this is how progress is made.

In any area of the unknown we must continue to imagine, conceive or visualize the pertinent conditions or circumstances based on all the available evidence, including, most importantly, our own experience and what we know about the experience of others. Based on our best judgment we should be able to develop some sort of mental model. It may be no more than an impression but it will have to do until we can obtain more information.

A great deal of this type of exercise involves the use of analogous reasoning, comparing this area or condition with another about which more may be known. This reasoning should attempt to bring out any similarities or dissimilarities. This involves recognition —something which is aided greatly by personal experience and an imaginative mind. Together, our own experience and imagination can bridge a gap from one small point to another and produce a continuous image in our mind's eye.

For instance, you recognize a friend's face, even when you haven't seen him for years. You recall his facial characteristics; your mind compares them to the image which your memory has called up and identifying the points of similarity—all this in the space of a few seconds. (Note that I haven't said anything about remembering his name.)

The ability to make comparisons is of great help in forming a reasonable mental image of geological conditions. It is regrettable that younger geologists no longer have as great an opportunity today to do surface geology and, thereby, plant in the mind, subject to recall, various structural and stratigraphic conditions.

On the other hand, however, we now have much more evidence to work with—which *should* make

"We try to fill in the gaps in our knowledge with our imagination. . . ."

our jobs much easier. For example, our better grasp of source rock considerations—the amount and type of organic content, the temperature, pressure, time and other factors influencing petroleum generation and expulsion—has enabled us to evaluate better the possibilities of petroleum occurrence. However, there are so many more factors we can now use as points on which to base an evaluation that our jobs have, in fact, become more complex.

I have stressed the necessity of

"New ideas have virtually revolutionized our methods of exploration. . . ."

trying to recognize points of similarity with an analogue, but points of *difference* are, of course, equally important. In both structural and stratigraphic studies we look for anomalies. Unusual structural attitude or position, stratigraphic changes, such as the sudden appearance of dolomite in a section usually limestone, breccia beds in such a series, sudden changes in thickness of strata and facies changes—we know all these are important. So we have both points of similarity and points of difference to consider.

After taking all present evidence, we try to fill in the gaps in our knowledge with our *imagination* and thus *create* a mental *image* or model. We can, of course, take overt action to discard this image or test it with experiments or further exploration, or we can passively await later developments. *(Cont'd)*

15

'Science...and any creative activity are essentially fun'

With the amazing rate of development in the science and technology of geology and geophysics which has occurred in the last 40 years, particularly in the last 20, we should continually go back to yesterday's unsolved problems and areas. We should review them in the light of recent developments and, with our present knowledge and capabilities, make the necessary changes in our models and concepts. This is the pattern of development and creativity, not only in our science but in others as well—a constant review and improvement in our concepts—building one, finding flaws and short-comings and, by inductive reasoning, searching out new evidence, altering concepts and trying again and again.

New ideas have virtually revolutionized our methods of exploration, our concepts and criteria. Consider the astounding developments in the acquisition, processing, display and accuracy of seismic data; the wealth of data obtained from deep-sea drilling; the progress in the field of organic chemistry.

Today, our accumulated knowledge and data are so great that fine opportunities exist to create new insights through imagination, visualization and intuition. However, we will hardly ever have sufficient data about any venture to apply a risk factor approaching zero, since our images are almost always hypothetical. So it's "back to the drawing board" time after time, failure after failure, to dream some more with a new set of observations and criteria, hopefully to create a new concept to explain and solve our problems. In this way, failure is truly the mother of success.

Creativity in petroleum exploration is not confined to envisioning stratigraphic and structural conditions under which oil and gas may occur. It is also needed in molding organizations and programs needed to carry on successful scientific and technological efforts.

Obviously, the more imaginative and inquisitive our explorationists are—both geologists and geophysicists—the more creative concepts they will produce. Both their heredity, and past environment, are strong influences but they are not the only factors. The type of organization in which they work and the encouragement and motivation they receive are great aids also.

Bringing together a group of gifted and experienced persons to discuss a common objective is usually productive. In this process, often called "brainstorming," they learn from each other, often breaking down old, fixed ideas and building altered concepts. But the actual creation of a concept from an individual's thoughts and imagination seems to occur at a later time, after such a meeting.

There seem to be periodic bursts of creative progress, often as a result of an increase in the amount and reliability of pertinent data. Sometimes this is based on pure research by universities, governments or foundations or comes from wartime efforts which are carried on beyond the bounds of economic profit. The greater availability of such data gives better opportunities to visualize, imagine and create new concepts.

Science and art and any creative activity are essentially fun. They are enjoyable to the practitioner. Exploration is like a game—the more difficult, often the more enjoyable and sweeter the success.

Since petroleum geology and exploration have many of the aspects of an art as well as a science, it is understandable why they do not flourish under a bureaucratic or repressive system. Small wonder that Russia and China as well as other centrally directed economies come to the United States for much of their needed science and technology. After all, this country has furnished the knowledge and methods that have found a majority of the world's petroleum. We must continue on the same path. ■

Above, Geologist Crandall in his "office" at home, taking a minute out for the cameraman. At right, three Chevron geologists are "going back to the rocks," doing surface geology on the spine of a ridge in northwestern Alaska.

This paper, based on remarks presented to the Southwestern Federation of Geological Societies in Fort Worth, Texas, on February 10, 1966, is printed by permission of the author.

Creativity—A Most Important Factor in Oil Finding

S. W. Totten
President, Chevron Oil Company
Standard Oil Company of Texas Division

Webster's big international dictionary defines "creativity" as an adverb derived from the verb "create," which means "to bring into being; to cause to exist—said especially of the creation of the world from nothing."

We all know, too, that the word "creativity" is generally associated with performance of the arts. In that context, names readily come to mind like Van Gogh, Beethoven, Hemingway, Cole Porter, even Walt Disney. All have been remarkably creative people, in the artistic sense.

So when we talk about creativity in oil finding, we're moving into some pretty fast company. Yet if you're not convinced of it, I'm going to try to persuade you today to agree that a great geologist is—and has to be—as creative as a great writer, composer, or artist.

That may sound a bit high-flown. After all, geology is one of the basic earth sciences and it *is* an earthy business. Most of us aren't given to wearing berets or beards; nor are we particularly temperamental, as creative artists, writers, and composers are popularly supposed to be.

Nevertheless, I believe our profession passes muster under the Webster definition of creativity. For to be successful we *must*, through our own individual efforts, cause something to exist that did not exist before.

I'm talking about an oil field, which is quite a work of creation in itself. It may not arouse emotions like a symphony by Brahms, but it can start a lot of other things humming.

I think it's tremendously important at this particular time to keep thinking about creativity in geology, because this is an era in which it is extremely easy to fall into the belief that advance technology and hardware can do our thinking for us.

This is the era of automation, of the computer, of electronic data processing. We fire astronauts into space and guide their every move by electronic equipment so advanced as to be almost beyond conception. From this we might logically conclude that all a geologist needs to find oil is a stable of IBM equipment.

Don't you believe it.

Let me emphasize first that I have nothing but admiration for the progress that has been achieved in applying EDP to geologic analysis. It has been a great support. There's no argument about that, and I want to make it clear that I'm not calling for a return to the use of witching rods.

My point is this: Geology is a basic science, with a fundamental body of knowledge. But this body of knowledge cannot be interpreted inflexibly, as if it were a periodic law or a table of logarithms.

There's also an element of art to our profession. For Mother Nature, being typically female, likes to surprise us. Our basic body of knowledge—and our computers—can take us only so far. Eventually, from somewhere within us, we have to rack up the physical evidence and make a judgment on it.

That's creativity in the purest sense. It is an act that cannot be delegated to a mysterious black box. It is, therefore, a specialized art.

Every judgment we make may not be sound, any more than every struggling painter comes up with a Mona Lisa. But neither can the black box guarantee us the right answer every time, or IBM would be writing hit musicals.

Experience has nicely demonstrated and confirmed for us that a geologist blessed with the intangible talent of creativity can beat the dickens out of just about any black box.

An outstanding example of this occurred just a few years ago in California—to the chagrin of a number of people. This was the discovery of the Asphalto field, a mile or less from two of the state's biggest fields—Elk Hills and Midway-Sunset.

The author of this prospect, a geologist, had no expensive tools or exotic equipment—only an idea. He inferred his prospect from two simple factors: first, the well-known southeasterly plunge of the Cymric anticline in the area; second, the need of a source for the Stevens sand production at nearby Elk Hills.

This geologist was not the first to relate these two factors. But he was evidently the most creative in his deductions. His thinking led to a discovery in what seemed to be a synclinal position, in between several dry holes.

The Asphalto field is but one example of a California discovery derived from an idea. There have been others, for instance, the oil fields located by core-hole exploration in downtown Los Angeles. What is significant about each of these is that it was achieved in an area where previous exploration had been intensive and knowledgeable.

What then brought about these discoveries? The human element of creativity. Men were creative enough to take another look at the data, and to have the personal faith in themselves and their colleagues to believe their interpretation was the right one, and that the previous appraisals were wrong.

There are many ways to generate creativity, many I'm sure that are already quite familiar to you. First, however, let's take a quick brush-up course on how *not* to encourage creativity.

Creativity is a peculiarly human talent—so is the ability to discourage it in others. Scratch a dull, unimaginative operation and you will uncover, more often than not, complacent, self-satisfied humans in charge of the management.

Typically, complacency sets in after an organization has had a run of success based both on ability and luck. It's like a pro football team nursing a three point lead with a few minutes to go. They figure they've got it made and there's no use trying anything tricky. You know what usually happens in such circumstances— the other team, with nothing to lose, throws the big bomb and wins.

So it is with the organization that thinks it has it made. It pays lip service to new ideas and methods, but somehow they never get past the talking stage. Eventually, the new ideas—the creative ideas—dry up. Often the creative people drift away, and decline sets in.

Hand-maiden to complacency, in submerging creativity, is another human trait—fear of the unknown. We can trace this one back to prehistoric times; doubtlessly the Neanderthal man who ventured more than a few miles from the family cave was considered pretty daring, if not some kind of nut.

In today's more sophisticated times this unreasoning fear takes more subtle forms. You may find a manager suspicious of a subordinate's ideas because they went to different engineering schools. Or because the younger man wears sport jackets instead of ivy league suits.

This kind of resistance to the ideas of others doesn't make sense. Nevertheless, it often occurs. So when we turn down an idea, we would do well to analyze our motives—are we sure that we are turning down the idea or its merits? Or is it because the idea or its originator are somehow alien to our experience or environment?

Be cautious when you evaluate ideas, but be certain that your caution originates in the mind, not the emotions.

Still another deterrent to creativity is our old friend the organization structure—or as it sometimes used to be called, "Everything in its place and a place for everything."

Naturally, any company, particularly a big one, has to have formal organization structures—lines of authority, channels of communication, uniform salary administration, and so forth.

And only experience can disclose whether a particular structure is flexible enough to accommodate and encourage the creative individual. You must observe and appraise that for yourself, in the light of day-to-day events.

I will turn now to more positive considerations— some specific "Do's" for stimulating creativity.

First, and most important of all, management ought to establish a climate of mutual trust—and, hopefully, admiration—between itself and its employees.

I mean by this that managers must demonstrate— by deeds and not merely words—that they will give every idea a fair hearing, give credit when it is due, and *not* pirate their subordinates' ideas to make themselves look good.

Next, I think it's vital to indoctrinate everyone to accept and welcome change. Actually, that should apply to everyone in the organization, beginning right at the top.

We live in an era of change and none of us can afford to overlook this simple fact. Change is often unsettling and frequently risky. But the greatest danger as a rule is not from change but from failing to recognize the need for change—particularly in preconceived ideas—until it is too late.

Another essential to a creative climate is adoption by management of a tolerant attitude toward half-baked ideas. I am not suggesting that you actually adopt them; merely listen politely, smile, and send the man back to his desk with an encouraging pat on the back.

If you don't, you're going to miss some *good* ideas.

I would sum it all up this way: honor the creative man, but don's shun the screwball—they may be one and the same man.

Speaking of honoring a man brings me to my next point—the vital requirement of recognition. As the Old Testament points out, "man does not live by bread alone." This applies particularly to geologists, who thrive on praise and professional satisfaction.

I am not one to discredit money as an incentive to creativity and superior performance. But you also have to feed that inner man—the psyche—to keep a person happy and productive.

Lastly, as a further stimulus to creativity, I believe management should maintain an attitude of strong faith in the capacity of its organization to explore successfully, no matter what the disappointments. If we display an air of faith and determination, it's bound to rub off on our people and affect *their* mental outlook. Maybe that sounds a bit old-fashioned in these sophisticated times, but it still makes a great deal of sense to me.

What do I mean by faith? My practical definition is simply being self-reliant and confident, to the exclusion of nagging doubts. Most of all, it is having optimism. And that is a factor in petroleum geology that is important beyond measure.

I strongly believe that in the normal course of events, excluding mechanical error, the difference between a dry hole and a producer can be traced to the degree of optimism incorporated in the prospect interpretation. By my standards, the natural born oil finder has to be an optimist and he has to have faith.

He is a scientist, of course, and he will take precautions to hold himself in check. And so his optimism will be tempered by the facts. But he will have some optimism, for without it he will be hard-pressed to find his share of oil.

This line of thought leads me to comment on what some people call luck. How often we hear the envious cry, "that lucky guy," or "he's loaded with luck," and so forth. I contend that a great deal of this so-called luck is something slightly different, called serendipity.

Webster defines serendipity as the "art of finding the unexpected." In the oil business some people label this as "drilling in the right place for the wrong reasons." Whatever the case, I believe that if a geologist studies, thinks creatively, and has faith—or, as I prefer to call it, reasonable optimism in his work—then he will be rewarded from time to time with a measure of serendipity. As the popular expression goes, "never underestimate the power of positive thinking." And oil discovered in the right place for the wrong reason still brings just as good a price in the marketplace.

I hope I've made my point—that our profession is an intensely human one.

Ours is a profession where gambling is a way of life, where the element of doubt can't be eliminated. True, we can slim down the odds with our field studies and our black boxes. But some of the gamble still remains. Frankly, I'm glad that gamble still remains; ours would be a pretty dull business without it. I believe we *can* win a surprisingly high percentage of our gambles if we encourage uninhibited creativity in our organizations.

Great technological progress has been achieved in our business. But computers are still incapable of thinking for themselves. Until they do, let's not forget that the best "black box" we have is still the one we carry around in our heads.

Reprinted by permission of the Permian Basin Graduate
Center from Permian Basin Graduate Center Publication 1- 84 (1984).

CREATIVITY IN OIL EXPLORATION
or
How to Find Oil and Gas Economically

Jack G. Elam

PREFACE

This book is a compilation of my lectures on creativity developed during the last decade in teaching various courses at the Permian Basin Graduate Center in Midland, Texas. These lectures focused on the generation, selling, and evaluation of prospects for geologists and non-geologists.

I specifically have recorded these ideas for a one-week course I taught at my old alma mater, U.C.L.A., in February, 1984. I taught the course in an abbreviated period of time, and I needed to spend most of my time working with the students on various mapping problems. I did not have the time to fully cover the subject by lecturing, hence the book. I have thought for sometime that it would be beneficial to record my musings for posterity, because it has taken me many years and lots of dry holes to develop this exploration philosophy.

I am an unabashed enthusiast about geology and exploring for oil and gas. It is my hobby as well as my vocation. I have been told that my enthusiasm and excitement turns off many of my peers. I guess they can't understand how I can maintain this optimism when there are so many disappointments in this business. Perhaps they think I have not had my share of dry holes and uncommercial wells. If I were to dwell on them, I couldn't stand it either. Many of my greatest geological successes have been commercial failures, at least to date. Only those who have spent many years in exploration can comprehend how many ways Mother Nature has to frustrate those who aspire to be right in this business.

Fortunately, wealth has been fairly far down on my list of priorities, and when I fail to develop that $10,000 per day income, as has happened more than once, I almost heave a sigh of relief. I don't have to contend with that money problem. Even so, I work far harder for Uncle Sam now than I do for myself. Although my wife and I live relatively modestly, we have the ultimate freedom to be able to live our lives as we choose. I never go to "work." You can't get any richer than that in my book!

I am very concerned about the near future of our industry. This is probably the most crucial time we have experienced to date. Almost all oil companies today are worth more dead than alive. However, cold logic will tell you that that doesn't make sense in the long run—something will have to change.

The real problem is that finding oil and gas economically is becoming much more of a challenge as the size of the prospective trap diminishes. The geologic profession has not been meeting the challenge. If the companies are not finding enough oil, it is because collectively we professionals are not doing our job. We have to figure out how to do it better. I hope this is one small contribution to that effort.

DISASTER IN THE OIL PATCH

A combination of factors has combined to wreak great havoc on the oil industry and, even more importantly, on our nation as a whole. These were 1) the low hiring rate from 1960-1973; 2) the boom from 1973-1983 that decimated the ranks of the corporations; 3) the tremendous influx of newcomers who did not receive proper training in oil exploration in school, and who were not able to get satisfactory on-the-job training with companies; and 4) the distortion of ethical and economic values during the super-boom. Although certain aspects of this impending tragedy are already apparent, few associated with our industry recognize the long-term implications. I am writing this paper to ring the alarm bell. Possibly if we all recognize the problem, we might collectively be able to make the right moves to mitigate its effects.

The symptoms are all around us. The takeover of Cities Service and the potential dismemberment of Gulf are cases in point. These are not isolated cases, and we are being promised more of the same in the future. While many of us know the direct reasons for these happenings, I don't believe most of us fully understand the basic reason why many companies are failing to replace their reserves.

Because of my particular experience and background, I am in a special position to be able to analyze the problem. I have been in exploration since 1946, and in contrast to most my age, I have remained a prospect generator to this day. Early in my career I recognized that I was not cut out for management, although I do run my own independent oil producing company now with a modest number of employees. While that background is not so unusual, I have combined it with a career of teaching. Since 1967 when we incorporated the Permian Basin Graduate Center under the sponsorship of the West Texas Geological Society, I have devoted about one-quarter of my time to the education of the professionals in our industry. In addition, I did a stint of teaching petroleum geology at Rensselaer Polytechnic Institute in Troy, New York. In 1960, I left the school because there was no demand for our graduates, and I thought my time too valuable to spend turning out an unwanted product.

For those who are too young to remember, I point out that from 1960 until 1973, there were very few employment opportunities for geologists, petroleum engineers, and other professionals who make our industry tick. Therein lies part of the problem. Our troubles started from the fact that we have a tremendous paucity of professionals in the 35 to 45-year-old age bracket. Those are the ones who should be replacing the reserves of Cities Service, Gulf, and other companies at a profit today. Many of the major companies have been in the process of depleting their reserves for some time, and that is why they are vulnerable and will either disappear or be forced to operate on a much reduced basis. As a competitor, I should look at this reduced competition with joy, but I cannot do so because, as an American, I know that there are many ultra-expensive prospect areas that will be exploited only if the majors are intact and viable.

Although I have never worked for a major (when I worked for Richfield, it was still a small company), that does not mean that I don't have tremendous respect for what majors can do that I, or Boone Pickens, or Mike Halbouty can never do. If the United States is to prosper in the energy business, we need to continue to have a good mix of company sizes, and that mix is being jeopardized. While Boone Pickens may be doing a great thing for the stockholders of Gulf, he could be hurting all of the rest of us. However, I am not assessing blame, because his contention is 100% correct. If corporate employees are liquidating the assets of their company solely for their own benefit, the company should be dismembered. When you don't replace your reserves, that is what you are doing.

I mentioned that the problem started with the hiatus in employment from 1960 to 1973, but that alone would not have caused disaster. It would have been inconvenient, but the 13-year gap could have been made up without too much difficulty if it had not been followed closely by the super-heated boom of the last decade. This combination is the prescription for disaster about which I am talking. What happened is that with the advent of the boom, the primary source of the geologists, geophysicists, petroleum engineers, and managers of all types was from the majors and large companies. Corporate employees with exploratory or management potential were offered the types of deals they could not refuse. It was impossible for public corporations to offer inducements sufficient to allow them to retain many of the personnel that should have made up top management today. I was personally close to several in middle management of Exxon at that time, and I saw it happen. Before the dust settled, almost everyone I knew who could have ended up as a top executive was gone. The companies all got high graded. I am not trying to say that no competent explorationist remained on the staff of any corporation, but those who remained are in such a minority that they don't have enough fingers to plug all the holes in the dikes.

The truth is, the companies are probably doing as well as could be expected under the conditions. The Dallas Cowboys would not be winners today if the USFL teams had had the money to hire away all Dallas' top players and Tom Landry was forced to play those who couldn't get a good enough offer. During the boom, the independents had the money and interests to offer, so they got the top players.

As bad as all this sounds, even this alone would not have caused the magnitude of disaster we are witnessing, because the oil is not discovered by management, but by the troops; and this is where my experience teaching the young professionals has given me an insight not available to most. As the boom progressed and new entries came into the industry, I was able to observe the collapse of our profession. The boom destroyed prospectives, and it became difficult to find any who even comprehended that they were expected to find oil or gas at a profit in order to justify their salaries. In one class of 85 registrants in 1982 on how to generate and sell a prospect, a show of hands indicated that only six in the class had ever had a prospect drilled, and only one had a discovery to his credit. These were geologists with an average of about five years' experience. With that type of productivity, few companies could make a profit.

Success depends more on the experience and training of the first six months on the job than on any other factor. As a geologist, if you work for or with a very competent, dedicated geologist, you will follow in his footsteps; but if you work with incompetent people who are only looking for the next opportunity to job hop, you will turn out the same way. This is well documented by research done by my friend Professor Al Shapero, formerly at The University of Texas and now at Ohio State. In one company after another it became fruit-basket-turn-over, and there was no one left to train the young geologist how to be a productive and valued employee. District geologists would have as little as one year with the company and were essentially untrained themselves, but that was only part of the problem. The lack of company loyalty and ethics hurt even more. Neither professional skills nor professional attitudes were being absorbed. At that time I taught a structural geology seminar, and I had seventeen employees from one major in the class. After they found out in the seminar they were really expected to read and give critiques on the papers that dealt with the origin of the producing structures they were supposed to be finding, thirteen out of seventeen went on strike and refused to come back to class. They would have had to do the preparation on their own time! That was bad enough, but they went to their training director to complain of my demands, and he wrote me a nasty letter saying that they were tender young people who did not take kindly to criticism. With managers like that, it is no wonder that that company has recently concluded that there is not enough potential here to keep a Midland exploration office open.

As I point out, we have had many problems in the oil business because of the boom, but that is not the only place where problems arose. Education of geologists also deteriorated, and I am convinced that part of our disaster lies on the doorstep of the universities and colleges during that period. There are more geology majors than there are working professionals, and many schools are obviously not qualified to turn out geologists trained to find oil and gas economically. My observations also lead me to believe that there was an understandable reaction by many faculty members to the fact that their former students a year or two out of school were making salaries two or three times what they were being paid. Many of the professors who were smart left and got jobs themselves, but many others reacted the other way and concentrated their teaching on subjects that would be of little value to an oil explorationist. All too often, I have to compile the basic rules of contouring for my young working Graduate Center students, because they have never been properly taught the most basic skill required of an explorationist. Where do you start if you can't contour?!

The second educational problem lies with the fact that a geologist needs spatial visualization to be a competent oil finder. If you don't see in 3-D, you are almost worthless as an explorationist. When I started at U.C.L.A. in 1939, that was the first aptitude they checked for; and if you couldn't handle descriptive geometry as a freshman, you were eased out of the department. Because that screening is not being done today, we have a tremendous number of young professionals who are finding out ten years too late that they have chosen the wrong field. We have to get those geologists out of exploration! The way to do it is to sharply increase the hiring and firing rate. We are fortunate we have such a vast reservoir of talent available in the universities. We have the luxury of doing the screening for talent at this late date. Fortunately, the boom is over, and those who find employment now are starting to appreciate their jobs. How do the companies survive this interim period until they have been able to rebuild their staffs up to an acceptable level? We are not going to be able to proceed ahead as usual.

In closing, I want to say that this is a period of unequalled opportunity in exploration for those who can take advantage of it. In my thirty-seven years in the business, this is the best year ever. As a subsurface geologist, I maintain that we are just now getting enough well control to use subsurface geology properly, and there are going to be many companies that will grow. If you have a staff of one hundred geologists and each one makes a discovery of one million barrels every year, that is just as effective as drilling ninety-nine dry holes and discovering one hundred million barrels with one good wildcat, *and it is more realistic!* It is easier to find a fifty-million-barrel field by chance when you are looking for five million, than it is by restricting your efforts to fifty-million-barrel prospects.

The industry has to learn to change with the times. The real question is whether top management will face reality. Without that, the prognosis is grim for the companies that cannot replace their reserves with the resources they have available. It is not a question of size. It is a question of exploratory talent. The oil and gas are there to be found, but it will only be found profitably by those who have competent personnel at all levels. It is a people problem, not a natural resource problem.

(Published in WORLD OIL, March 1984, under the title "Industry Problems Rooted in People, not Lack of Reserves.")

CREATIVITY AND THE EXPLORATION PROCESS

Musings of an Oil Finder

I have always been able to generate drillable oil and gas prospects. This gift has provided me with the very best life possible. There is no other profession in this world that can provide more freedom, joy, and satisfaction. Some writers and artists, such as James Michener or Pablo Picasso, might dispute this. But, overall, the pleasure of solving complex scientific problems and testing your theories provides fulfillment for your intellect, a sense of adventure, and an excitement that is seldom equalled elsewhere in this modern world. It has allowed me to be relatively indifferent to monetary success, and yet that has come, too. In this business, if you can find oil or gas, you cannot help but end up earning more money than you need for yourself or your family. It even allows you to be generous with your time, money, and ideas.

I do not know how or why the Lord gave me such a gift, but I feel very strongly that as I reach the apogee of my career, I have the obligation to pass it on to future generations if I can.

Over the past many years, I have tried to analyze why some geologists are so productive and others so barren. It is obvious that it has very little to do with I.Q., and there certainly is no correlation with grades. There may possibly be an inverse relationship between grades and economic success. I do not mean to imply

that it does not require intelligence, because it does; but it requires a type of thought process that generally is not taught or tested for in colleges and universities. In fact, as the geologic profession becomes more and more complex and specialized, it seems that we are moving away from the type of thinking required to be an explorationist. This is related to the fact that the bright new young geologist has so much more to learn and assimilate now, that one frequently does not have time to sit and think creatively like one used to. Some would say that in school we concentrate on filling the left brain with knowledge, but the creative process is a right brain function.

From my experience in teaching the new geologists in various courses at the Permian Basin Graduate Center in Midland, Texas and elsewhere, I have observed a marked decline in scientific curiosity and creativity in our field in the past few years. This may be the result of boom conditions in the oil business, which have tended to attract more opportunistic and less scientific geology graduates, and also because more unqualified departments are turning out geologists and geophysicists. However, it is more than that, because I have not noticed much correlation between performance and school, at least within my classes.

What makes the difference between the oil finder and the paper shuffler? Wallace Pratt, the original and most famous geologist to work for Humble Oil and Refining Company (now Exxon), put it most succinctly many years ago: "The qualities which make the individual oil finder are faith, persistence, the venture spirit, and vision. If he is informed and trained in the art of oil finding, so much the better, unless his knowledge makes him overconservative or obscures from him the fact that much remains always unknown to him. If his knowledge blinds him to the unknown, his discoveries will be fewer….

"What we learn, instead of illuminating what we still have to learn, sometimes casts a mental shadow over it, rendering it less discernible and impelling us to ignore it. Those who are trained and experienced in the art of oil finding appear to be particularly susceptible to this blindness which comes with the acquisition of a little knowledge." (Pratt, 1952; pp. 2234 & 2236.)

What Pratt observed in 1952 may even be more true today. In 1952 we knew that there was much in geology that we didn't understand, but today many of us think that we really do have a pretty good handle on the science. The fact is that we still have a long way to go. This has really been driven home to me recently in my research in structural geology, as I will attempt to show later.

Creativity and an Exploration Philosophy

I first began to lecture on this field ten years ago, and over the years I have continued to study and improve my understanding of the subject. As I progress in my studies, I find that my productivity has risen markedly. In my earlier years as a consultant, I found prospects quite difficult to come by. As I have improved my process of thinking creatively, my prospect generation rate has increased significantly. Part of this is related to experience, but I attribute a lot of my success to the fact that I have refined my exploration philosophy through thinking about it and by attempting to teach it to young geologists and geophysicists. I may not have imparted much to them, but the application of these principles has helped me immensely.

What is creativity? The definition of create is "to bring into existence, to produce through imaginative skill." There are two kinds of creativity—individual and synectic, or group creativity. From what I have observed, most geological creativity is individual; but this may be influenced by the fact I have worked for myself and by myself for the past three decades. Within companies where group geologic creativity sessions have been tried, the pecking order disturbs the process, and all too frequently it is the boss who has to be overcome in order for the process to be effective.

What is imaginative skill? Is it something we are born with, or is it something that can be learned? Years ago, listening to the radio on my way home from skiing in New Mexico, I heard the *University Explorer* describe an experiment done by the U.C.L.A. Psychology Department. This project really impressed me because it clearly demonstrated that creativity can be enhanced. And, of all things, it can be done by programmed learning. A series of experiments was conducted on fifth and sixth graders. All of the students were tested to measure their level of creativity and then separated into two groups with comparable scores. The first group had to solve a series of mysteries in a programmed manner. After a few weeks the two groups were tested again, and the psychologists found that there was a measurable improvement in creativity in those who had undergone the program of creativity enhancement. This type of mystery solution is particularly applicable to exploration geology, and I have used it frequently at the Graduate Center. All our problems are mysteries, and we never have enough control to solve them by using deductive reasoning. What we really do is induce a solution and then drill a well to test our hypothesis. If we are wrong, and wise, we then modify our solution and drill another wildcat. That should continue until we discover the field.

One of the reasons why I have worked for myself rather than for a company is that there the boss may get tired after a try or two, and that frequently prevents the explorationist from discovering the oil or gas field that he knows is there. As a consultant, I am free to get other investors in order to continue my exploratory effort. In looking back over my career, I now realize that my greatest single failure has been

that frequently I quit too soon and allowed someone else to make "my" discovery. If you can establish that there has to be "closure" in an area, there should always be a hydrocarbon accumulation. That has been driven home to me many times, and my pessimism has been my own worst enemy.

As subsurface explorationists, we are always keying off someone's dry hole. It should not make the prospect any less attractive because it happens to be *your own dry hole*. As a matter of fact, that is a plus. When you drill a wildcat, you are the first one who can take advantage of the new information. It may well mean that you will need to lease additional acreage, but that happens at the optimum time. Your dry hole will have cooled off the area. The landowners and leaseholders will be discouraged and it will be easier to deal with them.

It reminds me of an oil field story related to me by Charles Priddy, now chairman of the board of his own oil company in Midland, Texas. In his earlier years, he was a landman for the famous Texas wildcatter, Clarence Norsworthy. During the boom years of the 1950's in West Texas, competition was tough, leases were expensive and hard to come by, and the Permian Basin was in its early stage of development. There was a paucity of subsurface control. Norsworthy's geologist would locate a prospective area and Norsworthy would systematically drill around the fringes of the prospect (using other people's money). Finally, when his dry holes were many and his control was adequate, Norsworthy would be able to get the desired leases on reasonable terms and make the discovery. Although I was not aware of his exploration philosophy at the time, I competed with him on one prospect in Fisher County, Texas. He was able to outbid me because of his superior subsurface knowledge. He drilled the discovery well of the Ocho Juan Canyon and Strawn reef field in 1955. The name Ocho Juan came from the fact that it was the eighth well drilled there by his geologist, Dr. John Harrington. When the discovery was made, Norsworthy owned most of the working interest. His investors, too, had lost faith, and the profit accrued to him alone.

That discovery taught me another lesson that often is learned the hard way. When I lost the opportunity to drill the prospect, I resorted to the next best thing. I sent a landman out to buy the landowner's royalty prior to discovery. I mapped the reef crest immediately northeast of the wildcat, and I told the landman to restrict his purchase to the northeast section or square mile. The landowner had minerals under that section and the one-half section to the west. He insisted on selling under his spread rather than under the one tract. We were equally adamant, and he finally gave in and sold us what we wanted. The discovery fit my contour map perfectly, and I felt pretty good about my purchase. However, the reef took a sharp bend, and the crest of the reef was under the half section we turned down, and our tract was dry. Since that time, I always try to be under a spread of acreage, not just under my mapped crest of the prospect. In this business, it is easy to be "too smart" for your own good.

Getting back to the application of the findings of the U.C.L.A.'s creativity enhancement project—the solution of mysteries. In one example I recall, the students were shown a picture of a snow-covered clearing in the woods. In the clearing there were a number of animal tracks, some blood and feathers, and a little fur. They were first asked to describe in writing what they thought had happened. Then, in an accompanying story, a young girl and boy entered the same clearing and discussed what they thought had occurred. They then sought out their uncle, a detective, who came to the clearing and pointed out that they had overlooked this fact and that. The students were asked to rewrite what had happened in the clearing after having the benefit of the additional observations. This process was repeated several times, each time allowing the students to come closer to the answer. It was after five of these mysteries that the students were tested again and found to have enhanced their creative thinking ability.

What was being taught by this programmed learning process? They were learning the ability to observe and ask questions.

Creativity in geology is the ability to ask the right question. Testing has revealed that very young children are very creative, and that talent progressively lessens as they progress through school. Our school systems are teaching them the answers rather than the questions. As the field of knowledge expands, there are more and more answers to be learned, with less time for the questions. What I am trying to do is reverse that process so as to develop the traits required for success in exploration geology.

I could pose twenty questions to a group of geologists and give them a week to come up with the answers using any resource they chose. They would all come back with the answers. The real problem is, what is the question? With geologists working in the Permian Basin I often ask, "What question at present unanswered, would, if answered, help you the most in exploring for oil and gas in the Permian Basin?" Most look at me blankly. But the fact is, if geologists have not uncovered a number of unsolved mysteries as a result of their exploratory efforts, they are not thinking creatively. These geological mysteries are all around us, but you have to recognize them to solve them. If someone else has to point them out to you it does not stimulate your creative urge.

Let me cite an example I still remember vividly after many years. My master's thesis area was the Seminole Quadrangle, which contains Malibu Lake in the Santa Monica Mountains of southern California. I had mapped a channel deposit of coarse sand and conglomerate that looked like a stream channel; but it

was lying in the Miocene Modelo (Monterey) Shale, which even then I knew had been deposited in deep water. Dr. M. N. Bramlette, one of my thesis advisors at U.C.L.A., spent a day in the field with me in 1947. When he saw this unusual association of facies, he asked me for my explanation. I am still mortified that, although I had mapped it properly, I had not asked myself how the sand was deposited. I answered lamely that I guessed sea level rose and fell. Even before I got through with that explanation, I knew it couldn't be correct. I finally admitted I didn't know. We studied the outcrop below the channel and around the edge and onto the top, noting that there was absolutely no facies change in the shale. Yet, this sand body was intercalated, indicating a dramatic shift in depositional environment. Most young geologists today would not have any problem in explaining the outcrop. We were observing a turbidity current channel in a submarine fan. However in 1947, we didn't know anything about marine turbidity currents. I won't say I didn't know anything about turbidity currents, because I had read about the Lake Mead occurrence in Bramlette's sedimentology seminar; but we had read and believed that similar currents did not occur in the marine environment. Neither Dr. Bramlette nor I came up with the correct answer because we had previously been brainwashed. What we had learned had cast a mental shadow over it, as Pratt pointed out.

However, the point I am trying to make is that *I* had not been curious enough to try to solve the problem. Even though my professor asked the right question, it was not the same as my asking it. His question did not stimulate me and cause me to seek the answer.

The irony is that had I asked it properly, I could have solved it several years ahead of the crowd. I had the data required years before marine turbidity currents were recognized. I first heard about them from Dr. Maurice Ewing in 1950. In 1947, I was working as a subsurface geologist for Richfield Oil Corporation under Manley Natland, who had learned how to determine depositional depth from foraminiferal assemblages. I had been systematically mapping what I knew to be the deep water sands in Long Beach and other fields in the middle of the Los Angeles Basin. I remember concluding that "currents" were depositing the sands and discussed this fact with Natland, but I still did not ask the right question, "What was the motive force of these currents?" It was not until I heard a talk by Dr. Lewis Weeks (AAPG Distinguished Lecturer) about 1951, that all the data fell into place. What stimulated my answer was not Dr. Weeks' (then head of Esso Production Research) correct solution, but his incorrect statement that in the Los Angeles Basin, the Repetto (Pliocene) sands were deposited by shallow water processes, and I knew that was not true. Others' errors may stimulate creative thinking, as well as others' correct analyses.

What I think I have done in the intervening years is

to learn how to ask the right questions, and that has made a significant difference in my creative capability. I am convinced that in most geologists' lifetimes, they have opportunities to make major contributions to our science. When they fail to do it, it is because they did not ask the right questions at the right time. In my career, I have had at least five such opportunities (turbidity currents, hydrodynamics, continental collision, origin of secondary carbonate porosity, and the origin of "upthrust" structures). Only in the latter case have I pursued the problem to fruition as I should have, because I finally asked the right questions. Imagine the geological reputation I could have earned if I had really been creative! Of even more importance from an explorationist's point of view is the question, "How much additional oil could I have found if I had made a scientific breakthrough first?" Any one of these, answered correctly in a timely manner, could have made a tremendous difference in my exploration potential.

Skills Required

The basic skill required for the explorationist is the contouring skill. Unfortunately, this art is being ignored more and more as geological and geophysical knowledge expands. Few geological faculty members today have the slightest idea of how important contouring really is, because this craft is so mundane and "simple," and there are so many other "skills" that need to be taught. However, after thirty-seven years of searching for oil and gas, I can confidently state that a geologist who can contour creatively will be a prospect generator; and one who can't, will come up with naught. I am not trying to downplay the contributions of the other disciplines, but contouring is the integrator that ties a prospect together.

The reason that contouring is so important is that it is needed to display three dimensions upon a two dimensional surface. Explorationists must see in three dimensions (four really, because *time* of entrapment is so important). Exploration geology requires spatial visualization, an aptitude that is present in a small percentage of the populace according to tests. If the geology faculties are doing their job, those who are going to attempt to earn a living as explorationists will have survived a very tough 3D screening process. Unfortunately, it is not so essential for faculty members to have that propensity, and I find many of them reluctant to put demands upon the students that they have not had put upon themselves. I am not talking about what it takes to make a good geologist. I am just talking about what it takes to be a good explorationist. The vast majority of geology majors end up working in the oil industry. Geology faculties who don't recognize that fact, and fail to do the visual screening, are not fulfilling their responsibilities to their students.

I have a great deal of contact with the young professionals in their first years in the industry. Very commonly, they start out doing well-site work, etc., where the lack of these visual skills is not noticed. But, after about five years, they start to earn their pay by generating prospects. By this time, they are commonly married, have a couple of kids, and a large mortgage on a house. When they lack the ability to see in three dimensions, their exploratory efforts will be less than acceptable. It is one of the great tragedies of our profession that these young people, who have been led to believe that they are fully qualified, find out ten years too late that they have chosen the wrong field. That aptitude, or lack thereof, could have been determined in their freshman year. This problem seems to be increasing in magnitude and can no longer be swept under the rug. This lack of spatial ability is already sapping the economic strength of the oil and gas industry.

In 1939 when I started at U.C.L.A., there was not so much geology to be learned, and I was able to take many courses requiring contouring skills. We did our field mapping on U.S.G.S. topographic sheets, not air photos. Freshmen were required to take engineering drawing and descriptive geometry, and the latter demands spatial visualization. That course in particular should definitely be required today as the first screening course. In addition to the usual exposure to contour maps in physical geology, structural geology, and geomorphology, I had a year of surveying in the engineering department that included one whole semester mapping badlands topography on campus using a plane table. I even had time to take a cartography course in the geography department. By the time I received my A.B., contouring was old hat. That is a far cry from what I have encountered in recent years. Most geologists, regardless of degree or school, now have to be taught the basic rules of contouring before I can start to teach them how to find a prospect. It is true that I did not have optical mineralogy, petrography, carbonate sedimentology, environmental geology, volcanology, stratigraphy, oceanography, etc.—all courses now considered essential to make a well-rounded geologist at the A.B. level. I had most of those courses in graduate school, but collectively they have not contributed as much to my prospect generating ability as my earlier learned contouring skills. A broad geological knowledge is wonderful if you have the basic skill to hang it on. But without that skill, the breadth and depth of knowledge is of little value. I cannot overemphasize the importance of spatial visualization. It is the essence of the business.

Why is it so important? The chief problem we have in exploration geology is that we never have enough control. When I first came to the Permian Basin, the subsurface control was as little as one well per township (36 square miles); and yet, it was necessary to postulate what went on between the control points. Thirty-three years later, after many boom drilling years, we still are working with a well density of considerably less than one well per square mile. Even at one well per square mile, considering that a bore hole is less than one foot in diameter, the data covers less than one out of 27,878,400 square feet. It requires a great deal of imagination to be able to contour that uncontrolled part of the map. By contouring, in exploration, I don't mean just fitting the data points. Anyone can do that with adequate control. But you *never* have enough well data, and the contours must originate from the surface you *visualize in your mind*. When the subsurface data is not sufficient, a computer leaves that part of the contour map blank. An explorationist cannot do that, because the prospects often lie in the uncontrolled area. By forcing yourself to make the best interpretation possible where there is little or no control, you find the prospects that others have missed. The display tool is the contour map. What the contour map is illustrating is the sum total of all your geologic knowledge in that particular area. That is where your broad geological background is employed such as the structural, stratigraphic, sedimentological, geomorphic, and petrographic specialties. As I said previously, I am not downplaying the importance of these auxiliary tools; but the display is the contour, or isopachous map, or both.

It is extremely difficult to work with a small amount of data, and many geologists balk at using their imagination beyond a certain point. This is one reason why many academic-type geologists make poor explorationists, and also why many company men require so much seismic data before they can make a drilling recommendation. That is all fine to a degree, but it must be remembered that additional data does not come free. It is not at all uncommon to see a major company require so many seismic lines that they end up spending more on geophysics than they would to drill the well. No matter how much geophysics improves, it will never take the place of the wildcat. There will always be areas where geophysics is not usable because of velocity variations and other factors. Hydrocarbon discoveries there will be made by those who have the ability to visualize the subsurface and the courage to drill their convictions in the face of inadequate data.

The scientific method is to observe, infer, predict, and test. Exploration geology is one of the few sciences where we get to carry the scientific method to completion. That is why exploration geology is so exciting. We can use inductive reasoning, develop a hypothesis, test it, get a dry hole, modify it, and drill again. Few geologists have drilled as many dry holes in their lifetime as I have, but I have found more oil and gas fields than most, too. The issue is not the dry hole; rather, it is locating the hydrocarbon accumulation at a profit. Any other perspective is an aberration. A well is successful if it leads to additional drilling and, finally, the discovery. I often point out to my students that we are not in the scientific game or the geol-

ogy game. We are in the money game, and the tools are geologic knowledge, imagination, and courage. Some of the deepest wells in Texas have been drilled at my instigation, and they cost many millions of dollars. That money will be lost only if I don't have the persistence to stay with the play until commercial success is assured. Sometimes that takes many years. I have one area in New Mexico where I drilled the first wildcat in 1953, and I am still after the same oil I sought over thirty years ago. It should be tested successfully this year if my new interpretation is right.

Exploration Economics

Ted Bear, the new president of the AAPG, and I taught a course at the Permian Basin Graduate Center in 1982 on Generation and Selling of a Prospect. We had 85 registrants, most of whom were young; but there were a few with gray hairs, too. I would guess the average experience level was five years. At the start of class, I asked all those who had ever worked up a prospect and had it drilled to raise their hands. I only got six replies. Then I asked how many had made a discovery, and I got one hand. He sheepishly admitted that it was serendipity; his discovery was not in the zone he sought. I mention this to make everyone appreciate the cost of a prospect. Even if those six prospects were worked up in the past year, a quick calculation would indicate that those prospects cost their companies an average of $400,000 to $500,000 per prospect in salaries alone. The year 1982 was the height of the super boom, and it should be obvious to management that if those figures persist for long, the oil companies are doomed to bankruptcy. You cannot make money if you cannot generate prospects economically. If that happens, there will soon be few geologic positions available for petroleum geology majors. During the boom everyone forgot that the only reason to employ geologists was to make a profit from them. One year later, we began to harvest the fruits of that insanity, and many smaller companies are already folding.

The very long-term prospects of the oil and gas industry are grim indeed. Obviously we are going to reach the day when no company can continue seeking an ever depleting natural resource. Whether that happens in the near future or whether it occurs over 100 years from now depends on just one factor. That factor is our collective ability to make money exploring.

I want to interject that I am an eternal optimist, and I don't think we are near the end of the line yet. I have been able to increase my net worth twenty fold in the past dozen years, and I'll be disappointed if I don't double it again this year. My finding cost still is much less than my selling cost. When that reverses, I will have to shut down and hoard my remaining reserves. For an individual in his sixties, that is not disastrous; but for the U.S., it could mean the demise of the largest industry in the country.

I fear that the beginning of the end started at least ten years ago for many of the large oil corporations—almost without exception, their reserves are dwindling. This is partly political (loss of oil concessions, additional taxes, etc.) and partly economic. This debacle has been obscured by the fact that the selling price of oil has increased ten fold in the last decade, which makes the drop in reserves less obvious. Seventeen of the top twenty have failed to replace their reserves in the past five years (Nulty, 1984, p. 163). Even the mighty Gulf Oil Company, the fifth largest, is fighting for survival and may not make it because their exploratory efforts have come up short. The liquidated value of Gulf is greater than its survival value at this time. If they continue to explore for oil as they have in the recent past, they will use up their net worth unsuccessfully trying to replace their reserves. Some stockholders are now saying, "Give us the income now—stop exploring!" They are trying to prevent the employees from liquidating the company for their own benefit, and they have a very good point! If this challenge is successful, the stockholders of other companies will probably follow suit.

How did all this come to pass? It is the result of numerous factors, many of which were beyond the control of management. I am not trying to assess blame because, by and large, most major companies are run by good money managers. However, the early companies were run by oil finders; and now they have been generally shoved aside by lawyers, accountants, and financial people, who know "business" better. All majors are fighting the problem of size. Even a small operator finds that as he grows, it becomes a greater job just to replace the oil produced that year. However, this is offset by the fact that the more capital the explorationist has to work with, the easier it is to succeed in this business. So size alone is not the problem.

The second problem is common to all natural resources. When there is plenty, there is no reason to increase the price even if the cost of finding it rises. Finally, when a shortage develops, the price will rise disproportionately. Everyone will start exploring like crazy. After a few years of boom we will find too much, and then the price will drop disproportionately. Then everyone will start losing money. This is a cyclical business. I entered the oil business in 1946 right after World War II. These were years of shortages of manpower and exploration. No one could imagine at that time that we would ever have enough oil in the future, but by 1960 there was a great oversupply. From 1960 through 1973, eighty percent of those in the oil and gas business disappeared. For thirteen years, practically no geologists or petroleum engineers were able to find employment. All companies developed an age-skewed professional staff. With the oil embargo in 1973, we went from bust to boom overnight. We couldn't hire enough geologists, geophysicists, and petroleum engineers. The profession-

als with more than ten years' experience were in extreme demand, and it got so they could name their own terms. The shortage of explorationists in the prime age group was magnified. Exploration is a young man's game, by and large.

Because the large corporations could not offer geologists royalty or other interests in their discoveries, the professional staffs of most companies were decimated. There have never been enough true oil finders around at any time; but then, top geologists who had been making as little as $35,000 per year, were raking in as much as a million their first year out of Amoco or Exxon, etc. Morale was destroyed within the corporations. Not every oil finder quit, but there were not enough left to do the exploration required and to train the newly employed staff, too. On-the-job training, which had been the majors' strong point, became weak. Valuable young oil finders soon quit so they could make better use of their talents and learn more elsewhere, too. I know, because we were attempting to train this mass of new professionals at the Permian Basin Graduate Center. The registrants ended up using these courses to make contacts for better paying jobs with smaller companies and individuals. Major companies became afraid to enroll their staff in continuing education classes. Geologists lost sight of the fact that in order to earn their keep, they needed to be making a profit for their employers, too.

Not only have the companies been damaged, but many of the individuals who got caught up in this maelstrom were hurt, too. There is some question in my mind whether the damage to the industry will be permanent. Young professionals were wined and dined during recruiting and paid so much more than they were worth that they lost their perspective. They went to work for companies where their bosses had very little practical experience, and where there was little or no company loyalty or dedication to the job at any level of management. By opportune job hopping, many geologists made it all the way into top management without ever finding any oil. The way back to reality is going to be a long, hard one.

There are more geology majors now in school than there are working petroleum geologists. Because of the lack of the work ethic that developed during the boom, companies may have to clean house and start afresh with a whole new staff. But the fundamental problem still remains that there are very few professionals in their late thirties and forties available to form the nucleus of a strong exploratory staff.

Hope for the Future

Professor Al Shapero of Ohio State University, (personal communication) has found that the first six months on the job make or break most employees. If the young employee is fortunate enough to work first with or for a competent, dedicated, professional geologist who believes that he owes his company his allegiance, and who knows that he has to find oil commercially to prosper, he or she will be imbued with those qualities, too. However, the opposite is also true, and we have had a decade now where this type of professional role model has not been available to many geologists fresh out of school. I can vouch for Shapero's findings from my own experience. While still in graduate school at U.C.L.A., I had a chance to work under Ted Bear. It was the few months of being directed by Ted that made the difference all through my career, because Ted Bear is one of the foremost exploration geologists in the business. When I taught the prospect generation course with Ted, I found that our exploration philosophies were almost identical, even though he and I have gone our separate ways in different oil producing areas for over thirty years. Much of the exploration philosophy I cover in this paper came from Ted, directly or indirectly. Lest you think that this is a sampling of one, let me point out that shortly after I worked under Ted, he served as a consultant for Humble Oil and Refining Co. (now Exxon), and he proceeded to train a group of young geologists with that company. All of those who had this training have had outstanding professional records. Unfortunately for Exxon, very few of those trained by Ted are still with the company today.

How to Find a Prospect

Each prospect is different, but there still is an underlying framework that allows prospects to be located.

First, I need to define a prospect. In the industry this word has various meanings. But for me, a prospect is a geological lead on which you own an economic interest such as leases, royalty, or minerals. I do not consider a geological idea a prospect no matter how good the idea. If you come up with a strong lead, but the acreage is already leased to someone else, what good is that to you? A prospect is something that can make you or your company a profit.

Many geologists claim to have worked up numerous prospects but say their companies would not buy their ideas. If that is the case, they haven't worked up any prospects. This may have no bearing on their geological abilities, but if they can't sell their prospects to management, it does reflect on their ability to make a profit for their employer.

A poor claim to fame is for a geologist to brag that he found this field or that extension before it was drilled, but his company wouldn't buy it. My reaction is that the geologist really should be fired then because he failed to sell the prospect, and that cost his company a great deal of money. Being a good geologist is not nearly enough. You have to be able to pre-

sent it so that the boss can't say no. Remember, I am not writing about good geologists, but about good explorationists. When ideas don't sell, it is usually because the geologist trying to sell them has not done his or her homework and most probably doesn't fully believe in the prospect itself. It is always easy to claim how good a prospect was after someone else has discovered the oil or gas.

An even more futile exercise is to claim credit for keeping one's company out of a prospective area that turned out dry. When you consider that about 80 percent of the wildcats are dry, and four out of five discoveries are uncommercial, you can build a beautiful statistical record for accuracy by predicting that every well will be dry—but that does not make your company any money!

A good explorationist has to be a good salesman. The poor geologist, who is good at selling, will end up finding more oil than the good geologist, who can't sell. You never find oil without drilling. If you drill, you may end up with a discovery, even if your geology is wrong. In the Permian Basin with its multiple pays, probably more than half of the discoveries are accidental. I'll take one of them every time. I have often commented that if we only got to keep those that were geologically correct, all geologists would end up as failures.

It is important to be an unconventional thinker. Even if you are a good sound geologist and you approach the problem the same way as the thousands of other geologists in the basin, you will almost always find your prospects under lease. It is extremely difficult to be the first one there. I am a very unconventional thinker. Not surprisingly, an inordinate number of my prospects seem to have leases available on them. As I grow older and more unconventional, I find a greater percentage of prospects available. It matters not that everyone is out of step but me, *as long as I am making a profit*. Remember that when a prospect sells it is because the economic analysis indicates it is a risk worth taking. A low risk prospect on which you can only lease 160 acres may not sell, but a high risk prospect on which you can obtain 16,000 acres may turn easily. It *is not the degree of risk, but the potential return on risk that causes them to sell. This is one of the few professions where you can be wrong most of the time and still enjoy a good reputation.*

Any thoroughly studied area will yield geological leads. If that is the case, then the first priority should be to work in areas where the potential pay zones are very profitable. It is just as hard to locate a potential oil accumulation where the wells will make 20,000 barrels as it is where they will make 200,000 barrels. Before you start to work, you need to make some reserve calculations. You also need to examine the lease situation in your area of study. Fortunately, land maps are available in many areas. It is best to concentrate on areas where the leases are beginning to expire or will expire shortly. Fads are very common in oil exploration. An area can sit dormant for a long time, and then when someone makes a discovery the whole county will be leased up on new three or five-year leases. There is little opportunity to make a deal until the leases in that area get within a year or so of the expiration date. Then you have the luxury of either getting farmins and dry hole support from the various leaseowners, or you can wait and let the leases expire. Different areas work in different ways, but this method works well in the Permian Basin.

After you have ascertained that the reservoir quality is satisfactory and the leasehold situation is okay, you should be prepared to expend a major effort that may require a month or more of your time. The first thing I do is to construct a series of cross sections through the area. With the advent of copiers, these cross sections are easy to construct. On the first go-round, I make jammed correlation sections hung on a sea level datum. (By jammed, I mean to jam the logs as close together as possible). After constructing the cross section, I draw as many correlation lines as I possibly can. If there is any appreciable vertical relief, I will then reconstruct the cross sections using some datum other than present sea level in an attempt to decipher the stratigraphic relationships in past geologic time. One reason for making these cross sections is to check the accuracy of the correlations, and to begin to unravel the geological history. It is amazing how much geologic history you can reconstruct from a few cross sections. You can see the transgressions and regressions, the exposure cycles that develop the secondary porosity in carbonates, etc. Of extreme importance is the building of your conceptual depositional model so that you can be sure of the depositional dip and strike in times past. This is particularly important because look-alikes on the electric logs may mean similar depositional environments, not time equivalency. Correlation errors lead to poor maps and dry holes.

Until I make these cross sections, I can never be sure of my correlations, and if you are not sure of your correlations, your contour map is worthless. Even the major stratigraphic correlations within the Permian Basin are still in dispute, and when you go from major to minor correlations, the disagreements become even more acute. There is absolutely no substitute for the cross sections. Time and again I have correlated wells by sliding the logs alongside each other and then found that what I thought to be correct was in error after I made the cross sections. When I have located an area of interest, I construct a new set of cross sections that include every well in the area of the potential prospect. Cross sections also reveal errors in well elevations, at least within the Permian Basin where the shallow zones are relatively flat. If you find the vertical relief on your shallow marker is the same as on your deep marker for your prospect, watch out! Your whole mapping effort may be to no avail because of a surveyor's error, or even a figure transposition.

An additional point to look for on cross sections is the presence or absence of topographic relief on unconformity surfaces. If you are attempting to contour structure, and your contoured surface has been modified by stream erosion, your contours will be that of a composite feature. This can be handled by contouring the two surfaces independently. You can do this by removing the post-erosional tilting or folding and then make a paleotopographic map. After you construct that map, then you will be able to draw your structural contours, taking the topographic relief into consideration. Few geologists are capable of contouring composite surfaces adequately.

Cross sections will help define the structural complexities. Reverse faulting causes repeated section, and normal faults cut out section. Within the Permian Basin, drape over the basement faults is the most common structural feature; and steep dip will be evident as extended or stretched E-log sections. When the dip is greater than 15 degrees, the stratigraphic lengthening is measurable. The normal or isopach thickness divided by the stretched or isochore thickness is the cosine of the angle of dip, and every log should be checked for any abnormalities that can be attributed to steep dip or faulting. You cannot possibly contour an area properly without taking this detail into consideration.

Only after you have done all this detailed study are you ready to put your subsea points on the map and contour the area. The contour map is the composite reflection of your studies, and this is where you separate the oil finders from the also-rans. Contouring is an extremely demanding process that may take weeks and weeks to finish. No matter how much time you spend on a map, you can always improve it and come up with new ideas and prospects. If your work has been done properly, by this time you should have worked up four to six leads in your area of study. If you are lucky, you may even find that you can obtain leases on some of them. If you do not come up with several leads as a result of each detailed study, you will have a difficult time coming up with ten drillable prospects per year. That is a reasonable expectancy for a real pro. If you repeatedly make this kind of study and do not find prospects, you are doing something wrong. Very rarely do I study any area, regardless of the state or basin, without finding many places that need to be tested.

I remember that once when I was on the Executive Committee of the West Texas Geological Society, one of the officers brought up the issue that a major company geologist studied an area in the north Midland Basin for a whole year without coming up with a single prospect, and the company fired him. Some thought he had been unfairly treated, but I disagreed vociferously. As geologists, we have to recognize that our careers and our profession will only be viable as long as collectively we can produce commercial results. Every nonperformer in our profession weakens our collective record. If our basin develops the reputation of being uneconomic, each of our careers is endangered even though we personally may be able to find oil economically. In the late 1960's, Mobil Oil Corporation found that it was costing them $6.40 to find a barrel of oil in the Permian Basin when it was selling for $3.00. They promptly closed their office here, and now from Houston they are only marginally active in the Basin. One could point out now that they would have been glad to have that $6.40 oil when the price rose above $41 per barrel. The fact is, they have never moved back into active exploration in this Basin. There will always be a natural attrition rate as the less efficient companies reach the economic limit in any basin. Amoco, Chevron, Shell, Phillips, and now Texaco have also moved or are moving out and sharply reducing their efforts. Those of us who remain can argue that it leaves more for the rest of us, but it makes prospects more difficult to sell when others think a basin has reached its economic limit. There are too many who follow the leaders.

To keep the Permian Basin exploration viable is the main reason I helped start the Permian Basin Graduate Center and why I have freely donated my time trying to raise the exploratory capabilities of my peers and competitors. Many of my friends call me crazy for helping the competition, but I am crazy like a fox! There are many thousands of prospects yet to be developed in the basin. Exploration will end in the basin not when they are all found, but when we collectively conclude that we can no longer find oil and gas economically. I celebrate everyone's discovery and you should, too!

How to Contour

I have an obsession about contouring because I know from experience that my success has come from doing it well and differently from everyone else. Yet, technically I don't do anything but follow the rules of contouring, and these are included in the appendix.

A favorite oil-finding technique is to go to the literature and restudy fields that have obviously been miscontoured. Once these faux pas have been committed to print, their publication often causes a cessation of thinking about those fields. There it is; that is gospel! Yet, I have found very few instances where the contour maps and cross sections could not be improved upon markedly. In very many cases, the maps as drawn are impossible! I always wonder what the area would look like if it were contoured properly and with careful attention to detail. Mapping exercises of this type often produce good leads or prospects.

If the map is wrong, it may be wrong for many reasons; but there are a number of commonplace errors that I look for first. The most frequent error is that the contours do not fit the points. I was recently asked to evaluate a new manual on contouring written by a local explorationist, and even his contour maps frequently failed to observe this cardinal rule. My explanation is that many think we are working in an approximate science, and therefore it is all right to contour approximately. The opposite is true. We never have enough subsurface data, and we need to milk the last drop of information out of each data point. First and foremost the points must fit your contour map *exactly!* Many will argue that the rate of dip is unknown and variable, and we can never know what a surface looks like. That has nothing to do with the discussion. After you have drawn your contours, you then have an *exact* interpretation of what you think the surface looks like, and you can calculate very accurately where each data point should fall on your map. If the mapped and actual points don't agree, something is wrong!

The smaller the map and the larger the contour interval, the more likely that errors will be present. To check this hypothesis, enlarge any published contour map several-fold and drop the contour interval from 100' to 25' or less. The map will seldom look the same, and you often find it radically different. The advantage of the change in scale and contour interval is that even though it will not eliminate the misfits, the errors will be smaller in magnitude. You can't be off more than a few feet when you use 10' contour spacing, but many published maps are off 50' or more with 100' contour spacing. Theoretically, if you could eliminate all your errors, contouring on a smaller interval would not help. But that is the only way I can minimize my errors even when I try to be careful.

I recently restudied an area that I had contoured about 14 years ago. In the interim period one hundred and thirty-four wells had been drilled to the contour horizon, and it was an opportunity to see how my contouring checked out. I had previously mapped several tilted fault block structures, and my gross structure was reasonably correct. However, there was a lot of detail on the structures that my map didn't explain. My old base map had a 5000'-1" scale, and I had used 100' contour intervals. I enlarged the map and dropped to a 25' contour interval. All of a sudden, the details fell into place. Although the tilted fault block interpretation was correct, I was mapping on an unconformity surface. Superimposed on these structures was a low relief topographic surface. The oil was trapped on the preexisting topographic ridges. The structure was of secondary importance as a trapping mechanism. Wells drilled on the crest of the structures but in topographic valleys were dry because the porosity zone had been removed by erosion. A true scale cross section showed very modest relief on the unconformity (Fig. 1). I had missed it completely using the 100' contours, as had the author who pub-

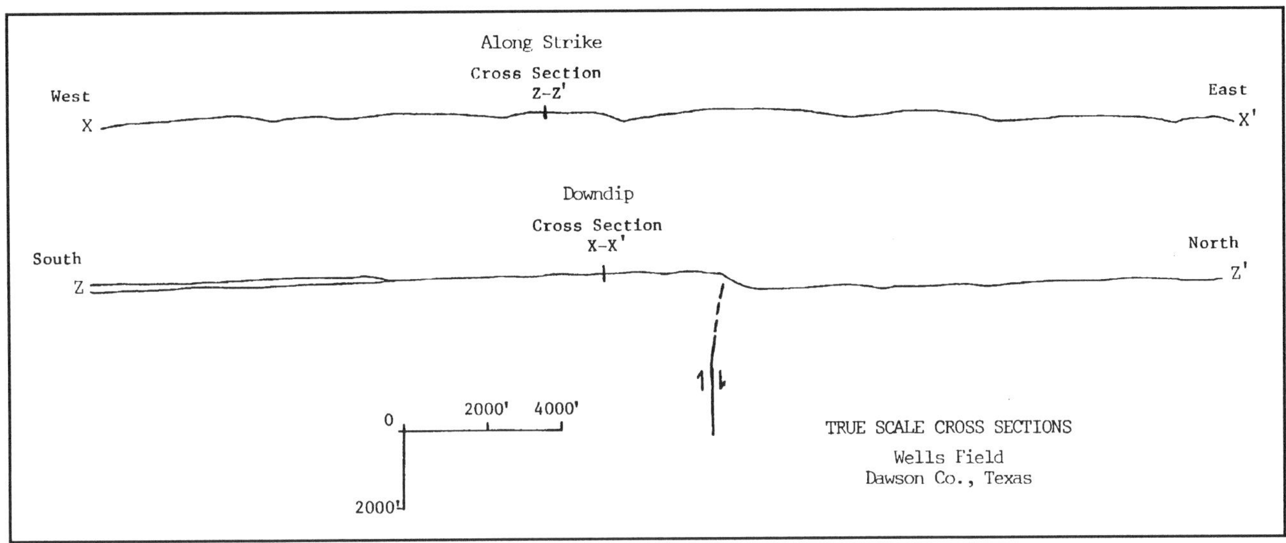

Fig. 1. The east-west cross section illustrates the low magnitude of relief on the Siluro-Devonian unconformity, but the production is limited to the ridges. On 100′ contour interval, this controlling factor was missed, and I believe the tilted fault block structure (Z-Z′) is only partially developed.

lished the map of the field. The pay was at a depth of 12,000′. Deep prospects usually require seismic confirmation but the seismic record here was awful. Only two out of the one hundred thirty-four wildcats drilled since 1970 made commercial discoveries from that zone. I assume that almost all operators used seismic to spot their wells. These features lie under a thick Pennsylvanian reef section, and the velocity variations within the reef distort the seismic. Because of the poor seismic statistics, I was able to sell the deal on the basis of good subsurface control. Why use seismic if it doesn't work?! This prospect was uncovered merely because I changed the scale and contour interval. Considering that our objective has perhaps a $100,000,000 potential, it was well worth the time invested to contour it carefully! So often it is the little changes in technique that separate the oil finder from the failure.

A common error made by geologists is to think that the contour map is one of the first things you construct. You cannot draw contours until you know what kind of a surface you are contouring. Topographic contours do not look like structural contours. This is why the computer will never replace hand contouring, at least at the prospect level. Contouring has to be done with some bias, and the bias has to be based on knowledge. There is no such thing as unbiased contouring because to use no bias is the bias of no bias. You may choose first to make an unbiased map, but it should be an interim effort, or you can use a computer to get an overview of the surface. The computer is good at many things in exploration such as trend surface analysis, etc.; but the fact that it contours the same, regardless of the surface, indicates its liabilities. I have

been told that some new software can induce a bias but I have yet to see the results. Artificial intelligence can never replace human intelligence.

If contouring cannot be done correctly without some bias, then where does the bias come from? The ultimate objective in any area is to work out the geological history from start to finish as I pointed out earlier in this paper. If you are dealing with structural geology, you need to know what the producing structures look like, how they were formed, and when. Structures formed after the oil has migrated will often be barren. Reading what others say about how the structures were formed or what they look like may be a liability rather than an asset. If your informant is wrong, you will be wrong, too, and besides you will be following a beaten path. In exploration you need to develop the attitude that *everything is wrong* unless proven right by your own work. Only after you have some knowledge can you afford to examine the literature. I believe that those who really know the answers don't publish them because they are too valuable! For example, even as bland as this paper is, I am reluctant to publish it at this time. I am still not willing to give exploration secrets away. Methodology, yes; secrets, no!

The best way to learn how to map undiscovered structures is to do detailed mapping in existing structural fields. If you go antelope hunting, you need to know what an antelope looks like, and it is a lot better to map it out in detail than have someone try to describe it to you. In many oil companies exploitation and exploration geology are separate departments, and the twain never meet. You have one group of geologists knowing what it looks like but not looking for it, and the other group looking for it but not knowing what it looks like. Hardly a prescription for success. For the past decade, I have concentrated on structural accumulations, and I spend 75 percent of my time mapping existing fields and only 25 percent looking for analogs in the wildcat realm.

Another common mapping error is to draw faults on the contour map prematurely. Even if you know that a fault cuts a well, you usually don't know the strike. Even if you have faults cutting two wells, you frequently don't know if they are the same fault. Seismologists are particularly prone to make this error. As my structural professor (the great Dr. James Gilluly) pounded home, "Don't put a fault on the map unless you are absolutely positive it is correctly located." I often see faults drawn on contour maps and when I question the mapper, he defends it by saying that he put it there because the contours were getting too close together and hard to draw. That would mean that he thinks the presence or absence of a fault is a function of contour interval and map scale alone. Remember that a single misspotted fault will usually invalidate an entire map. If you contour very carefully, and a fault does cut through your mapped area; it will show up as closely-spaced contours. The anomaly then will be defined by the subsurface control. When you finish contouring, you might then be tempted to replace the steep dips with a fault; but you don't need to, even if you are sure it's faulted. The closely spaced contours will illustrate the same phenomena. In the Permian Basin, the brittle basement is faulted, but the overlying plastic sediments on which we map flex over the basement faults; and the basement fault plane dies out rapidly upward. I generally never put faults on the map, at least until I prepare the sales brochure. At that point it may simplify the drafting, but you may be illustrating knowledge that you don't have. It is very embarrassing to have to admit you have drawn a fault whose existence you cannot prove. In the Permian Basin, there is little agreement on the stress system that caused the faults, and arguments about faults are often self-defeating.

How to Contour Creatively

Up until now, I have been discussing the rules of contouring, but contouring creatively requires a whole additional mindset. What is involved philosophically is excellently brought out in Dr. Gordon Rittenhouse's excellent 1959 SEPM presidential address, "There Is A Reason." Study what he has to say very carefully. It is probably one of the most important exploration papers ever written. I attempt to employ his philosophy. I was fortunate enough to hear him deliver it, and it has had a profound effect on me ever since.

"Mr. Chairman, fellow members of S.E.P.M. and A.A.P.G., and guests, if I were a preacher instead of a geologist, I might take for my text today the words 'There is a reason.' I am not a preacher, of course, but as some of you know, I have been somewhat of a missionary within our profession for a way of thinking about rocks—or if you wish, a philosophy about how to look at rocks. This philosophy or way of thinking is not new—certainly it is not original with me but antedates by far my entrance into the field of geology some 30 years ago.

"What is this way of thinking—this philosophy? It is based on the assumption that rocks as they occur today did not just happen, but that there is a reason for their being what they are and where they are. What rocks are and where they are are due to a combination of physical, chemical, and biological factors that have acted through time according to definite laws. These factors have left their imprints on the rock in varying degrees and, from these imprints or clues, the history of the rock may be reconstructed with varying degrees of success. This reconstruction of the total history of the rock is to me of paramount importance. To do it one needs not only to observe and ask 'why'; one must believe 'There is a reason.'

"This may seem quite simple—something that we all do—and therefore you may be wondering why I am talking about it today. It is simple—but it certainly is *not* something we all do.

"Let me illustrate this with two examples. I keep on my desk a piece of core that I call my 'Separator of Men from Boys.' I have given many budding geologists fresh from more than a score of universities a hand lens and knife and asked them to study this core for a few minutes and then to describe it, to give the rock a name, and to indicate what geological history can be deduced from it. I have found few 'Men' and many 'Boys.' Less than half recognized that information about the relief, the climate, and the kind and age of rocks in the source area can be deduced from the kind and size of pebbles, or that size, sorting, and rounding of the pebbles suggest a relatively short distance of transport and deposition under fluvial conditions, probably in an alluvial fan.

"Fewer than one in ten recognized that the orientation of the pebbles indicates that the beds are almost vertical, or that low porosity, close packing, and the kinds and number of contacts between pebbles indicate extensive post-depositional modification. Most of those young geologists were not asking themselves 'What properties does this rock have that will tell where it came from, how and under what environmental conditions was it formed, and how has it been changed since it was deposited?' Nor were they saying 'There is a reason' why it is like it is today.

"As a second example, I might point to descriptions of measured sections. In publications these commonly read 'sandstone, brown, fine-grained, cross-bedded, 14 feet,' and I have no reason to believe the descriptions are more extensive in many field notebooks. Here we have major lithology, color, size of grain, and the presence of a primary structure. For distinguishing this bed from those above and below, or for comparing it or the beds in a sequence, with beds in another measured section this may appear to be adequate, in fact, may appear to be a first very necessary step. But is it adequate? From the terms 'sandstone' and 'fine-grained' it may be inferred that the agent of deposition was a current of moderate velocity. What kind of current and what was its direction of flow? I submit that these are important questions to answer, since the answers may bear on selection of where to measure the next section or how the lithologic variations between sections are to be explained. In our example, cross-bedding may be a useful clue. But what do we know from just the word 'cross-bedded?' Our answer may come from the direction of dip of the cross-beds, their inclination, their thickness and lateral extent, and their relation to beds above and below.

"I might follow this example further, using the lack of information on composition conveyed by the broad term 'sandstone' and the lack of information on how much induration has occurred and what caused the induration. I hope, however, that I have made my point; namely, that it is common to overlook many clues that could be used in reconstructing the history of rocks because we are not wondering what they mean or assuming that 'There is a reason.'

"In looking at any rock we have some purpose—and a limited time in which to achieve that purpose. Therefore we can not describe everything—we must be selective, observing or measuring those features of the rock that will permit us to accomplish our purpose. In making this selection we are guided by our experience— and by our way of thinking about rocks.

"When our objective is limited, such as in identifying a subsurface structural datum, our choice may be easy, because our experience or that of others may indicate what to look for. Generally little thought is required to identify the datum—although much thought may be needed to interpret what the position of the datum means.

"When our objective is broader, the choice may be more difficult, requiring a combination of experience and a way of thinking about rocks. Thus in our 'sandstone, brown, fine-grained, cross-bedded, 14 feet' example, reliance is being placed on experience but this time with the *hope* of accomplishing the purpose, experience having shown that useful geological interpretations may be made in some cases if vertical and lateral variations in lithology are known. Outlining lithologic geometries may be a powerful tool, particularly in subsurface work—provided enough control points are available to define the geometry and provided we know what the geometry means when we have it. Commonly our control is not adequate, and how we interpolate between control points to outline the geometry will depend on our way of thinking about rocks.

"All too often, experience is substituted for thought—not made its partner. Because certain types of observations or measurements have contributed to the successful solution of problems in the past, they are selected and applied indiscriminately to new problems. Here we have one type of 'shot gun' approach, based on a philosophy which, in effect, says 'If we can make enough observations on enough rocks, put them in a machine and turn the crank, something useful may come out.' To some, this is the 'modern statistical approach' to geological problems. This maligns statistical methods—which can be valuable in geology—though not as a substitute for thinking. Since this approach does not reach the objective in a minimum of time, if at all, it is wasteful of time, manpower, and money. I prefer the 'There is a reason' philosophy.

"In selection of features that may be significant in rocks, we are faced with ever increasing knowledge in geology and related sciences, and ever increasing specialization. In looking at rocks and at problems involving rocks there is an increasing tendency to look at specialized aspects of the rock, rather than at the rock as a whole. The sands 'belong' to one specialist; the clays, the carbonates, the isotopes, the trace elements and various fossil groups, to others. Specialization has great potentialities for our science, since it can lead to a better understanding of the physical, chemical, and biological factors that have made rocks what they are today. But it has great hazards as well. Let's not forget how the elephant appeared to the three blind men—like a wall, like a post, like a snake. Just as we need to consider the entire elephant, we need to consider the entire rock—putting each type of observation and deduction in its proper perspective.

"With an increasing number of things to observe or measure, it becomes increasingly important to make the proper choice—the choice that will permit us to reach our objective in the time available. To me the obvious procedure is to go from the simple to the complex, from the rapid to the slow, from the less expensive to the more expensive. In general, but not always, this will mean going from the large to the small, from the visible to the invisible, from the direct to the indirect. There is no magic formula, no 'gimmick,' that will indicate what to look for or when. This will depend on the problem, the purpose, the time, the material available, and the way one thinks about rocks. Because I am firmly convinced that what rocks are and where they are has resulted from a combination of physical, chemical, and biologic factors that have acted through time according to definite laws, and that those factors have left their imprints on the rock in varying degrees, I believe that as we study rocks we should observe, ask 'why', and remember 'There is a reason.'..." (Rittenhouse, 1959)

If creativity is related to asking the right question, the more questions you ask the more likely you are to ask the right one. That particularly applies when you are contouring. Asking the simplest questions is more fruitful than asking the complex ones. The reason is that if the question is too difficult, you probably won't answer it correctly.

In order to contour creatively you first have to be an optimist. There is nothing more damaging to your productivity than the feeling that every field has already been found, and that even if one is present, someone has probably leased it already. *You have to believe that you are the best there is!* However, you can't kid yourself; and if you do careless, sloppy work, there is no way you can convince yourself that you are the best. I have run a number of exploration games over the years, and it is amazing how well a contestant does after he or she starts winning. Usually fewer than five will get rich, and the other thirty or so will end up bankrupt. However, when I run games in sequence, the results in the second and third games may be radically different. Every time an explorationist gains confidence, his or her results are better. It has also happened to me in real life. I have drilled six or seven discoveries in a row and begun to feel I couldn't miss. Then my luck has changed. After my first dry hole, I have run through so many dry holes that I began to feel that I couldn't do anything right. When you get this mindset, it sure is hard to find a prospect. Many attribute these streaks to luck, but it is far more than that—it is also a state of mind. There is always a chance that a geologist's effectiveness may be destroyed forever. You really have to be tough to survive when you are wrong most of the time. This is why there is such a small percentage of good old exploration geologists. Slowly, year after year, the disappointments destroy an explorationist's capability.

If you have difficulty remaining optimistic, you probably should not be an explorationist. Dry holes are so traumatic that you will need all the moral support you can get. A most important source of support should come from your spouse if you are married.

There I speak from personal experience because I have had both types of wives. Geologists should expect to be wrong most of the time and they may learn to accept that, but spouses who do not understand the probabilities often don't comprehend why this is so. You really don't need someone harping on your stupidity. My success as an independent can be correlated with my success in finding an understanding wife. I have known many good explorationists who gave up their independence because their spouses could not stand the disappointments. However, dry holes are not as hard to take as the uncommercial discoveries—four out of five discoveries will not make a profit. When that $10,000 a month royalty income drops off to $49.75, some wives (and geologists) go berserk.

Asking the Right Question

When you ask the right question, it can be any kind of a geological question that will lead to a prospect. My primary major was sedimentology and much of my early creative work was in that field; but in the last decade or so, I have concentrated more on structural geology. However, many of my most creative questions throughout my career have been geomorphic. In exploration you cannot afford to be a specialist. My greatest creative insights recently have been in seismology, where I have had little formal training. It is easy to ask lots of questions, particularly if you don't know much about the subject. Again, often those are the best because they are simple.

The one question I have asked, which has led to more prospects than any other, was one that was so obvious that most geologists did not even bother to ask it. I had worked in the Permian Basin nearly fifteen years before I asked it myself! It was the question, "How were the structures in the Basin formed?" I had mapped them and read theories on their origin and seen published maps of them, but I had not bothered to ask the most critical question of all. They are anticlines and everyone "knows" that anticlines are formed by compression. I did not realize it early in my career, but I know now that they are all tilted fault blocks, cored by basement, and bounded by nearly vertical faults. In 1965, I had the good fortune of attending a structural seminar by Dr. John Prucha, formerly with Shell Research, but since that time at Syracuse University. He described the structures in the Wyoming Province of the Rocky Mountain foreland, and after realizing they appeared identical to those I had mapped in the Permian Basin, I asked him how they were formed. I clearly remember his answer was, "I haven't the foggiest idea." That really got my creative juices flowing. Although this was his verbal answer, he speculated on their origin in his paper (Prucha et al., 1965). My subsequent literature search and examination of text books indicated that this particular structural style has been consistently ignored or its importance downplayed. This is amazing because from a hydrocarbon accumulation point of view, this style could well be as important as any other structural style.

Part of the reason this style had been ignored was because at that time, we did not have a correct model of the earth. Until the modern plate tectonics theory evolved, we did not have the means of fully understanding how the structures could have been formed.

After I worked on this problem for nine years and hopefully solved it in 1974 with the help of the great plate tectonicist, Dr. John Dewey, I found that all I had done was to reinvent the wheel with some important modifications. If you don't know the problem, how do you search the literature? Dr. Hans Cloos (1930, 1931, 1932) had used a similar solution in the Rhine graben fifty years previously, but it had evoked so much controversy that his major contribution has been almost completely lost to the profession. My work has provoked the same controversy. My structural model has now been tested for ten years, and the validity has been confirmed again and again in the borehole. Others may think me wrong, but as long as my income continues to rise through its use, I couldn't care less. As I said earlier, it is a distinct advantage to be an unconventional thinker. At the end of this paper, I will ask you to solve this structural problem. When I finally thought to ask the right question, the answer came quite easily. My only problem was that I did not ask it early in my research. And so far as I can tell, no one else working on the problem ever thought to ask this most critical question.

I have mentioned the importance of geomorphology in exploration, and this is particularly important where carbonate reservoirs occur. Most carbonate porosity is secondary and related to exposure and subaerial leaching and erosion. The easiest, most common, and productive contouring horizons are these unconformity surfaces. The main reason that paleotopographic mapping lends itself to creative contouring is that the sine curve is a natural function, and erosional ridges and valleys tend to follow some variation of a sine curve. Thus, once you have a few control points to identify the periodicity of the curve, it permits you to project highs into that part of the map where there is little or no subsurface control. Three of the most productive carbonate horizons in the Permian Basin—the Ordovician Ellenburger, the Siluro-Devonian carbonates underlying the Woodford Shale, and the "Canyon" reef of Pennsylvania age—have unconformities at the top of the formation. The first is a karst surface and the latter two exhibit ridge and valley topography. Prospects can frequently be found by working out the periodicity by careful contouring, and by projecting it into the area where the control is meager (Fig. 2 & 3).

In some high structural areas such as the Central Basin Platform of the Permian Basin, the interbedded carbonates and shales are eroded deeply enough so that the anticlines are breached. The deep seated anticlines are often hidden because there is velocity attenuation in valley fill at the crest, and the seismologists tend to concentrate on drilling the flanks. Prolific deeper horizons often are never tested. Working with problems of this type has made me appreciate the value of the geomorphology courses I took at U.C.L.A. under Dr. William Putman. His favorite exercise was to throw you a U.S.G.S. topo sheet and ask you to map the underlying structure from the topographic contours alone. Thanks to Putman, I got pretty good at that, and at least one-quarter of my prospects have come from using his methods. Apparently, not many students are exposed to this type of geomorphology anymore. It too has become the domain of specialists.

Few geologists can handle the contouring of a composite surface, so we need to eliminate as many variables as possible. I recommend an exercise to make you appreciate my point. Take a relatively simple topographic map and develop a set of data points that will allow you to contour it with confidence. Then tilt the surface in any direction, adjust the same data points for the tilting, and then try to recontour the map. You will find it an almost impossible task even if you know what the surface originally looked like. Prior to contouring tilted paleogeomorphic surfaces, you must restore the surface to its original position. In competing with other explorationists, you will find that even minor improvements or changes in attacking a problem of this type will yield great dividends.

One way of working with paleogeomorphic surfaces is to isopach the overlying unit, but I find it more difficult to do creative isopaching than to do creative contouring. In Saskatchewan I had occasion to work an area mapped by the late Dr. Rudolph Martin, an AAPG distinguished lecturer on paleogeomorphology (Martin, 1966). He had advocated isopaching the overlying Watrous red bed series to locate Mississippian cuesta closures in the northern part of the Williston Basin. I came up with a technique using a computer to remove the regional tilt and restore the erosion surface to its original position. I was able to locate subtle closures in many locales that were not apparent previously. I developed and sold six prospects the first month that I worked in the Williston Basin, and the first test resulted in a flowing well. It was a most rewarding geological exercise. Unfortunately the Saskatchewan provincial government, good socialists that they are, decided to take more than 100% of the profits soon afterwards. It got so there was not enough left to pay for the operating costs, let alone explore. I understand there has been some modest improvement in recent years. Political actions of this type reduce the demand for petroleum geologists throughout the world.

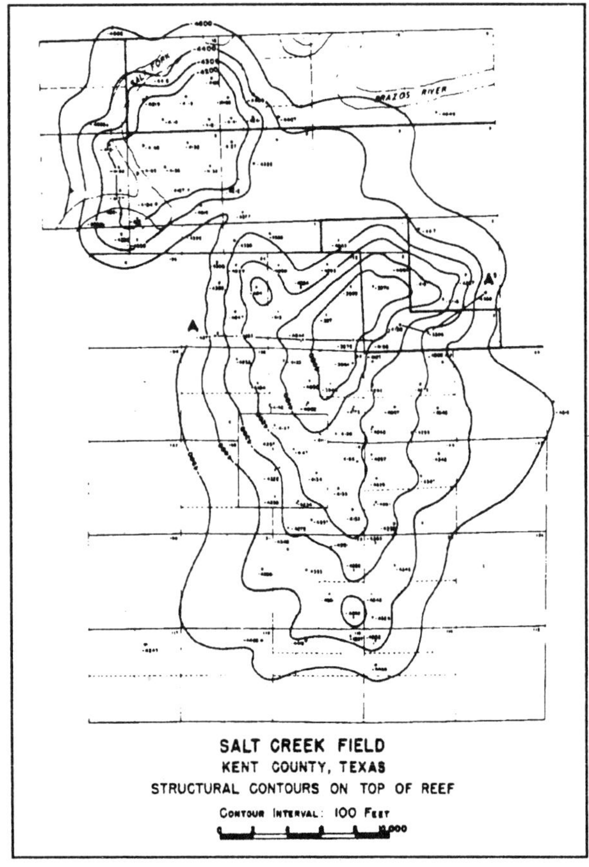

Fig. 2. Salt Creek Unit Map, 1962
This contour map illustrates typical contouring of a "Canyon" reef. Two things are wrong with these contours. The first is that the actual points poorly fit the contours throughout the map. The second is that the reef here is exceedingly porous and permeable and has to have been eroded. It was not until the mid-1960's that we came to realization that reefs are normally dense unless they have been exposed to meteoric waters that selectively leach the aragonitic fossils, and the porosity is secondary.

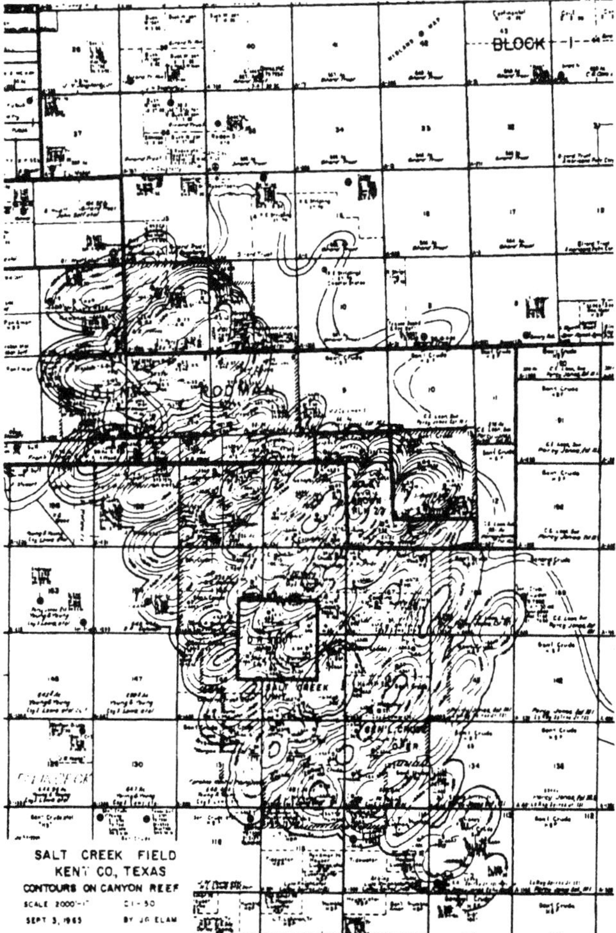

Fig. 3. Salt Creek "Topographic" Map, 1965
This map was contoured using the points from the unit map, but the scale was enlarged, and the contouring interval was reduced. It was not until this contouring was done that it was apparent that this was a paleotopographic surface.

I have worked many basins in the United States and Canada, and I have frequently found that with a fresh approach I can commonly find prospects that the locals missed because they were too close to the problem. On the other hand, there is no substitute for experience. In my later years, as I have gained experience, I have tended to stay in the areas I know best. That has improved my success ratio. As always, the more you know, the more you need to know.

One way to overcome your familiarity with an area is to repeatedly challenge your own ideas and concepts. Very commonly I force myself to recontour an area using a bias other than the one I have been using. Structures in the Permian Basin commonly run northwest-southeast, but other directions occur with some frequency. Sometimes I have deliberately contoured the structures using various orientations, and every once in a while, I am startled to find out that some direction other than the one I thought was right fits the data points better. Experience has taught me that whatever I postulate will probably be wrong. When you experience that repeatedly, it is not so difficult to force yourself to consider other alternatives. We almost never have enough control to "know" anything for sure in exploration geology.

Anatomy of a Prospect or Prospects

To portray one example of the type of thought processes that go into the making of a prospect, I cite a recent case in point. Things are tough in the profession right now, and there are many geologists who have lost their positions because of the cutback in exploratory activity. At the behest of my wife, I offered a young friend of hers a small retainer with the intention of showing her how she could make it on her own by generating salable prospects. This was a good learning experience for me because I could catalogue my thought processes and record them for this paper.

As a condition of employment, I asked her to present three prospects that she had recommended to the management of the several companies she had deigned to work for over the past few years. She had been a job hopper. Although she had been exploring for five years, she had no proven exploratory record. She submitted three geological ideas that had merit, and we chose the one that offered the greatest potential for developing salable prospects. It happened to lie in a geological sub-province I had never studied, so I had little knowledge to offer. However, I do have a methodology that yields prospects, and I was able to apply it here. Whether it yields any actual prospects still remains to be seen, but it has proved to be an interesting application of my exploration techniques.

In the deep Delaware Basin, one of the sub-basins of the greater Permian Basin, the final clastic phase prior to the terminal evaporite cycle was composed of deep water sands deposited by density currents, turbidity currents, etc. (Fig. 4 & 5). The accepted trapping mechanism is that porous and permeable sands were deposited in channels or submarine canyons and they are productive because they are cleaner. The trapping mechanism is hydrodynamics, or fluid flow, and/or capillarity, so these must be considered stratigraphic traps. Structure is not involved.

Fig. 4. Permian tectonic elements, West Texas-Southeast New Mexico (Williamson, 1979; p. 40).

Although I had never worked in this geological sub-province, I knew enough to wonder why the regional geology did not seem to fit with the detailed explanations or conventional wisdom. One always has to be careful that the apparent spatial relationships have not been distorted by regional tilting or deformation. However, my quick restoration indicated that the tilting would adversely affect regional contours and make the generally accepted solution even

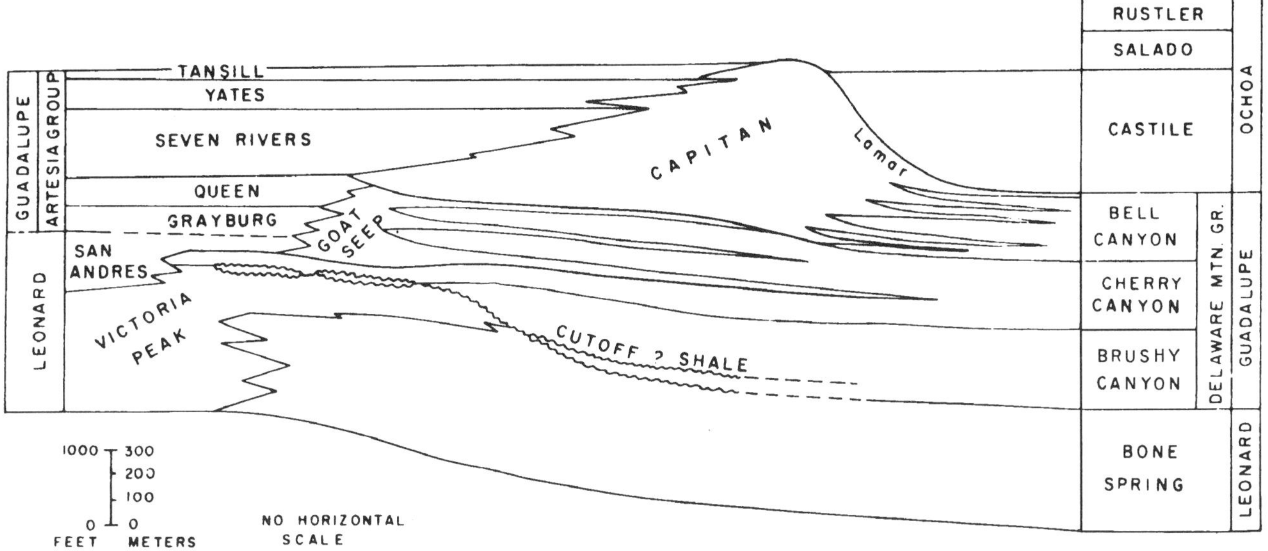

Fig. 5. Stratigraphic relations for Permian rocks of the Delaware Basin, Texas-New Mexico (Williamson, 1979; p. 40).

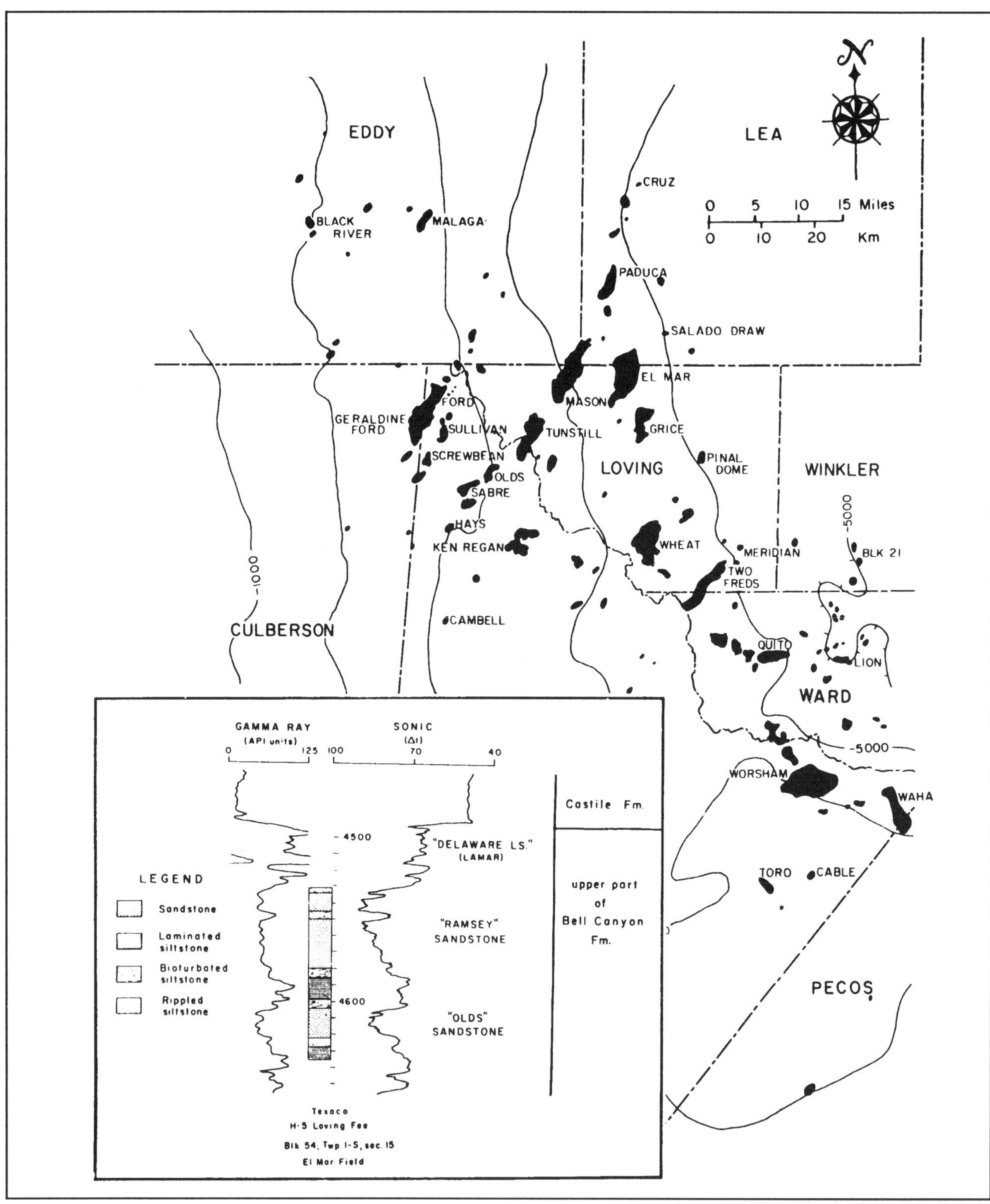

Fig. 6. Map of Delaware basin showing principal Bell Canyon oil fields and structural contours on top of "Delaware Sand." Representative Gamma-ray-Sonic log, upper Bell Canyon Formation-lower Castile Formation, showing informal subsurface stratigraphic nomenclature and log response to lithology (Williamson, 1979; p. 48).
NOTE THAT THE ELONGATE FIELDS ALONG THE TEXAS-NEW MEXICO BORDER ARE ALIGNED IN A SOUTHWEST DIRECTION. IF THE SAND SOURCE IS TO THE NORTHEAST, THE "SCOUR" CHANNELS WOULD HAVE BEEN CARVED BY CURRENTS FLOWING UPDIP.

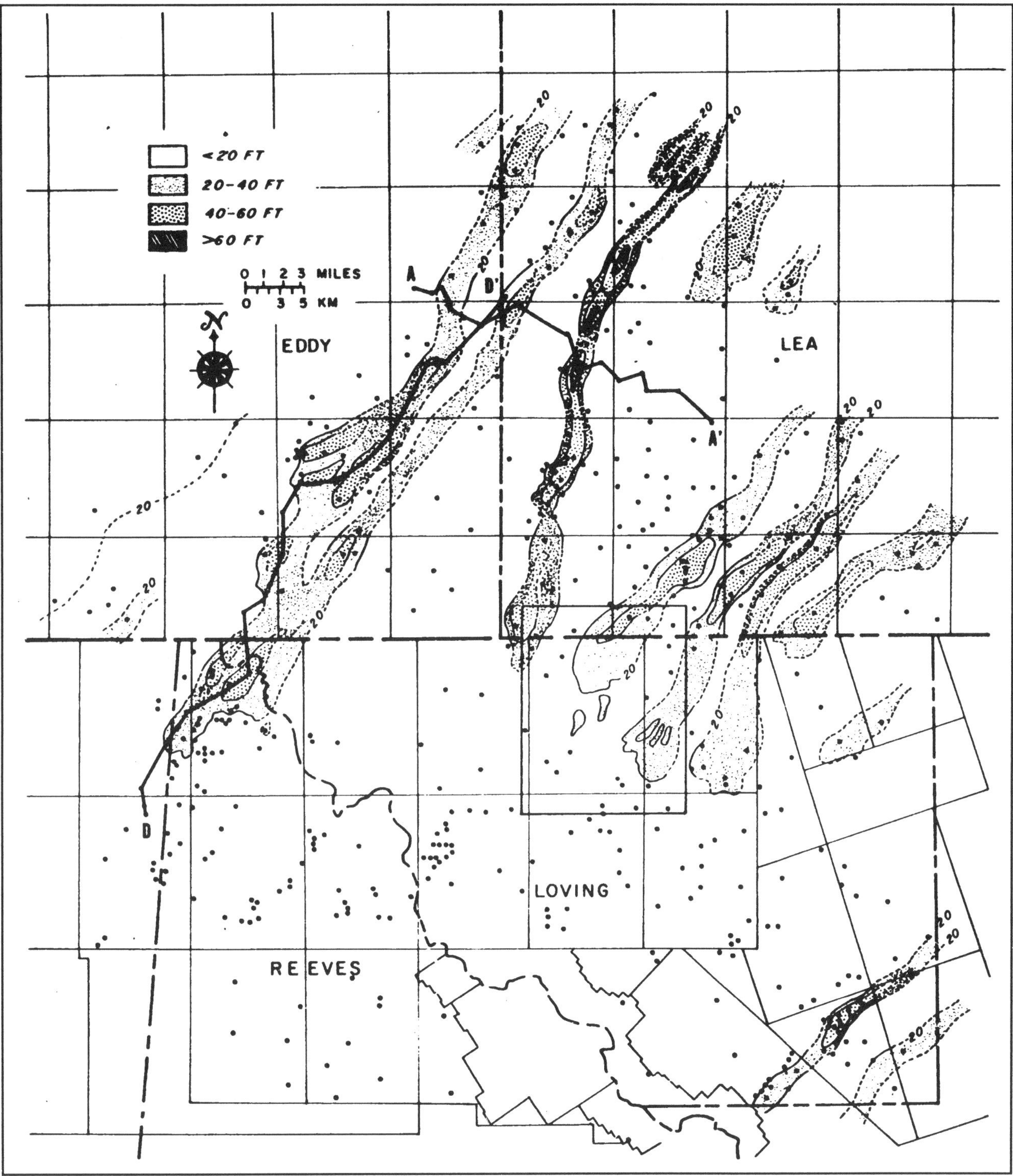

Fig. 7. Regional sandstone isolith map of the "Ramsey" sandstone emphasizing interpretation of maximum northeast-southwest continuity of major submarine channels. This map is contoured with the same data as Figure 8 (Williamson, 1979; p. 50).

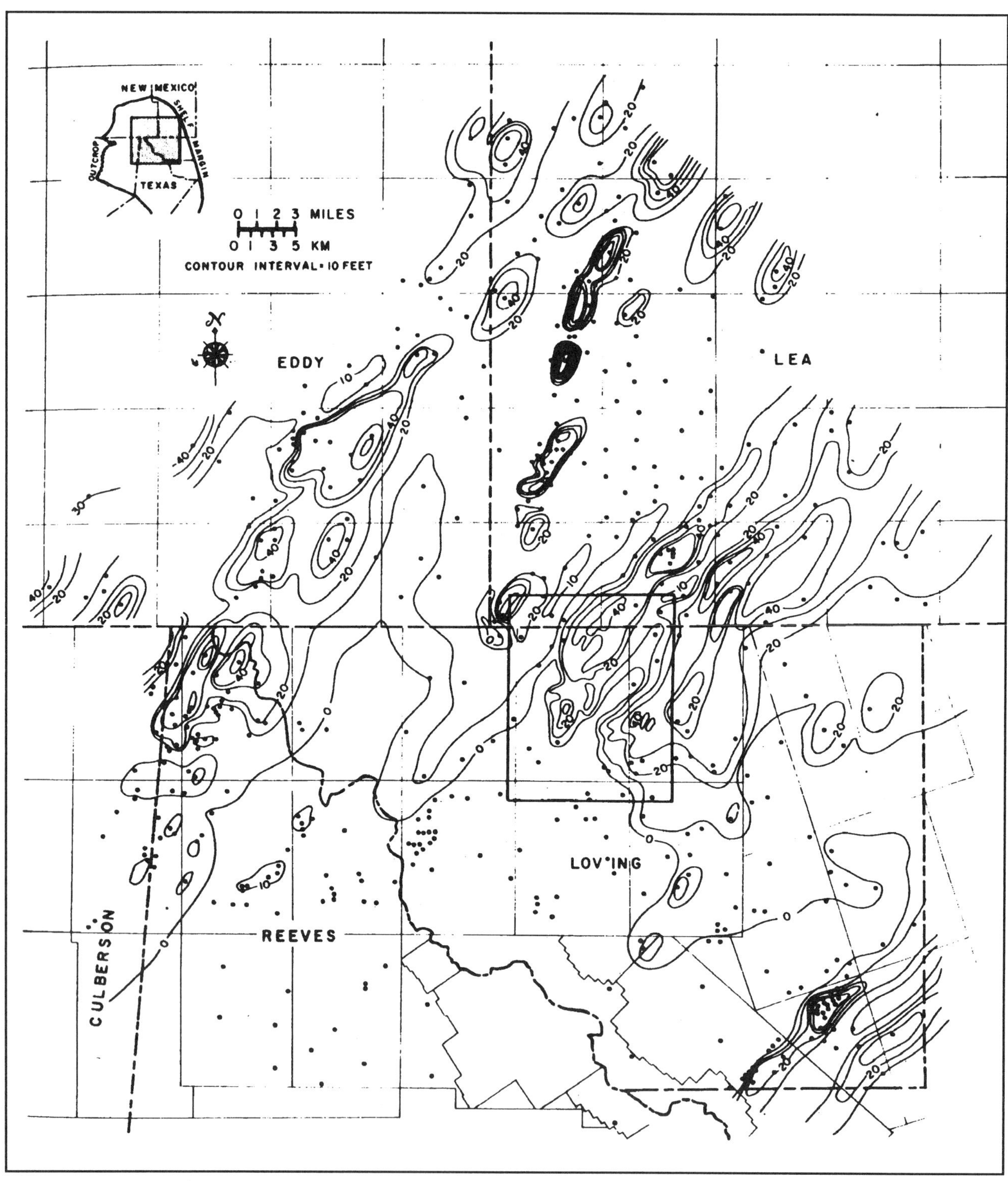

Fig. 8. Regional sandstone isolith map of the "Ramsey" sandstone, upper part of the Bell Canyon Formation (Williamson, 1979; p. 49).
ASIDE FROM THE PROBLEM OF HAVING THESE SCOUR CHANNELS FLOWING APPARENTLY UPDIP, THERE IS A PROBLEM OF EXPLAINING HOW THE CHANNELS CAN SCOUR LOCAL HOLES IN THE CHANNEL BOTTOM. THIS IMPLIES RADICALLY DIFFERENT RATES OF FLOW. WILLIAMSON SUGGESTED THAT THESE HOLES WERE CAUSED BY A MEANDERING THALWEG. DOES THIS LOOK LOGICAL?

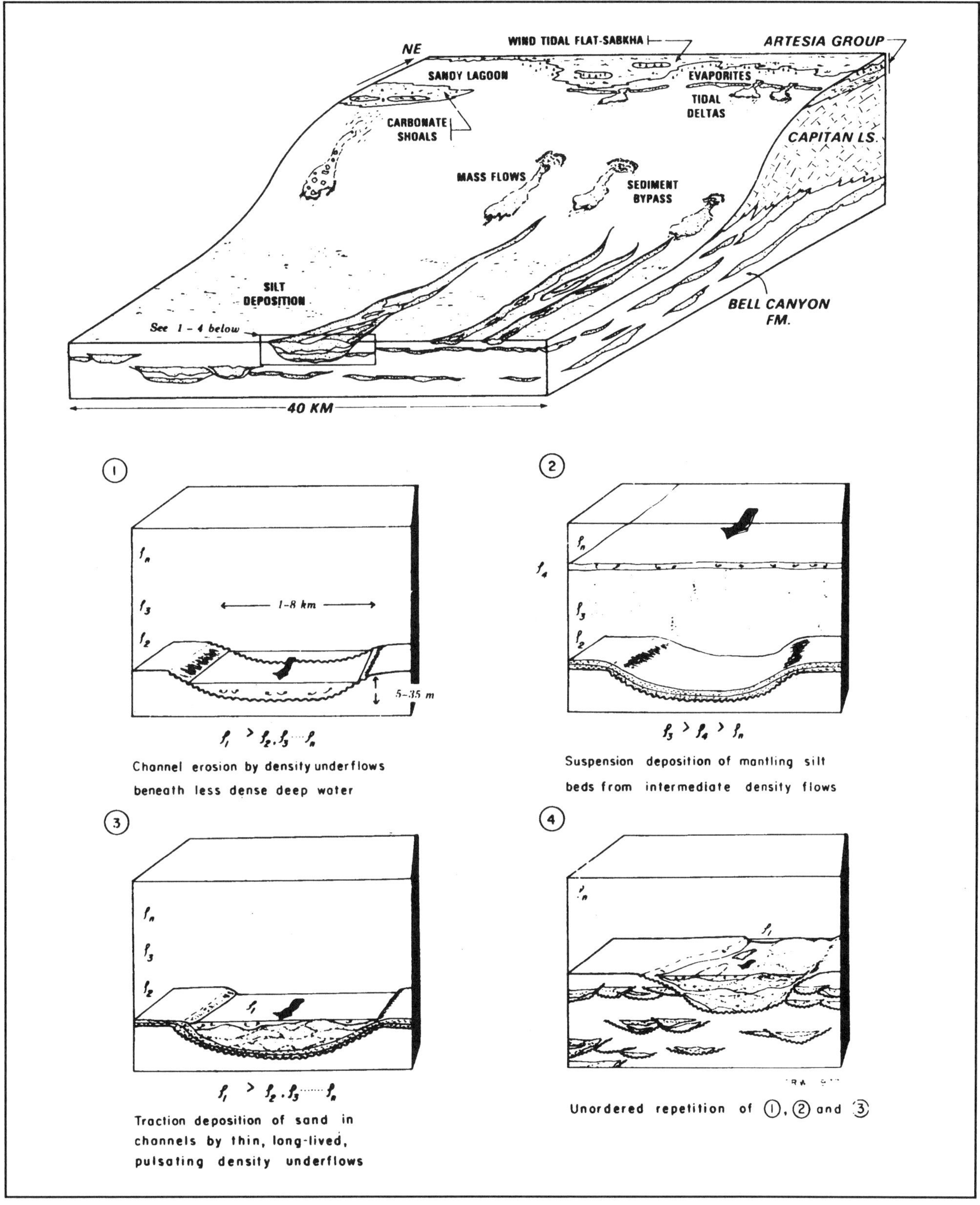

Fig. 9. Depositional model for upper Bell Canyon Formation and inferred sequence of events leading to channel filling (Williamson, 1979; p. 67).
NOTE THAT THIS MODEL INDICATES THAT THE SCOUR CHANNELS DO FLOW DOWNSLOPE. ALSO NOTE THE SLUMP SCARS WHERE HE SHOWS MASS FLOWS.

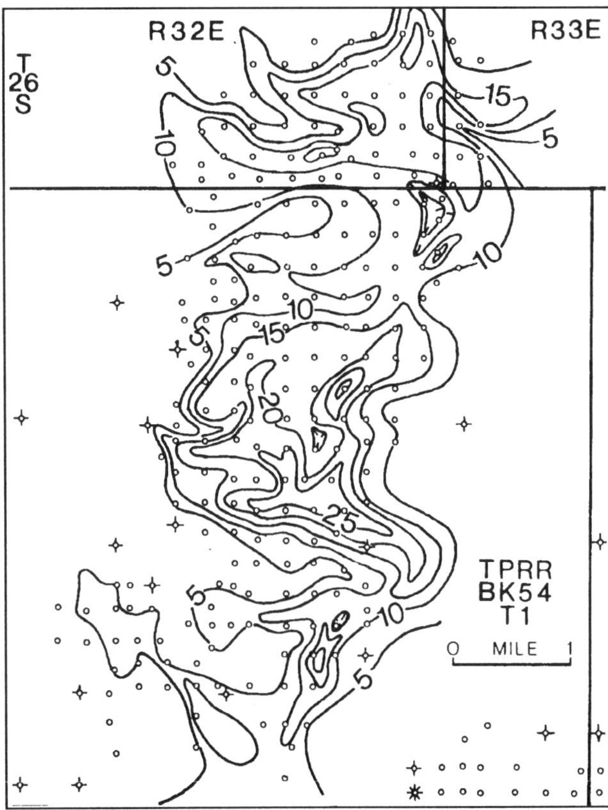

Fig. 10. Ramsey net lower sand map of El Mar Field (Weinmeister, 1978; p. 55). ALTHOUGH THIS MAP WAS HIGHGRADED TO BEST ILLUSTRATE THE APPARENT **NW-SE** ORIENTATION OF THE SAND THICKS AND THINS, DETAILED ISOPACHING OF THE SAME DATA ON OTHER MAPS WITHOUT THE BIAS OF SOUTHWESTERLY FLOW AND SCOUR FREQUENTLY SHOWS THE SAME PHENOMENA.

less probable. Most of the Delaware Sand oil fields on the upper paleoslope are elongate parallel to the contour or even trend updip—improbable flow directions for scour channels created by density flows that originated in the hypersaline lagoons and passed through breaks in the reef (Fig. 6, 7, 8, 9).

A second point also bothered me. The only available sand source was from the Pedernal Uplift that was situated northwest of the Delaware Basin. However, the orientation of the proposed scour channels required that the sands enter the basin from the northeast side of the basin. Whereas the sands could possibly be carried laterally along the back-reef lagoon for forty to fifty miles before they spilled through the reef into the Basin, it is certainly not the simplest solution. Beware when it is not the simplest solution.

The first order of study was to carefully map the nearest production analog. You never accept conventional wisdom, no matter how plausible it sounds. If you can possibly come up with another explanation

from the same evidence, you may have the ingredients for generating many prospects. Avoid having to generate prospects one at a time, which is a hard way to make a living.

After examining a number of cross sections and studying the published papers, it became obvious that a map at the base of the channel would describe the morphology of the channel immediately prior to the deposition of the productive sand body. Although previous workers said these channels were scoured by a current that flowed in from the northeast, the detailed contour map on the channel base suggested a series of ridges and valleys oriented almost at right angles to that alleged "scour" channel. An isopachous map of the productive sand seemed to confirm that orientation. That new apparent scour direction is nearly parallel to the present dip of the strata (Fig. 10). The economic implications of that one finding could be immense because this is a prolific producing area at a shallow depth. If the depositional environment has been wrongly interpreted, many of the fields may not be properly exploited and there could be *many* exploratory and field extension prospects.

My young geologist pointed out that the "channel" geometry was distinctive in that the updip bank always appears to be steeper than the downdip bank. That is not a normal channel shape, and whenever anything is out of the ordinary, it demands an explanation. Those who thought the channels were carved by density currents that were traveling so fast to the southwest that inertia kept them flowing and scouring for miles parallel to contour or flowing updip would not be bothered by that shape. However, it should be apparent that density flows will turn and flow downslope soon after they enter the basin regardless of inertia.

Isopaching the productive sand suggested that although the downslope scour orientation was clearly evident, there were an inordinate number of thick sand isochores. That statistic also needs to be explained. The producers encountered thick sands about ten times as frequently as thin ones, but the distribution didn't seem to make sense on the contour map. We never have enough subsurface control, but we make up for that by using statistics. What kind of an underlying depositional surface would yield ten low points (sand thicks) for every high point (sand thins)? A sine curve would give equal numbers. Induce a solution! Draw different types of curves until you have one that will give you low vs. high points in the right percentages. In this case, the erosion surface appears to be scalloped ∿∿∿. That morphology is what you would expect if each of these cross channels were cut by density or turbidity current flows of small and finite dimensions, or by broad flows exhibiting numerous density and velocity variations internally.

When your geologic concept is right, all the data will fall in place. Then you will know you have mastered the problem. Until you have that moment of revelation, you need to continue your research. I think what we have been recording here with these depositional data are different phenomena possibly related to slumping on a paleoslope. The shape of the "channel", with a steep updip slope appears to be that of a slump scar. After the slump scar is formed, density currents flowing down the paleoslope, more or less in equilibrium, will accelerate as they flow into the holes created by slumping. The steep slope and nearby bottom of the slump scar will be scoured creating the lesser channels drawn earlier. As the flow starts to come out of the slump scar, the slope lessens, the scouring ceases, and the overloaded current starts depositing. In time the scar will be backfilled as the loci of sand deposition shifts upslope. Eventually, the slump scar will be filled back up with sands that are cleaner, more porous and permeable than those that were there before the slumping took place. The shape of the sand bodies is a combination of the slump and scour channel morphology, and that is what the sand isopachs illustrate. This depositional environment alone would not create the oil trap by itself. The oil is trapped because there is hydrodynamic flow west to east into the basin or down the regional tilt. Oil is trapped under hydrodynamic flow conditions when there is a measurable change from less porous and permeable sands to more porous and permeable sands in the flow direction. Each one of these pods of clean sands may reflect the local change in depositional environment caused originally by the slumping.

If you examine the contour and isopachous maps out of context and without having worked out the geological history, you fail to understand the trap. You cannot solve exploration problems out of context as many are wont to do. This error is illustrated by the Ph.D. dissertation and papers by Charles R. Williamson (1978, 1979, 1980) from which many of my illustrations were obtained. He adjusted his maps and concepts to coincide with the conventional wisdom. The fact that his thesis was approved by world renowned academic specialists in sand deposition indicates how dangerous it is to tackle an exploratory thesis problem without having a strong background in exploration. He was looking for the answer, when he should have been looking for the question. None of his advisors asked the right question either. Not until you have spent many years drilling dry holes because you ignored this little detail or that, do you ever learn that *you cannot afford to leave a single question unanswered*. Most geologists do not work in enough detail, and their bosses do not ask enough questions either.

Only after we had developed our depositional concepts were we finally ready to contour the map. All the original contouring and isopaching, prior to the conceptual development of the proper model, needs to be discarded because you cannot contour or isopach correctly without the concept. *It is extremely*

important at this point to start with a fresh map. There ain't no free lunch.

The reason my young geologist was not able to sell her prospect to management was because she did not understand the prospect well enough to sell it. I repeat, a good geologist who can't sell his ideas is of no more value than a stupid geologist when it comes to finding oil. The way to sell is to understand the history of entrapment better than anyone else. Actually, after we built the model properly, we found that she did not have a prospect where she thought it should lie, but there were many other prospective areas nearby.

A Word of Caution:

My methodology may well yield prospects. However, you should be warned that just because I have come up with an attractive alternative working hypothesis, it does not necessarily mean that I am right; or that if I am right, I am the first one to correctly solve this mystery.

I mentioned before the tremendous value of any new working hypothesis. Its value here could easily be measured in millions of dollars. You can be assured that if I knew I was right, and first, I would not include this hypothesis even in this paper. This is why you cannot go to the literature on exploration and find much of real value. *Who publishes economically valuable ideas knowingly?*

If my depositional model is correct, others probably have already beaten me to it. When I try to put a prospect together, I may be too late. I have no way of knowing that until I have been able to zero in on specific prospects.

This methodology has only led to a working hypothesis at this time. We are now in the stage of trying to prove it wrong by very careful field checking. It will be at least another month before we can begin to validate or disprove the hypothesis. It is also highly probable that as we proceed, we will modify it and possibly alter it substantially. Thus the finished product may be far different from what I discuss now. It will take several test wells to check the hypothesis to my satisfaction. I emphasize the point that only when an idea has been tested does it have validity.

Even so, I want to emphasize that the opportunity to come up with a novel and unique idea in exploration probably will happen to you only a few times in your lifetime *if you are creative*. For many explorationists, it may not ever happen. I am writing this paper to increase the probability of it happening by teaching you to ask the right questions and not to be satisfied until you have the right answer. I am not trying to give you the answers.

The Problem—a Challenge

This is going to be an unusual treatment for teaching geology because I am going to give you the question, not the answer. I believe it has a very simple answer, but it is not available in any structural geology textbook today, so far as I know.

The most difficult problems to solve are ones for which there is no correct answer using conventional wisdom. It is very difficult for most of us to conclude that there is some body of knowledge that remains unknown to us, especially when we are attempting to solve what seems to be a simple problem. This is not to say that this problem has never been solved, but that any previously correct solution has not been accepted within our body of knowledge that constitutes the gospel.

I mention this because I have presented the answer to this problem many times to petroleum geologists interested in the Permian Basin, and I have encountered exactly the same arguments that were used fifty years ago to counter Hans Cloos' (1930, 1931, 1932) similar solution. Although I have had my model intact for ten years, and I have spent this last decade testing it with the borehole and not found it wanting, there are very few of my peers who are willing to make use of it to find oil and gas even today.

Inasmuch as it is an extremely valuable exploration concept, my feelings are not hurt by the lack of competition. Many geologists have told me they will wait until I put it in print before they will test it. That has caused me to hold up on my paper that was written and sent out for critiquing two years ago. For me, it is wonderful that most geologists don't believe anything until they see it in print, because who needs more competition! I only care if my solution works. It doesn't matter if others don't accept it. That is the great luxury of being independent. Fortunately, I have tenure with my own company!

The problem is very simple. As I said earlier, every structure in the Permian Basin is a tilted fault block, and there is every indication that the bounding fault is vertical or nearly vertical in the basement. Every structure is basement cored, and the plastic sediments drape over the tilted basement block. Enclosed are a few illustrations of typical structural oil and gas fields in this area (Fig. 11, 12, 13, 14).

Many challenge the vertical orientation of the fault. However, we do have some pretty good field evidence such as the bore hole cutting the same fault several times. A normal borehole will corkscrew down through the sediments, particularly if they are nearly flat lying. Thus, there may be many intercepts of the same fault.

Many geophysicists interpret the same faults on seismic sections as having dips in the 60-degree range, even where I think they are vertical. If 60-degree faults are the norm, then our bore holes should cut many faults. Normal faults cut out stratigraphic intervals. The vast majority of the 158 or so faults that have been encountered in the 50,000+ exploratory wells are reverse in nature. Steep dips are much more common than faulting of any kind. The statistics alone should

suggest that the only way the bore holes could have avoided cutting a greater number of faults would be if the faults are more or less parallel to the bore hole or they are not present. You have to stick with statistics.

Before I ask the right question, I would like to point out another problem with conventional wisdom. If conventional wisdom is wrong, and you accept it, you are really in trouble. In 1942, Anderson pointed out that all deformation could be explained with three orientations of principal stresses (Fig. 15). In fact, they were the only three directions he could think of so he thought they should suffice to explain all faulting. The maximum principal stress can either be vertical or horizontal. The intermediate principal stress can also only be vertical or horizontal. The angle of sheer failure is approximately 30 degrees to the maximum principal stress and parallel to the intermediate principal stress. In his example, you get low-angle thrusting by compression, high-angle normal faults by extension, and wrench faulting where the intermediate principal stress is vertical.

If Anderson's structural concepts are complete, there is only one possible origin for these tilted fault-block structures. They must have been formed by wrenching, and that interpretation is commonly accepted by many exploration departments of companies in the Permian Basin today. So far as I can tell, there are no hard data whatsoever to support the wrench interpretation other than that the fault is vertical. However, if a vertical fault is by definition a wrench fault, do you need any other support? (Only if it is wrong!) This is the hazard of accepting what anyone else says without checking the data yourself. I know that very, very seldom can you check out anyone else's work in exploration and validate his conclusions. There is too much wishful thinking in all of us.

Another group of geologists, led by Dr. Robert Berg of Texas A&M and Bill Brown, formerly structural specialist for Chevron but now at Baylor, is convinced that these structures, at least in the Rockies, are not bound by vertical faults, but are folded thrust structures caused by compression (Fig. 16). Even when the faults are vertical on surface exposure, they believe they curve and flatten downwards. Recent geophysical interpretations, particularly by Cornell's COCORP group, have tended to support their interpretation (Smithson et al., 1978). I say "tended to support" because these structures lie in front of the overthrust belt, and there certainly is plenty of evidence for some element of compression in the Wyoming Province. The recent discovery of production under low-angle thrusts in some parts of Wyoming have given a tremendous boost to Berg's model, even causing many to discard field evidence to the contrary by Stearns (1978), (Fig. 17); Prucha et al. (1965); and others that indicate that there was no lateral transport on many of the important faults. One could say that if only one period of deformation was to be permitted,

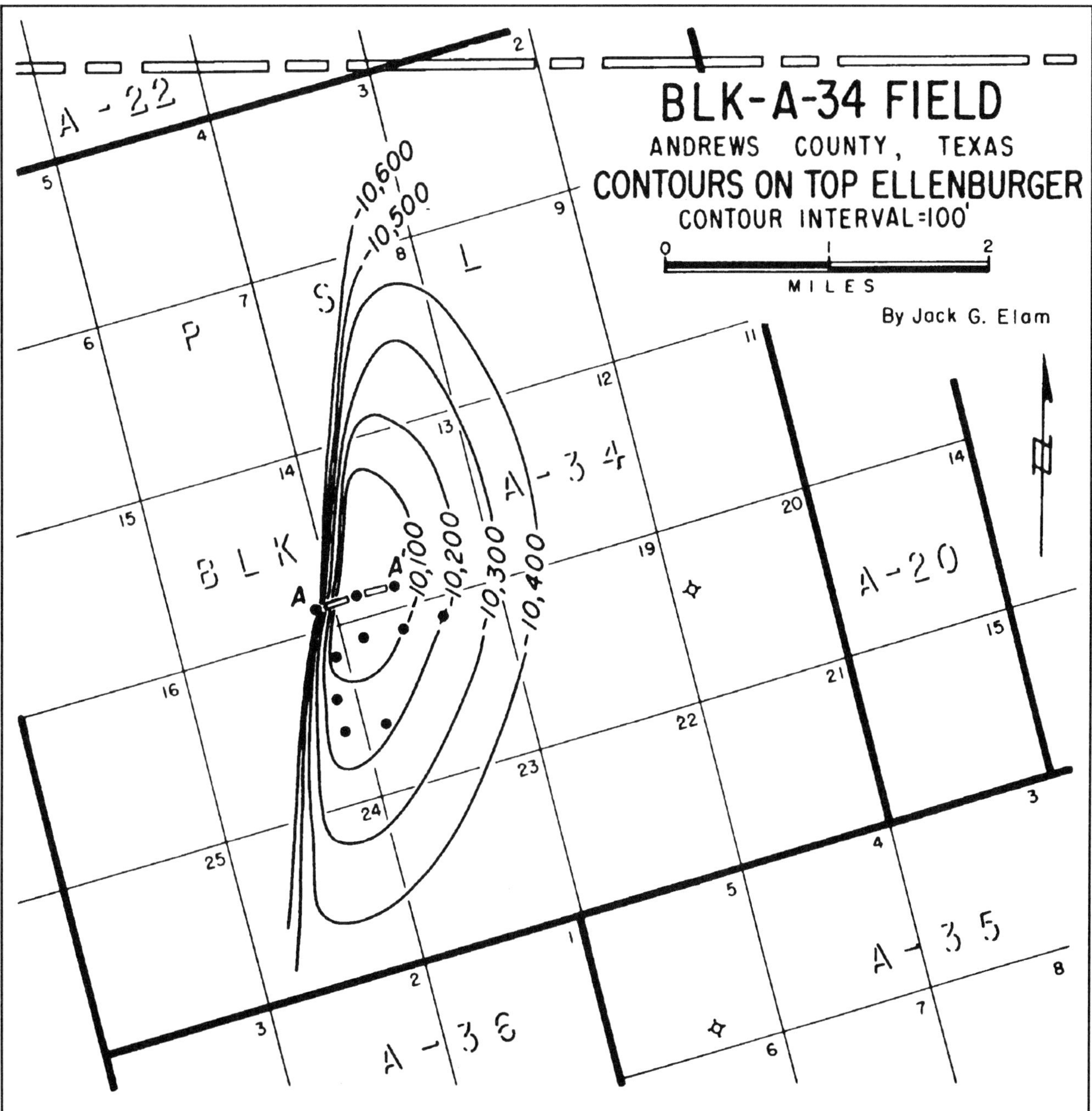

Fig. 11. (Elam, 1969; p. 62).

then Berg, Brown, et al. are right. However, there is nothing that places a limit on the number of periods of deformation. If the Wyoming Province were allowed to have two periods of deformation—one that caused the "upthrust" structures, and a later period that caused low-angle thrusting—there would be no problem explaining all the field data that has been reported from both camps (Fig. 18).

One point always must be kept in mind in geology. If you explain 100 percent of the data, your solution might be right, but if you only explain 99.9 percent of the data, your solution is wrong! That dictum is awfully hard for many geologists to accept. I fully appreciate it because almost every time I drill a dry hole, it's always that little nagging fact that I didn't explain that reached up and grabbed me. This is the reason we are wrong so much of the time. It is almost impossible to explain everything completely and adequately, and we drill lots of dry holes.

Fortunately, I had occasion to test my vertically

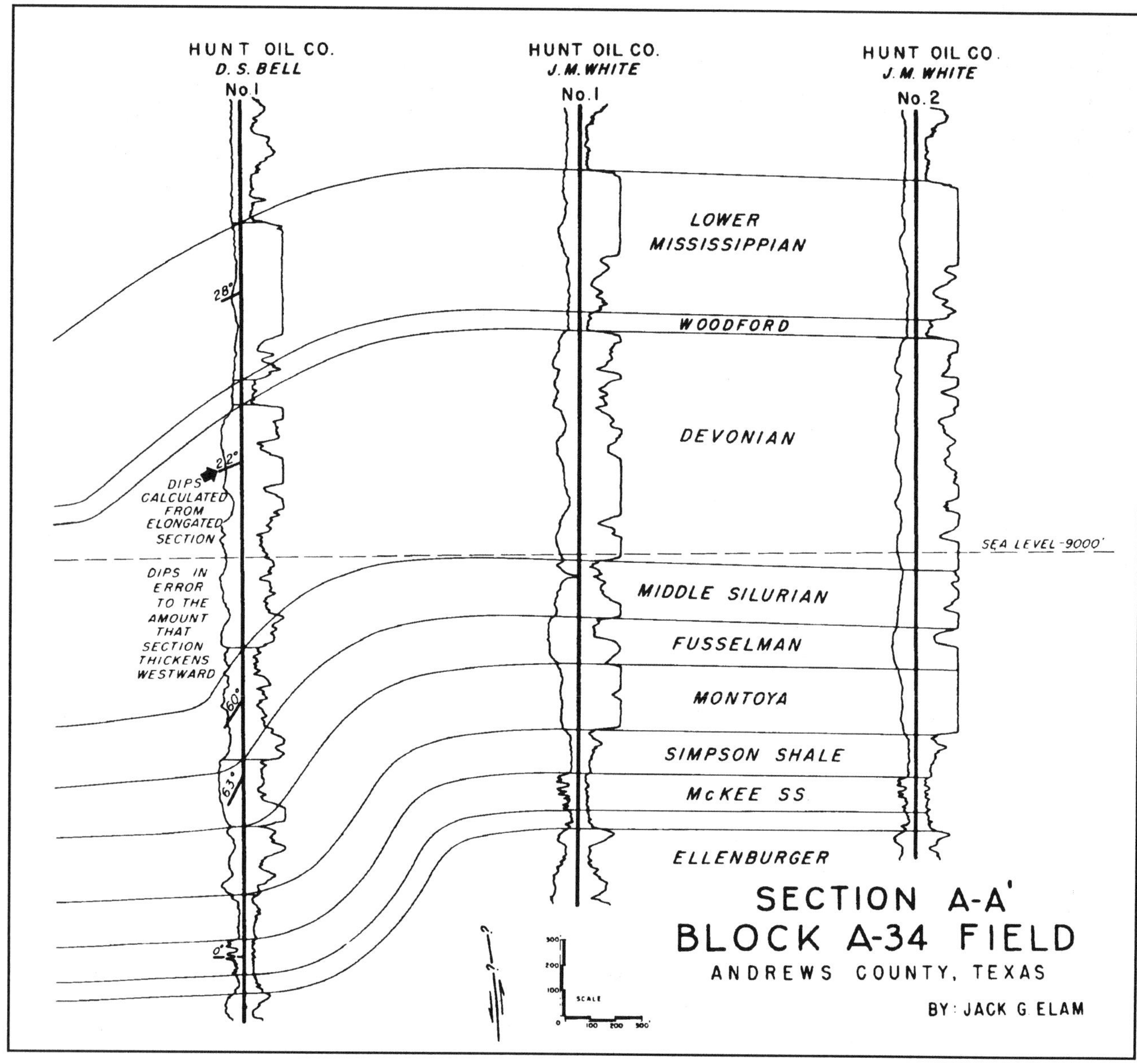

Fig. 12. Cross section across Block A-34 Field illustrating typical deformation overlying basement block fault (Elam, 1969; p. 63).

faulted model vs. the low-angle thrust model in the Permian Basin two years ago. Chevron drilled a well first, and thought they found a low-angle thrust. Unfortunately, it was very low and dry at a prospect cost of $29,000,000. I got a farmout from Chevron on the same block after they were through. The Chevron well cut 11,000 feet of granite before it crossed the reverse fault and went back into Mississippian sediments at a depth of 23,500 feet. We did not encounter their large fault in our well even though we drilled a little over 2500 feet in front of their well. Our well penetrated a normal section little disturbed by lateral compression. Although their fault was a reverse fault, it had the steep dip I had postulated. My structural interpretation was essentially unchanged after drilling, except that our well was even higher than I expected. It was full of gas, but it was pore plugged and too tight to produce commercially. It was a great geological success!

I mention this last episode to point out that I do have a lot of field evidence to support my contention that the faults are high angle, but not necessarily perfectly vertical.

I am ready to ask you the question that will allow

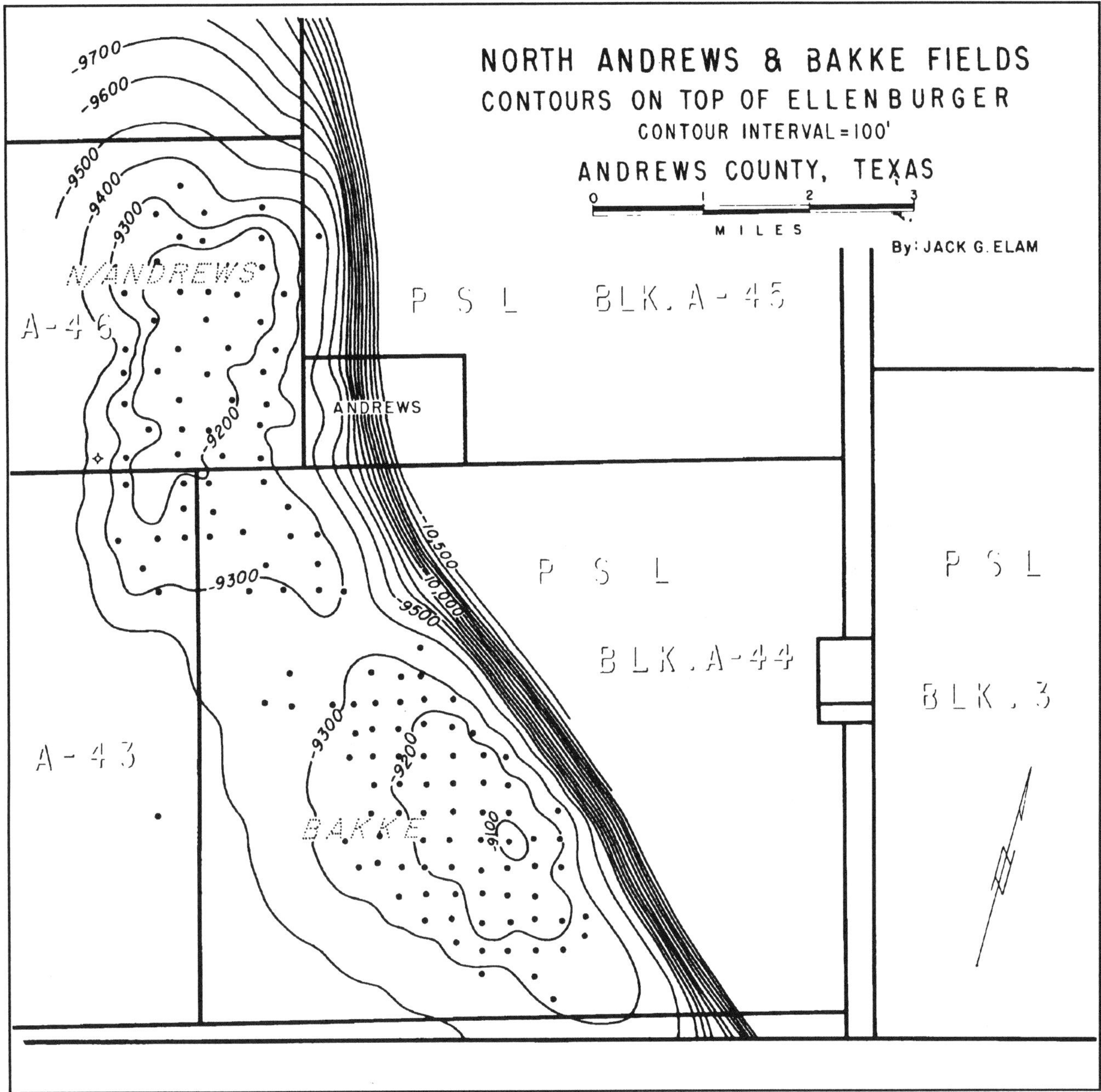

Fig. 13. (Elam, 1969; p. 61).

you to solve the problem. It is simply, *how do you tilt the block*? This is the question Prucha et al. (1965) never addressed, and the failure to do so cost his team a lot of answers (Fig. 19, 20, 21, 22). You have to remember that the block has a third dimension. We know the basement is brittle down to a depth of at least fifteen kilometers, so you have to liken the problem of how you would be able to tilt a bunch of rectangular 4 × 4's that are several feet long and in juxtaposition. When you determine how to tilt the 4 × 4 blocks, you will have lots of answers. The reason no one has solved the problem is because no one bothered to ask it. Almost any researcher could have answered it.

I have one solution, but I don't know that it is unique. Very seldom do we have enough data to render a completely unique solution. I would be very interested in hearing yours. So far as I can tell, you will not find the answer in the literature. This is what makes it fascinating, because possibly 25 percent of all the oil and gas in the world is trapped in structures created by this stress system, and yet it is ignored or

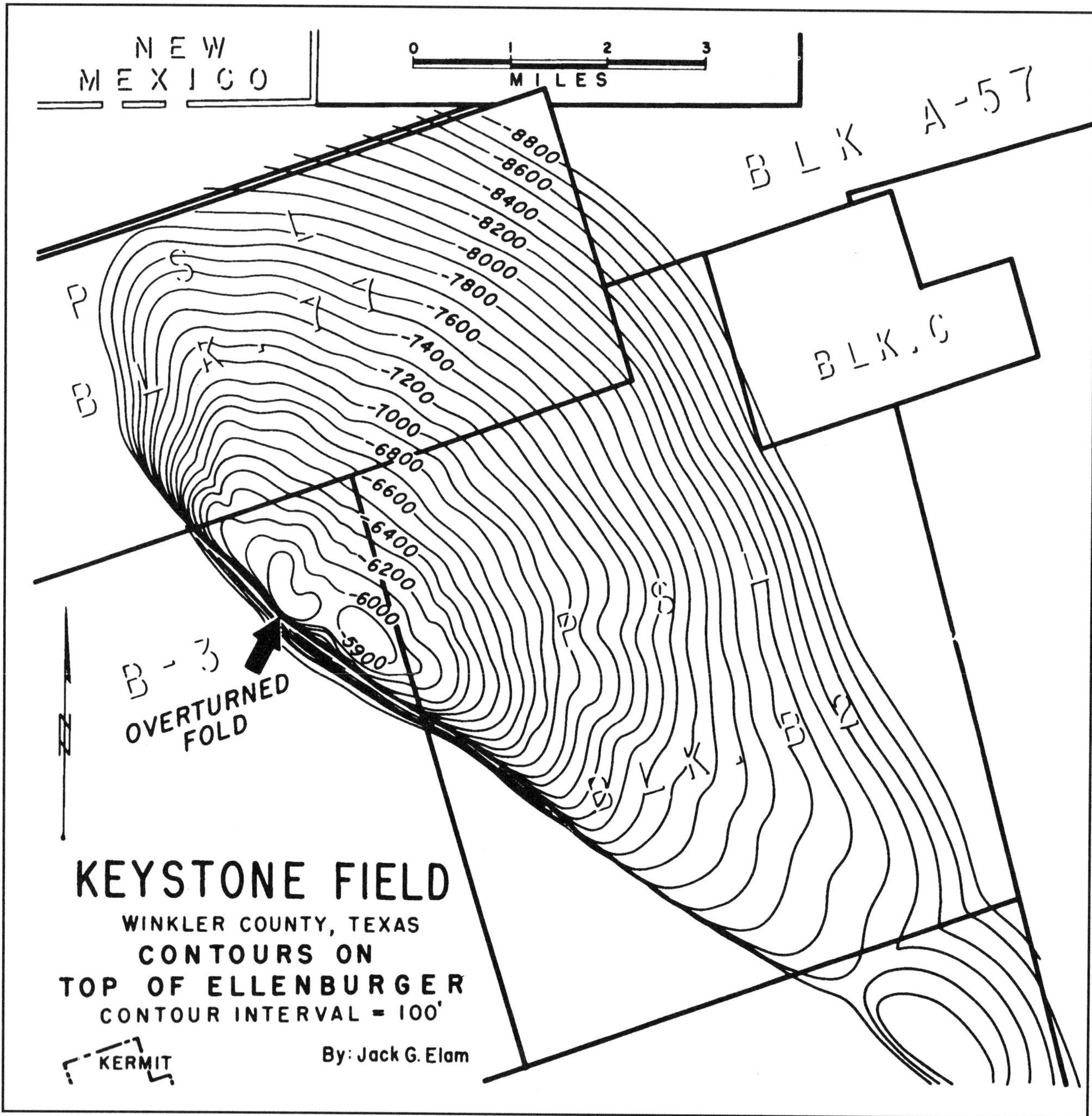

Fig. 14. The Keystone Field is almost always shown as faulted on the west side. No wells actually penetrate the fault plane, but overturned beds are encountered. If the basement fault does not extend upwards through the Ellenburger, however, considerable tectonic thinning would be required (Elam, 1969; p. 65).

downplayed in structural geology textbooks. I call it the fourth structural style.

Summary and Conclusions

There are lots of oil and gas fields out there to be found. However, the prospects are getting smaller and will require an increasing amount of geological skill and creativity to find them. As the average discovery decreases in size with each passing year, it is offset by the fact that our skills, knowledge, and control are increasing. There has been a tremendous improvement in geophysics recently. Some will maintain that this tool will become so good in the future

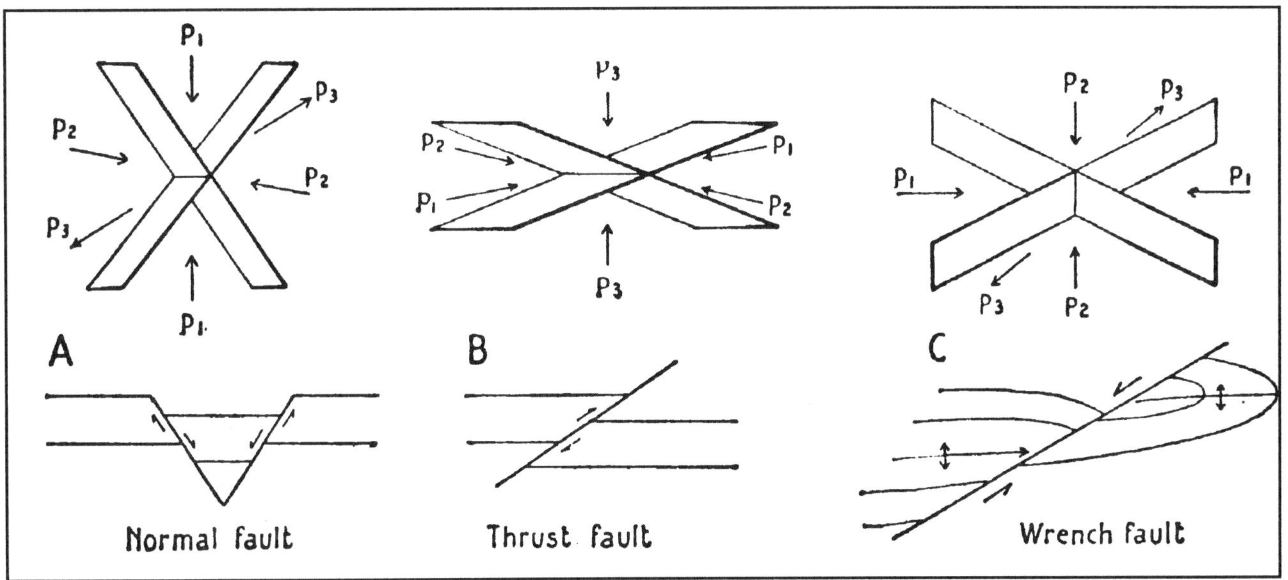

Fig. 15. The origin of faults. P_1 = largest, P_2 = median, and P_3 = smallest principal stress (DeSitter, 1956; p. 119).

that geologists will become redundant, but there never will be a time when a creative geologist will not be an important factor in exploration.

I have tried to emphasize the importance of creativity in the process, because we are continually faced with mysteries that require solutions. Wm. J. J. Gordon (1961) defined creativity as problem stating, problem solving. He analyzed the psychological states of creativeness. In interviewing a great number of creative people in all types of fields, he found that they followed a pattern, be they artists, composers, inventors, or writers. After studying his conclusions, it became obvious that creative geologists should also be included. His psychological states are as follows:

(1) *Detachment*—work without specific objective. Map for understanding.

(2) *Involvement*—when you are contouring a surface, imagine that you are a drop of oil and you are going to migrate upwards until you are trapped, or that you are walking around that surface looking at the trees, etc., or snorkeling around the reef.

(3) *Deferment*—don't jump to conclusions. This is one of the greatest failures of eager young geologists. It is very dangerous to latch onto the first solution that comes to mind. You have to keep an open mind until you have considered all possibilities.

(4) *Speculation*—think of all possible solutions. Develop numerous working hypotheses. Attempt to prove each wrong. Never try to prove you are right. If you can't prove your ideas are wrong, then and only then are they usable; but that still doesn't make them right.

(5) *Autonomy of object*—all falls in place. This happens if your working hypothesis is really valid. All of a sudden, in a flash, all the loose ends fall into place

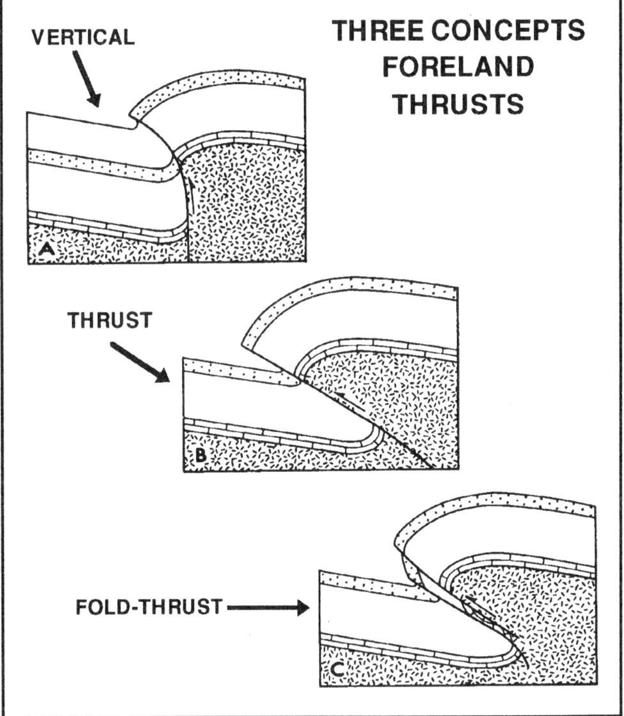

Fig. 16. Diagrams illustrating hypotheses of mountain flank deformation: A—block uplift, B—thrust uplift, C—fold-thrust uplift (Berg, 1962; p. 2028).

and you are overcome with a sense of elation. Once you have it happen, you will really know it. You will have generated a real prospect, and your discovery

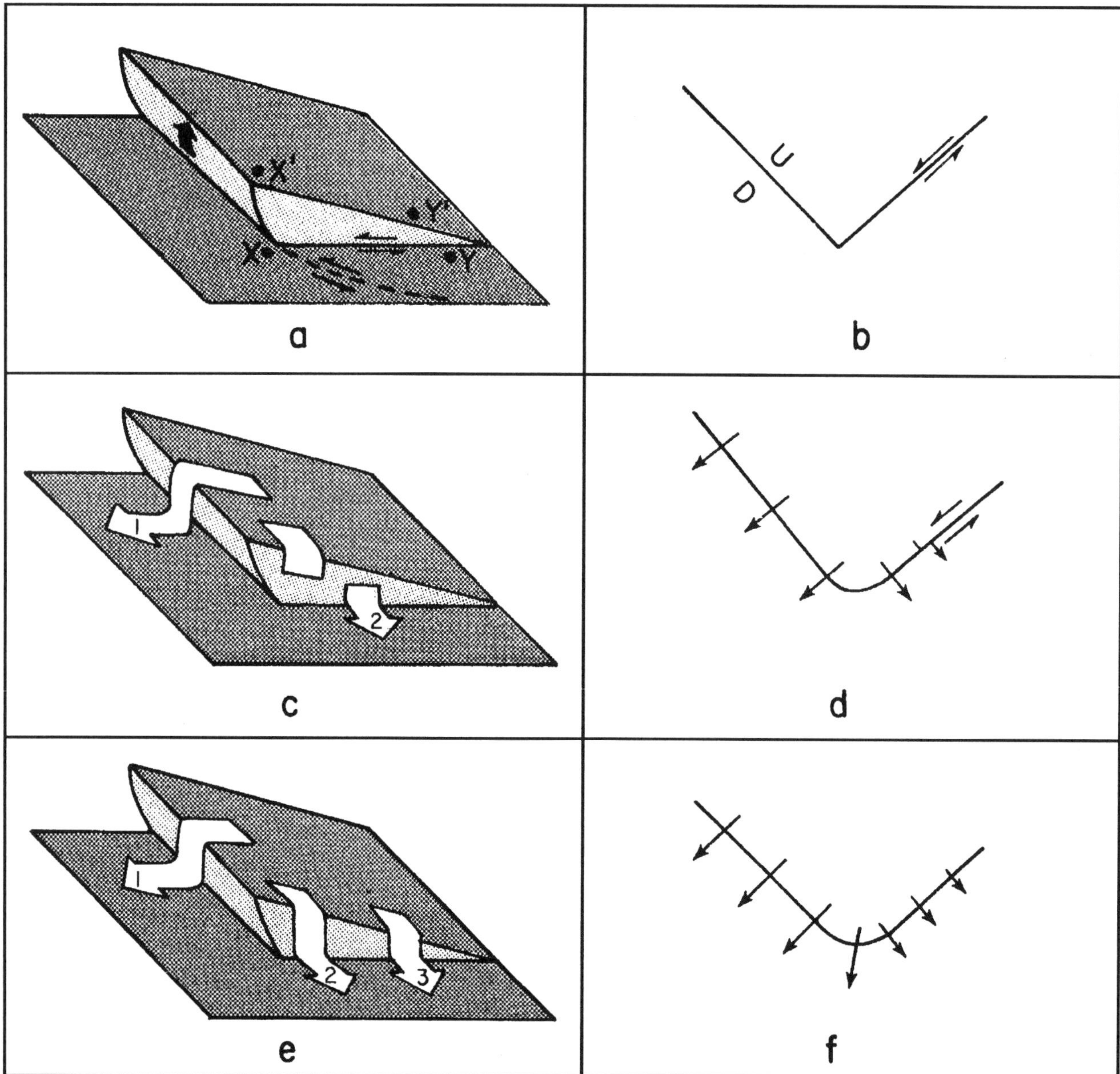

Fig. 17. Schematic diagrams showing block configurations if they are underlain by a low-angle thrust fault (a-d) opposed to what actually occurs (e and f); b, d, and f are simplified maps of the block configurations shown in a, c, and 3, respectively (Stearns, 1978; p. 14).
IN THE PERMIAN BASIN, THERE IS AMPLE WELL CONTROL ON THESE RIGHT ANGLE FAULT INTERSECTIONS THAT SHOW SIMILAR DRAPE OF THE OVERLYING SEDIMENTS OVER EACH FAULT SEGMENT. IF THERE IS LATERAL TRANSPORT, THE PATTERNS OF DEFORMATION WOULD BE DIFFERENT BECAUSE THE STRESS SYSTEMS WOULD DIFFER.

rate will be much above the industry average. In many cases, there will not be enough data initially to reach that point with your first well on the prospect; but if you remain convinced and have courage enough to drill for information, you often will reach that point after you have paid the price to obtain the data required. In many cases, you have to recommend drilling locations that you know will probably be dry, but that are justified if your objective is of sufficient size to make a risk of this type economic.

Many years ago, a former president of the AAPG, Frank Morgan (Vice-President, Richfield Oil Corp.), recommended that we set up a special classification for wells drilled for information alone, so that they

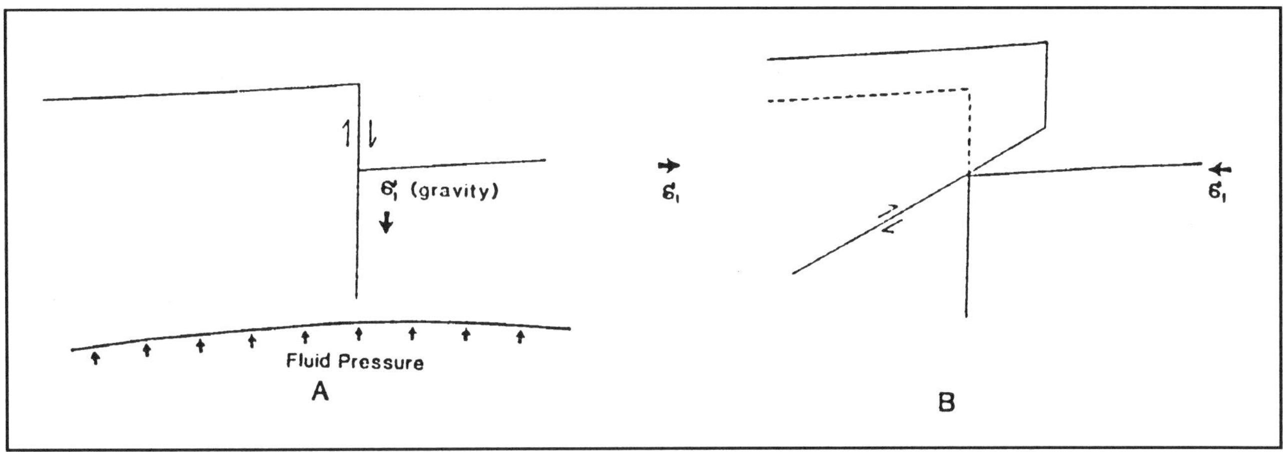

Fig. 18. Composite structure. Vertical "upthrusting" followed by compressed shear failure.

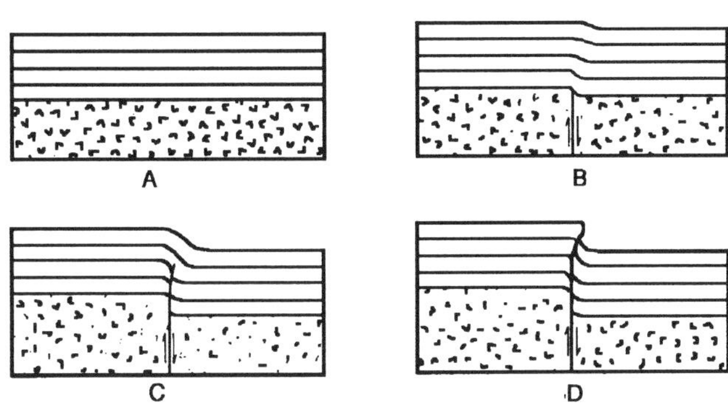

Fig. 19. Postulated sequence in deformation of sedimentary rock beds overlying basement fault blocks. (A) Before faulting in basement. (B) Initial stage of faulting in basement. Sedimentary beds adjust by bending without faulting. (C) Intermediate stage in basement faulting. Lower sedimentary beds have exceeded critical degree of bending and are faulted; upper sedimentary beds are bent but not faulted. (D) Advanced stage of basement faulting. Faulting extends through sedimentary beds to surface (Prucha et al., 1965; p. 983).
THIS MODEL WAS ACTUALLY CONSTRUCTED WITH TWO WOODEN BLOCKS SAWED VERTICALLY AND OVERLAIN WITH MULTICOLORED SAND. THE BLOCK WAS PUSHED UP BY HAND. THIS IS TRUE "VERTICAL UPLIFT."

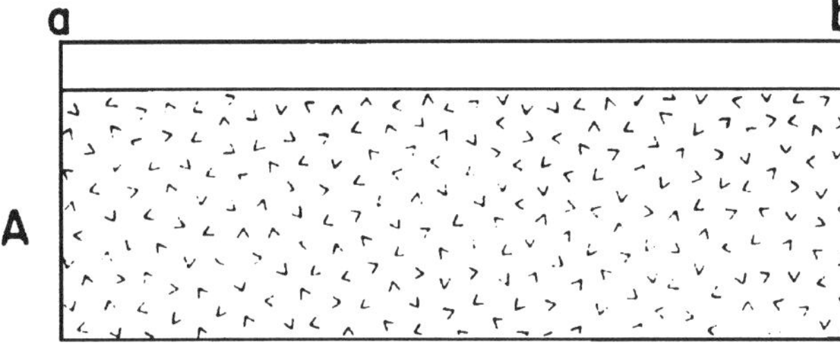

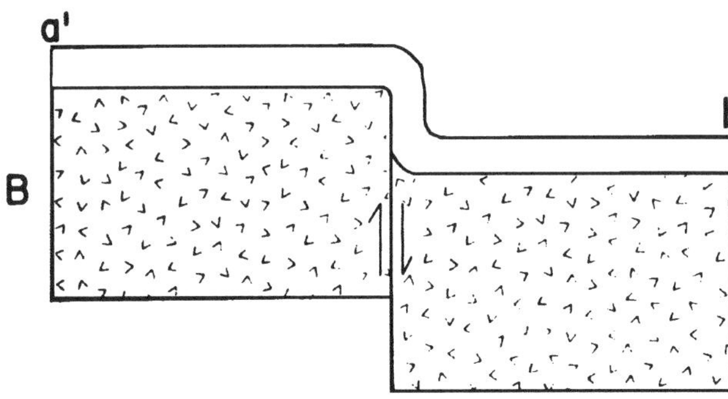

Fig. 20. Draping of sedimentary beds over edge of a basement fault block. In A, basement is not faulted; in B, basement is faulted, and sedimentary beds are draped over edge of upthrown block. a'b'>ab (Prucha et al., 1965; p. 983).
THE ONE THING THIS MODEL FAILS TO EXPLAIN IS THE PROCESS BY WHICH a-b IS LENGTHENED TO a'-b'. IN THE PERMIAN BASIN, THE PLASTIC SEDIMENTS ARE STRETCHED AND THINNED.

would not be included in the wildcat statistics. This was to avoid the stigma of a dry hole. The nearest we have come to that are some of the COST wells drilled offshore under the auspices of the U.S.G.S., but paid for by the companies interested in leasing the area. These wells are deliberately drilled off structure as defined by geophysics so no one expects them to have any oil or gas. Surprises do show up sometimes because hydrocarbons are trapped off structure, too. However, no one feels bad when they are dry.

Very few geologists have the courage to recommend wildcat locations that they expect to be non-productive, but I have found that if you level with your investors, you can commonly sell this type of risk. It is

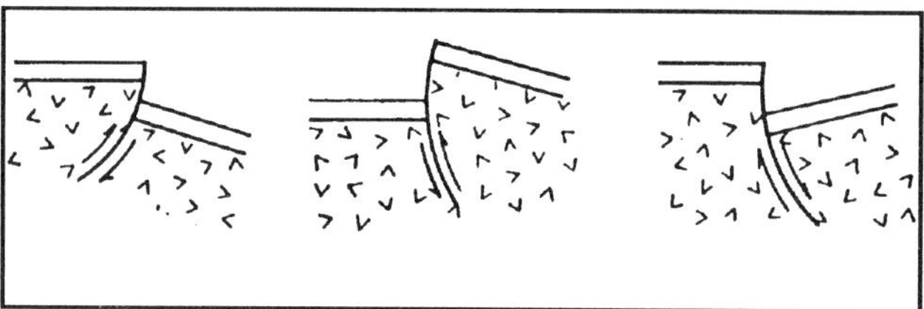

Fig. 21. Geometric relations of differently tilted fault blocks explained by curving faults at boundaries between blocks (Prucha et al., 1965; p. 982).
THIS IS ONE WAY TO GET A TILTED BLOCK, AND TILTED BLOCKS MAY BE FORMED THAT WAY IN THE ROCKIES, BUT IN THE PERMIAN BASIN, NO RANDOMLY TILTED BLOCKS HAVE BEEN MAPPED.

one way to make major discoveries. Human nature being what it is, though, seldom do all of your investors stay with you throughout the project. Times change, and investor economics change, too. The real losers are those who drop out prematurely, because deals of this type usually would not have passed economic muster on the first wildcat alone. The real winners are those who have the opportunity to get in on the deals after the first or second dry hole has been drilled!

In closing, I want to point out that successful exploration is done by those who have the best feel for the risk involved. Any wildcat, no matter how risky, is worth drilling if it promises to give a satisfactory return. If you are looking for $1,000,000,000 worth of oil or gas on a prospect, you can drill a very risky wildcat. The real key to success is to drill in areas where you are relatively certain that the trap has to lie in there somewhere, even though you are not exactly sure where. It is amazing how much additional light will be shed on the specific location of the trap by a single test well. If your hypothesis is not disproven, hang in there until you find the field! The scatter gun approach is probably the least economic method of playing the oil game. You drill for knowledge. The money will follow. Even to this day, there is no comparable economic opportunity in other lines of endeavor. As I said earlier, it is the most exciting game in town. You have to be careful that it doesn't consume you.

BIBLIOGRAPHY

Anderson, E.M., 1942, The Dynamics of Faulting: Oliver and Boyd, London, 183 p.

Berg, Robert R., 1962, Mountain flank thrusting in Rocky Mountain foreland, Wyoming and Colorado: AAPG Bulletin, v. 46, p. 2019-2032.

Cloos, Hans, 1930, 1931, Zur Experimenteller Tektonik I & II Dic Naturwissenschaften, 18 (34) p. 741-7; 19 (11).

Cloos, Hans, 1932, Zur Mechanik grosser Bruche und Graben Centr bl. f. Min., etc., p. 273-86.

Dallmus, K. F., 1958, Mechanics of basin evolution and its relation to the habitat of oil in the basin, in Habitat of oil: Tulsa, Okla., AAPG, p. 883-931.

De Sitter, L.U., 1956, Structural geology: McGraw-Hill, New York, 532 p.

Elam, Jack G., 1969, Tectonic style in the Permian basin and its relationship to cyclicity, in Chuber, S. and Elam, J., eds., Cyclic sedimentation in the Permian basin: West Texas Geologic Society, pub., p. 56-69.

Elam, Jack G., 1984, Industry problems rooted in people, not lack of reserves, in World Oil, March 1984.

Gordon, Wm. J.J., 1961, Synectics—The Development of Creative Capacity: Collier Books, Collier-MacMillan, Ltd., London.

Hefner, W., 1951, Stress distribution and faulting: GSA Bulletin, v. 62, p. 373-398.

Martin, Rudolf, 1966, Paleogeomorphology and its application to exploration for oil and gas: AAPG Bulletin, v. 50, p. 2277-2311.

Nulty, Peter, 1984, Big oil's top explorer takes to the high seas: Fortune, v. 109, n. 2, p. 162-164.

Pratt, Wallace E., 1952, Toward a philosophy of oil finding: AAPG Bulletin, v. 36, p. 2231-36.

Prucha, John James, John A. Graham and Richard P. Nicholson, 1965, Basement controlled deformation in Wyoming Province of Rocky Mountains foreland: AAPG Bulletin, v. 49, p. 966-992.

Rittenhouse, Gordon, 1959, There is a reason: AAPG Bulletin, v. 43, no. 7, p. 1500-1502.

Sanford, A. R., 1959, Analytical and experimental study of simple geologic structures: GSA Bulletin, v. 70, p. 19-52.

Smithson, Scott B., Jon Brewer, S. Kaufman, Jack Oliver, and Charles Hurich, 1978, Nature of the Wind River thrust, from COCORP deep-reflection data and from gravity data: Geology, v. 6, p. 648-652.

Stearns, David W., 1978, Faulting and Forced Folding in the Rocky Mountain Foreland, GSA Memoir 151, p. 1-37.

Weinmeister, Marcus Paul, 1978, Origin of upper Bell Canyon reservoir sandstones (Guadalupian) El Mar and Paduca fields, southeast New Mexico and west Texas: Texas A&M master's thesis, 95 p.

Williamson, Charles R., 1978, Depositional processes, diagenesis, and reservoir properties of Permian deep sea sandstones, Bell Canyon formation, Texas-New Mexico: Texas Petroleum Research Committee, Report No. UT 78-2, 260 p.

Williamson, Charles R., 1979, Deep-sea sedimentation and stratigraphic traps, Bell Canyon formation (Permian) Delaware basin: SEPM (Permian section), Symposium and Field Conference Guidebook, Publ. 79-18, P. 39-74.

Williamson, Charles R., 1980, Sedimentology of Guadalupian deep-water clastic facies, Delaware basin, New Mexico and West Texas: N.M. Geol. Soc. Guidebook, 31st Field Conference, Trans Pecos region, p. 195-203.

APPENDIX

Rules of Contouring

1. Contours never end.

2. Contours never cross (except on overhangs or overturns).

3. Contours never bifurcate.

4. When you go down through a contour, you have to go back through the same contour again before you can go upstructure.

5. Contours tend to be parallel.

6. Lacking any other data, contours are equally spaced.

7. It is impossible to contour without a concept. The concept of "no concept" is still a concept. Computers contour with a very limited concept.

8. Where do you develop the concept? From cross sections. You must have a good grasp of what you are contouring before you can do an adequate job of contouring.

9. The contouring ability is the most important play-making ability a geologist can have.

10. In order to contour properly, you must have a good conceptual model in your mind. Contouring is an art form.

11. Contouring should always be as optimistic as possible. Every wiggle you put on the map should have a reason. As a result, contour maps should be averaging maps. By being overzealous in making them appear "real," you may eliminate some acreage from the play that really should be bought.

12. For picking a location, you should use the most pessimistic contouring possible.

13. In order to really evaluate risk, you need to contour the area every way possible. The more amenable it is to multiple interpretations, the greater the risk.

14. Your contours should always account for the oil shows (unless you know the shows are stratigraphic).

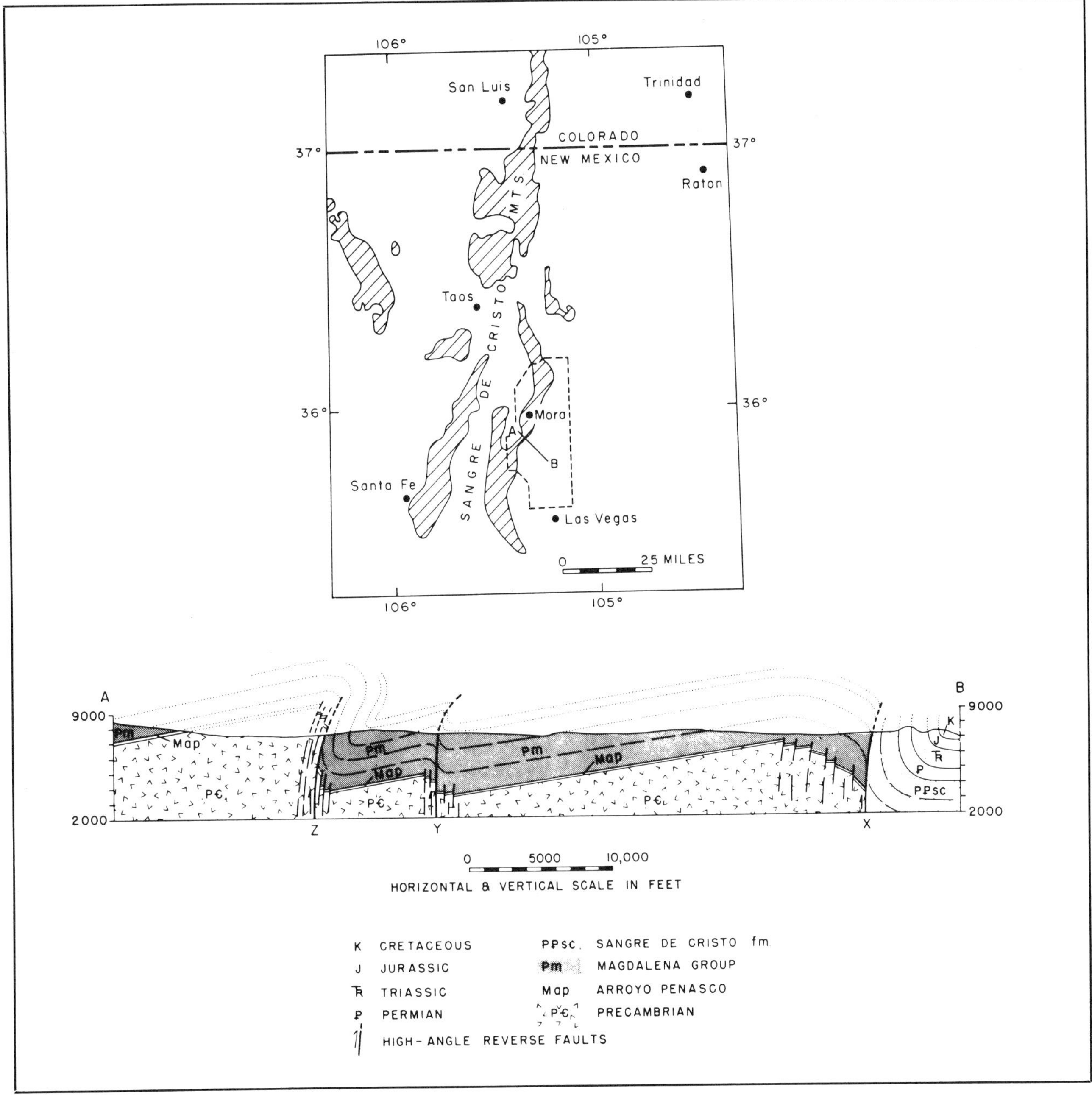

Fig. 22. Basement block faulting and related structures in southeastern part of Sangre de Cristo Uplift near Mora, New Mexico. Index map of Mora area and cross section A-B showing relation between basement block faulting and asymmetrical folds in overlying sedimentary formations (Prucha et al., 1965; p. 977).
THIS IS THE NORMAL STYLE OF TILTING IN BOTH THE ROCKIES AND IN THE PERMIAN BASIN. THIS SUGGESTS THAT THE TILTED BLOCKS ARE PART OF A SYSTEM. THE KEY IS TO DETERMINE WHAT SYSTEM WILL CAUSE THIS GEOMETRIC RELATIONSHIP.

Volume 27   Number 7

BULLETIN
of the
AMERICAN ASSOCIATION OF
PETROLEUM GEOLOGISTS
JULY, 1943

DISCOVERY THINKING[1]

A. I. LEVORSEN[2]
Tulsa, Oklahoma

Every man, woman, and child in the United States had a stake in the oil industry during times of peace, but now that we are at war, the stake is infinitely greater, for, as top-ranking government officials say, "petroleum is one of the most essential of war industries."[3] This stake is expressed in many ways: through ownership in the industry; in utilization of its products as fuel, power or raw materials; in an appreciation of the many taxes paid by the oil industry; and in the applications of petroleum and its products to war. The affairs of the petroleum industry have outgrown the simple business of producing, manufacturing, and selling petroleum and its products; they are now of universal military, economic, social, and political importance. As a matter of fact, the public is in the oil industry and probably here to stay.

Because of this public interest in petroleum, society, in planning and thinking of its future, has the right continually to ask "How long can we depend on a supply of petroleum adequate for our needs?" "What about the future reserves?" "Can we continue to build an economy based on an abundance of raw materials in the form of petroleum and gas?" And we in the oil industry, as the custodians of this national resource, should continually be prepared to give answer so that the time does not come when the public finds that it has been misled; when, instead of scarcity there is plenty; or when, instead of plenty there is scarcity. Much planning—economic, military and political—will be affected by the answer and it becomes a public responsibility that the answer be as accurate and as up-to-date as possible.

In approaching such a problem as this, it is well first to get some idea of its size and proportions. The problem from a national viewpoint is not the same as it

---

[1] Read before the Association at Fort Worth, April 7, 1943. Manuscript received, April 14, 1943.

[2] Consulting geologist, 221 Woodward Boulevard. Chairman, Association research committee.

[3] Harold L. Ickes, Petroleum Administrator for War, and Paul V. McNutt, Chairman, War Manpower Commission, *Chicago Journal of Commerce* (February 10, 1943), p. 13.

A. I. LEVORSEN

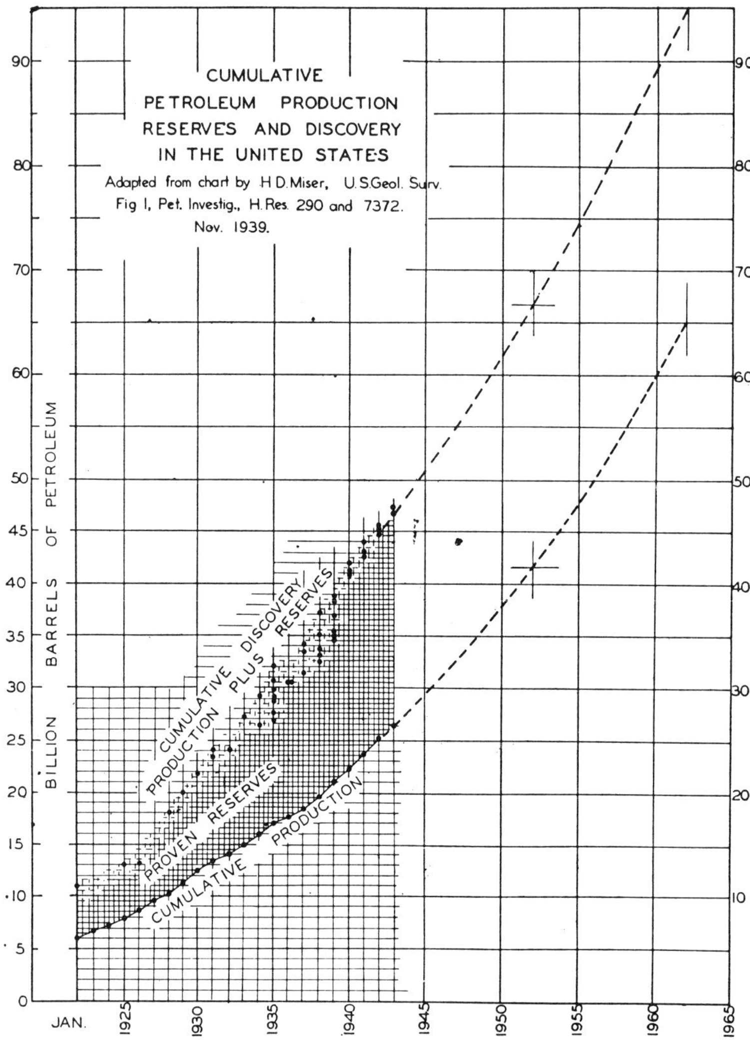

FIG. 1.—Chart showing, through use of cumulative curves, past trends in petroleum production and reserves in United States. When projected into the future, dashed lines indicate magnitude of discovery problem if present trends continue. Upper black dots indicate estimates of reserves by various authorities in different years.

is from the viewpoint of the individual or even that of a major oil company—the individual or company needs are infinitesimal by comparison. Figure 1 is designed to indicate the future needs by the projection of our past record of growing supply and demand. Here we see that during the next 20 years, or by 1963, if we are to maintain the same rate of increase in production as in the last 20 years, and if we are to continue to have as a discovered reserve 13 or 14 times the annual production, it will be necessary to discover the staggering amount of 50 billion barrels of new oil. This is a greater amount than has been discovered by all methods since the beginning of the oil industry 80 years ago. The present trends, when projected into the future, indicate a steadily increasing tempo of discovery, production, and consumption. We are talking about a lot of oil and in the light of our current decline in rates of discovery in the face of steadily increasing war needs, we have something deserving the most serious and careful attention.

Experience has demonstrated that the uses of petroleum have expanded with the expanding supply. When there has been an extra amount of petroleum, industry has developed new uses to take up the surplus. We have recently seen how a local shortage—due it is true to a shortage in transportation, but in effect a shortage in petroleum, nevertheless—quickly caused a shifting to other sources of energy and heat. This may well be the clue as to what will happen at any future time if a shortage of petroleum should develop. Since the supply of petroleum is directly related to discoveries of new oil pools, it is quite obvious that the key to the whole question of future supply rests in the ability of the industry to make the necessary discoveries. Failing to discover new pools, the entire economy dependent on petroleum is affected.

Since no one can know in any exact or even approximate sense what the discoveries of a year, or even a month in advance may be, another problem is to decide what sort of an approach can be made toward some reasonable idea of the amount and location of the future petroleum reserves. It seems to me that the only answer available to us is that the approach must be made through the reason and philosophy which result from past experience and which may be applied to the future. The best we can hope to do is to continue to add to our present small supply of philosophy concerning the relations of petroleum to geology, then to apply all such understanding to our knowledge of geological conditions. From that we may decide whether or not there is reason to believe that there are adequate reserves remaining to be discovered, and also how the discoveries may be made. Our answer will be a function of our philosophy, of our theories, and of our thinking.

This has been most effectively stated by Pratt,[4] in discussing the lack of successful discovery in many of the foreign countries.

[4] Wallace E. Pratt, "Geology in the Petroleum Industry," *Bull. Amer. Assoc. Petrol. Geol.*, Vol. 24, No. 7 (July, 1940), pp. 1212–13.

... The suspicion grows that physical conditions within the earth's crust impose fewer and less formidable obstacles to the development of commercial oil fields over much of the land area of the globe than are raised up by our prevailing mental and social climates of opinion. The limitations put upon the discovery of oil on earth to-day are more mental and social than physical. Gold is where you find it, as the old adage has it, but oil must be sought first, at least, in the mind ...

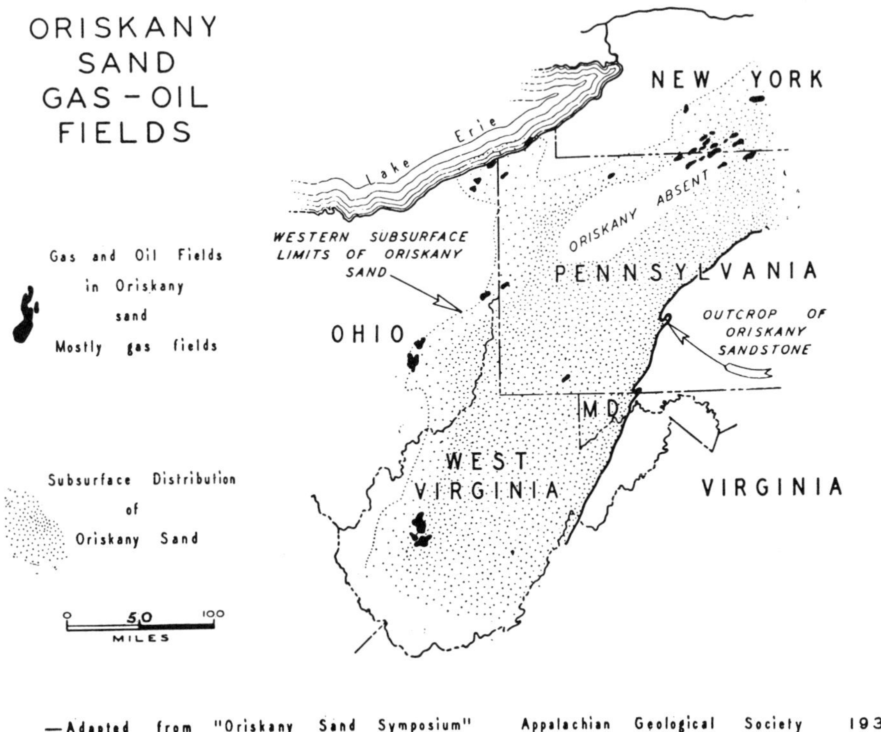

FIG. 2.—Map showing subsurface distribution of Oriskany sand (Lower Devonian). Sand thins out updip northwest. Along this belt of wedging-out of porosity, gas and oil fields producing from Oriskany sand are found. Many fields are in local anticlinal folds.

Much oil has been discovered by mere random drilling, and much will most assuredly be discovered through such means in the future, but the only rational way to make available to society the oil and gas which now lies underground, as the undiscovered reserve, is through the application of scientific methods and reasoning. It is the one approach which offers reproducible results. A correct scientific principle applied to the accumulation and occurrence of oil and gas can be applied repeatedly to Oklahoma, to the United States, or throughout the world—this year, next year, or a hundred years from now. It becomes a foundation upon which we can build indefinitely into the future.

The occurrence and distribution of oil and gas deposits are intimately related to the occurrence and distribution of varying kinds of geological conditions, and because of this relation, the problem of where to search for oil pools and what might be expected to result from this search is fundamentally geological in nature. Within the oil industry, the petroleum geologist is the one to provide the answer about

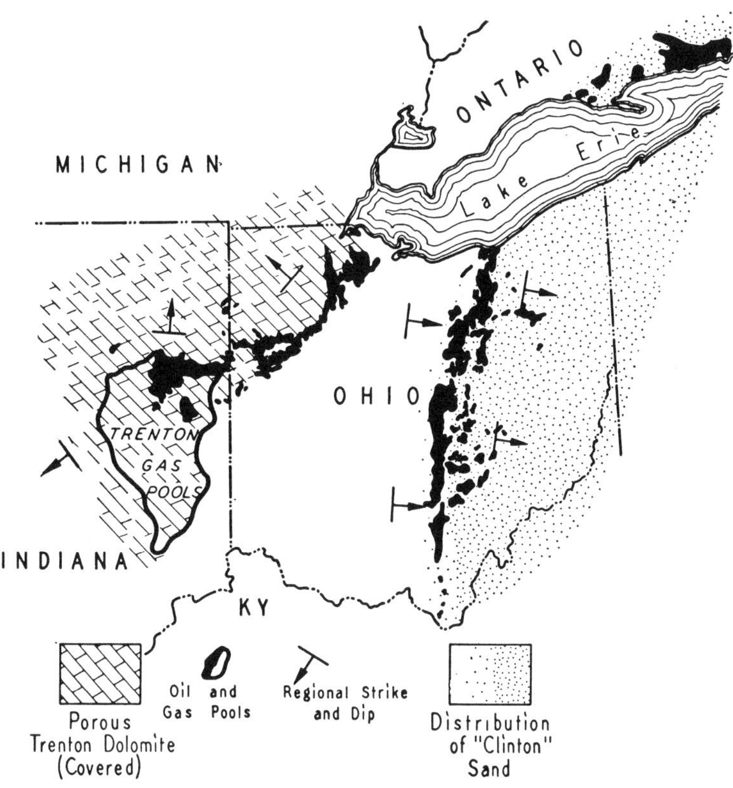

Fig. 3.—Map showing subsurface distribution of Clinton sand (Silurian) with gas and oil fields associated with its thinning-out updip westward, and subsurface distribution of porous dolomite part of Trenton formation (Ordovician) with oil and gas fields associated with updip edge of porosity. Individual pools in both provinces are in part on local structure but mostly associated with local porosity.

the adequacy of our petroleum reserves and about the methods by which they may be discovered. A petroleum geologist is one who is engaged in the interpretation of the relations of the rocks to the occurrence and production of petroleum, whether he began as a paleontologist, a mineralogist, a geophysicist, a geochemist, or as an academic geologist. This problem of the discovery of an adequate oil supply becomes our responsibility to the industry and in turn to the nation. Since the petroleum geologist is the one whose primary purpose is to guide the discovery effort, it can readily be seen that he occupies a most important post in the national economy—in fact he may well be considered as the keystone in the arch of that part of our future national economy which is dependent on petroleum whether as a source of power, as a fuel, or as a raw material. Truly, geology in all of its history has never reached such a plane of useful application to humanity. The size of the problem gives one real cause to approach an answer in a spirit of deep humility.

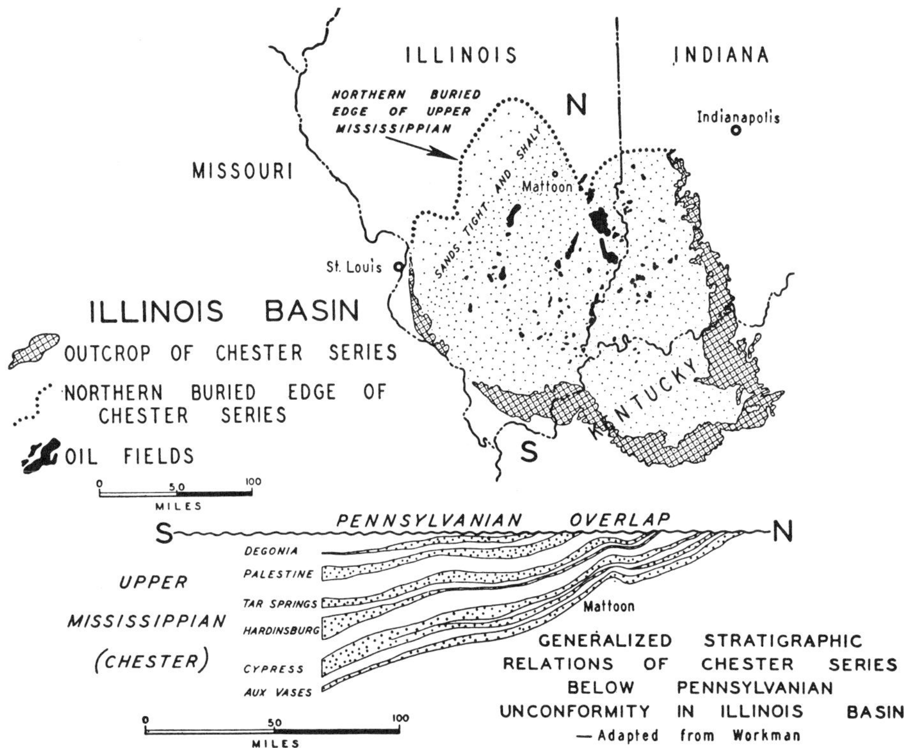

Fig. 4.—Pools of Illinois basin region are in region of updip wedging-out of numerous sands and reservoir rocks of Chester or upper Mississippian series. Area now underlain by Chester rocks in Illinois basin is only small segment of area originally part of wedge belt. Chester rocks extended southeast into Arkansas and Oklahoma, east into Appalachian states, and south into Alabama. See A. I. Levorsen, "Pennsylvanian Overlap in United States," *Bull. Amer. Assoc. Petrol. Geol.*, Vol. 15, No. 2 (February, 1931), pp. 118–48; and "Studies in Paleogeology," *ibid.*, Vol. 17, No. 9 (September, 1933), pp. 1107–32, particularly Fig. 2.

As we have seen, the normal future expansion of the petroleum industry calls for discoveries on a scale which to many is almost, if not, impossible of achievement. Some feel that the decline in discovery rate is because of the approaching exhaustion of our ultimate reserves—that we are scraping the bottom of the barrel, in other words. If this be true the situation is indeed serious. On the other hand the discovery decline may be due to an insufficient discovery effort, in which case something can be done about it.

The first part of what follows is intended as a contribution to our foundation of geological principles which apply to the search for new reserves in the form of new oil provinces. At best it can show only a few aspects of an infinitely large and broad subject—but a subject to which we must give continual attention if we are to utilize and apply the ever-increasing mass of new data being discovered daily throughout the oil areas. Principles, without data on which to apply them,

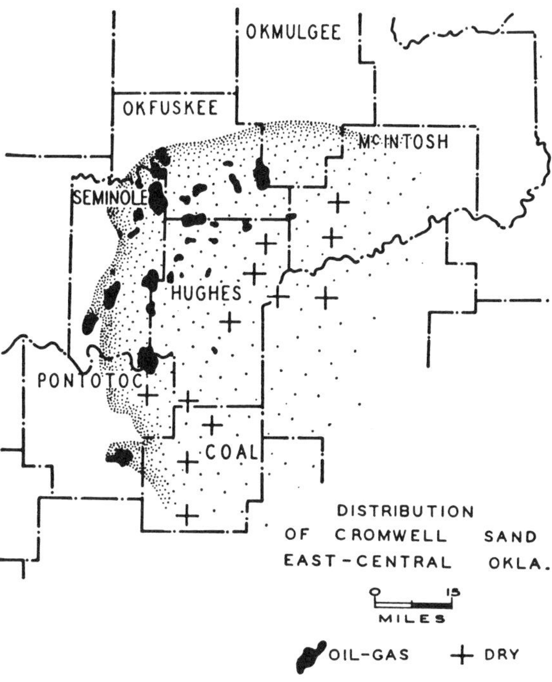

FIG. 5.—Map showing subsurface distribution of Cromwell sand (Pennsylvanian) in east-central Oklahoma. Oil and gas fields, most of them on local anticlines, shown in black; crosses mark equally good or better anticlines which are dry in same formation. Association of oil fields with updip wedging-out of sand is evident.

are of academic interest only, but data without principles by which they might be applied, are useless, costly, and wasteful.

The second part is concerned with some ideas for improving the discovery effort from a strictly geological viewpoint.

UNDISCOVERED RESERVES

Whether the declining discovery is the result of lack of oil to be discovered or lack of effort, there is one factor which enters into the over-all problem and that is, that from a national viewpoint we are too often thinking in too small terms. By far the greatest part of our effort is in searching for extensions to pools already discovered, in searching for deeper sands, and in working and reworking progressively smaller and smaller areas within the developed provinces. We finally get to think almost exclusively in terms of individual oil fields, when we should be thinking in terms of new oil provinces; in discoveries of millions of barrels, when we should be thinking in terms of billions of barrels; and in acres of new production when we should be thinking in square miles. If you agree that our work is essential to the

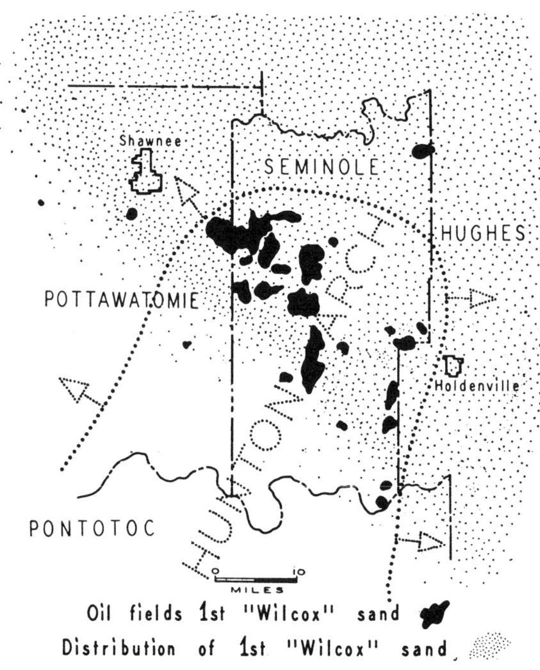

Fig. 6.—First "Wilcox" sand (Ordovician) grades out into dolomite and limestone toward the southwest. Hunton arch crosses edge of porosity as it pitches northeast, and in area where arch crosses wedging-out of sand are found rich pools of Seminole district in Oklahoma. Individual pools are localized by upfolds and local deformation, and province is localized by combination of favorable stratigraphy and regional arching.

future oil supply of the nation, it means that we must turn an increasing part of our thinking into terms of not one but many wholly new oil provinces. It means that we must attempt to think in terms of a steady procession of major oil discoveries into the future. It means that we must think in terms of "big" geology so that we may be assured of "big" results.

In order to reduce this discussion to more concrete terms and to everyday experience, two types of geological concepts or principles are presented to show the kind and the scale of thinking that seems to be called for, in an attempt to formulate a better idea of the nature and the extent of the undiscovered reserves of petroleum in the United States. The layman may say, with a wave of his hand, "There are plenty of reserves, we have always found enough, why worry?", but we, who are technically trained, must attempt at least to put the answer into terms which can be justified by geological reasoning based on experience and applied to known geological conditions.

The first of these principles has to do with updip wedge belts of porosity in reservoir rocks and the second has to do with the layers of geology which are sepa-

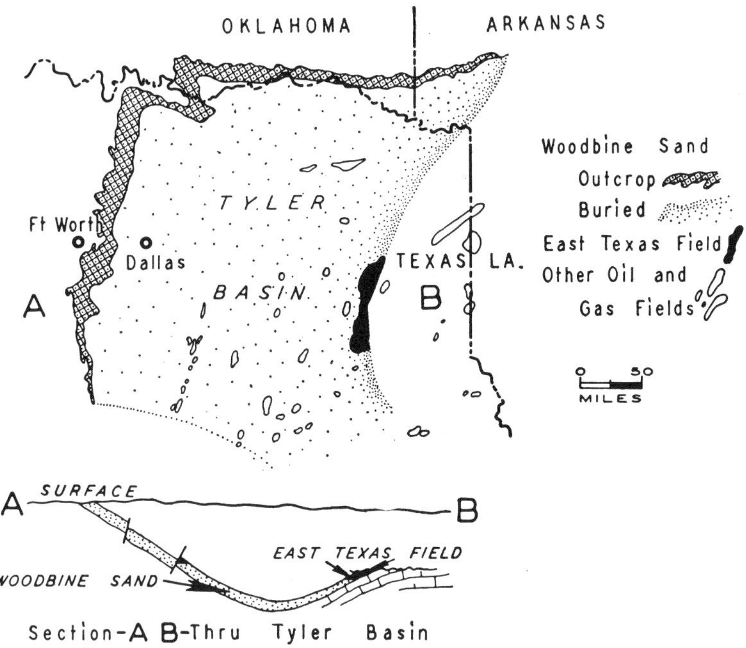

FIG. 7.—Map showing outcrop and subsurface distribution of Woodbine sand (Upper Cretaceous) in northeastern Texas, with East Texas oil field located along eastern, updip wedge edge of porosity. In this case, wedge edge is result of uplift and truncation of Woodbine sand with subsequent overlap of younger formations.

rated by unconformities. They may be considered as principles bearing on the occurrence of oil provinces, both discovered and undiscovered, and it is hoped that an understanding of them will enlarge and clarify your concepts of the occurrence of oil as they have mine.

WEDGE BELTS OF POROSITY

From Pennsylvania to California and from the Ordovician to the Tertiary, we find that oil province after oil province is located in close association with the updip wedging-out of the porosity in the reservoir rock. This association is not coincidental or occasional, as will be seen from the familiar examples which follow, but is a principle which is fundamental to many or probably most of the oil provinces of the United States. The implications of such a philosophy are far-reaching.

Beginning in the eastern United States and moving over the country to California, we find many examples of this relationship and those that follow are selected to show some of the many variations in geological conditions in which the principle is expressed.

These wedge belts of porosity are only a part of those that can be shown to be related to the present distribution of oil and gas pools. Other familiar examples

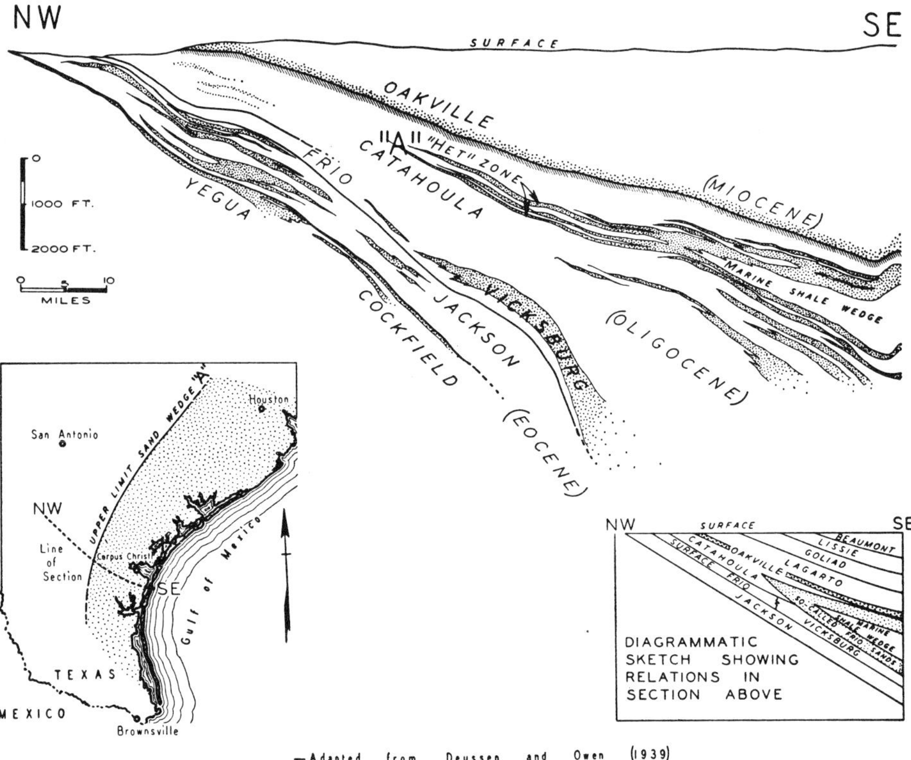

Fig. 8.—Characteristic section along Gulf Coast showing updip wedging-out of most producing sands.

are the Appalachian fields in the Mississippian and Devonian sands; the "Corniferous" production flanking the Cincinnati arch in Kentucky; the Simpson, Bartlesville, and Burbank sand production of northeastern Oklahoma; many of the pools of north-central Texas producing from sands of Pennsylvanian age; many pools in the Laredo district of South Texas; the Lower Cretaceous production of Arkansas and Louisiana; and pools in the Santa Maria Valley, the Eocene and Miocene production in the vicinity of the Coalinga nose, and much of the "East Side" production of the San Joaquin Valley in California. The areas mentioned are shown in Figure 12.

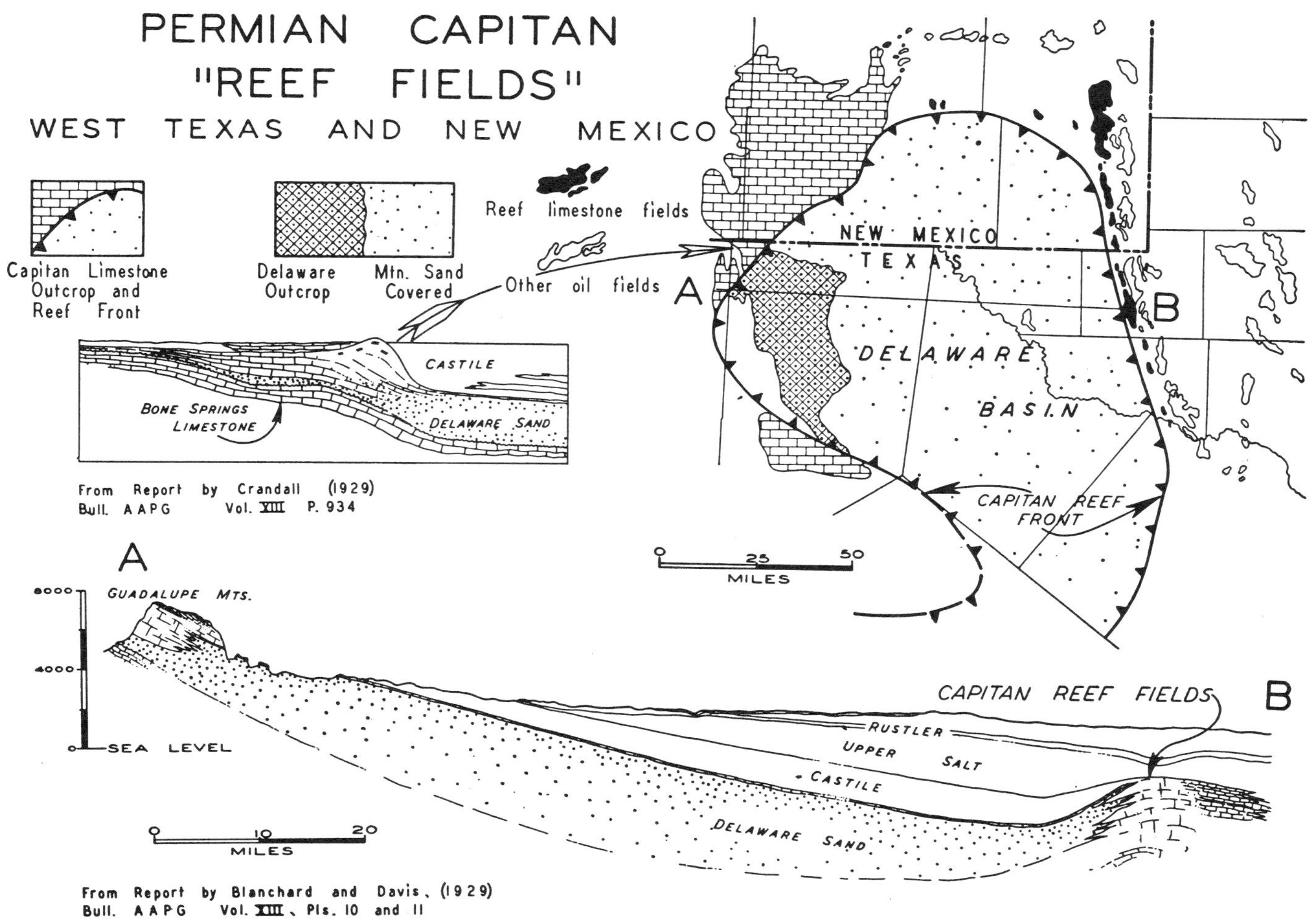

Fig. 9.—Map showing "reef fields" of West Texas and southeastern New Mexico. Fields shown in black are located along fringing reef limestone which occurs along edge of Delaware sandstone (Permian). Porous Delaware sand grades into, and is replaced by, porous reef limestone which in turn becomes dense and non-porous eastward. This edge of porosity marks eastern edge of fields in this province. Individual pools are on local folds which occur along reef front.

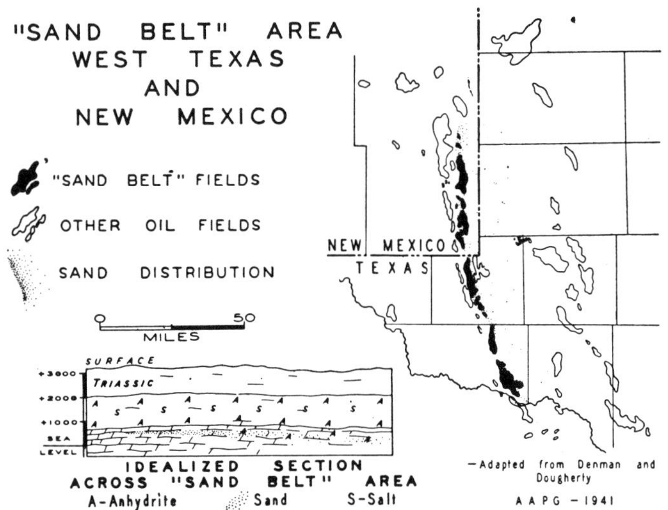

FIG. 10.—Sand Belt pools of West Texas are located along western edge of sand belt in upper part of Permian limestone on back or lagoonal side of fringing reefs.

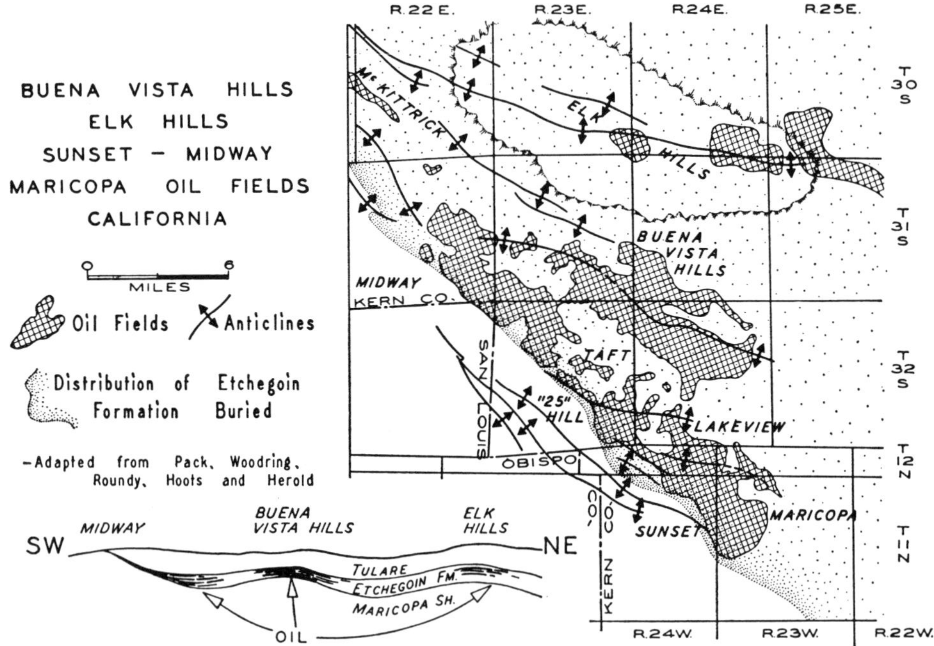

FIG. 11.—Production in Midway-Sunset-Buena Vista province in California occurs in vicinity of updip wedge edge of Etchegoin formation (Pliocene). Sands go out southwest where they are buttressed against rising underlying Maricopa shales along edge of depositional basin in Etchegoin time. Individual pools are associated with local folding and with variable sand conditions.

The origin of a porosity wedge seems to be of less importance than the fact that there is a porosity wedge present. Thus some wedges are the result of a lateral change of facies from sand to shale; others a change from porous limestone or dolomite to non-porous limestone or dolomite; while others are the result of up-lifts, truncation, and overlap. The physical presence of an updip wedging-out of porosity, irrespective of origin, seems to be the important factor rather than the nature of its origin.

If we grant the fact that many or even most oil provinces are associated with the wedge edge of porosity of the reservoir rock, whether the accumulation is in the form of single large oil fields, such as the East Texas field, or is in a number of separate pools, each in a local structural trap yet grouped in the vicinity of the edge of porosity, then the question "Why is this the fact?" may well be asked. Three possible explanations are the following.

1. Wedge belts of porosity may be favorable areas of origin of oil for some unknown reason. Such an explanation may readily account for oil located in wedge belts which are the result of primary environmental conditions such as shore lines, sand bars, or areas of facies change, but it is more difficult to see how this would apply to those wedge belts which are due to secondary conditions, such as uplift, erosion and later overlap.

2. Another explanation is that within the wedge belts of porosity, pressure gradient factors are set up which may be instrumental in extracting the oil and gas from the surrounding shales, limestones, and other possible oil sources. In the regional history of almost any reservoir rock, from the time of its deposition to the present, it has experienced many effects of uplift, faulting, tilting, burial, erosion, or in other terms, the pressures within it, whether hydrostatic pressures or rock pressures, have fluctuated almost as a bellows due to its geologic history, diastrophism, and temperature variations. The wedge edge, because it is effectively sealed in one direction, offers resistance to these pressure changes, and the operation and effect of pressures in such areas may be different from what they are in the center of the basin where circulation is free and there is relief from pressure in all directions. There may then result, in the vicinity of the wedge edge, a suction effect on the surrounding rocks when the pressures are lowered in the neighboring regions due to erosion, lowering of hydrostatic head, or diastrophism. This suction effect may "pull" the hydrocarbons out of the surrounding formations and they would thus replace the water which was "pulled" to areas of lower pressures deeper in the basin.

3. The third explanation is that wedge belts of porosity become favorable areas for the accumulation of oil which may have had its origin elsewhere. It depends on the acceptance of the theory of the free migration of oil and gas. Briefly, a wedge belt of porosity is such a favorable area of accumulation because it is the oldest trap in the reservoir rock, commonly being contemporaneous with the reservoir rock in origin. This means then that it is the trap which has the longest period of time in which to accumulate whatever oil and gas are in the reservoir

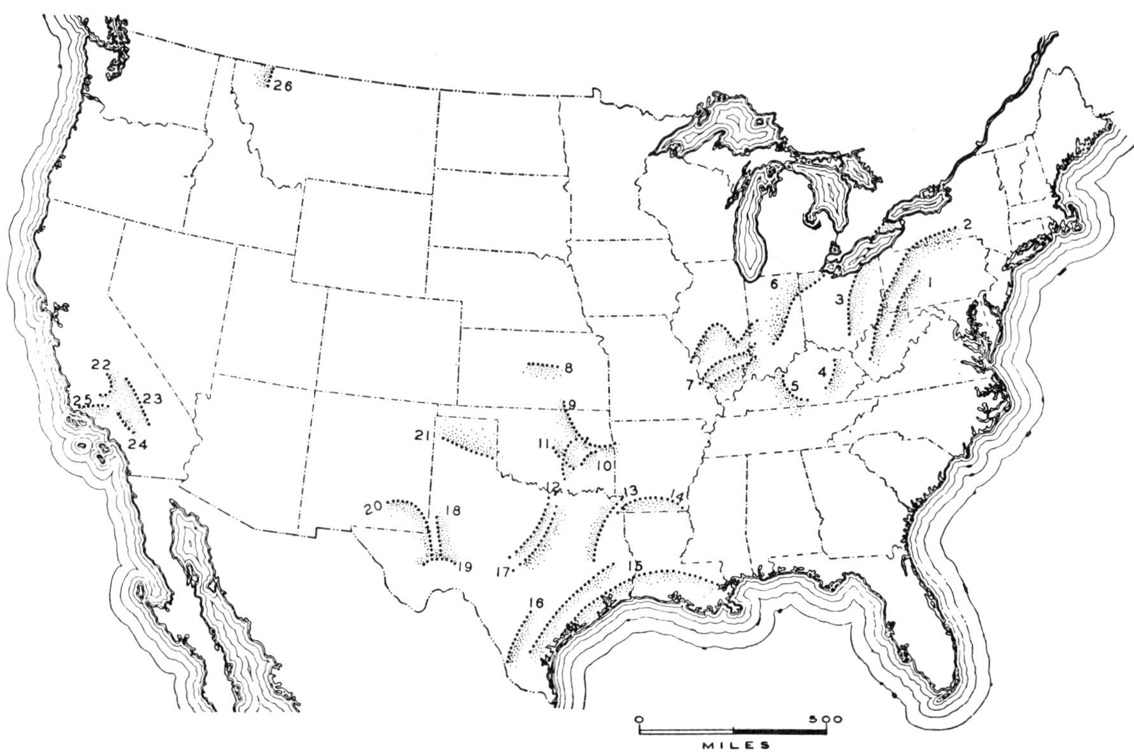

Fig. 12.—Outline map of United States showing some of the petroleum provinces which have in their early history an up-dip wedging-out of porosity in reservoir rock.

1. Numerous lenticular sands, wedging out westward in Mississippian and Devonian of Appalachian region.
2. Oriskany sand as shown in Figure 2.
3. Clinton sand as shown in Figure 3.
4. "Corniferous" province in eastern Kentucky.
5. "Corniferous" production in central Kentucky.
6. Trenton province shown in Figure 3.
7. Several sand wedges of Chester series of upper Mississippian age shown in Figure 4. In addition there are many sand-wedge belts in overlying Pennsylvanian rocks, particularly in old Illinois producing area on LaSalle anticline.
8. Hunton of central Kansas.
9. Simpson sands which wedge out due to overlap of Mississippian rocks around southwest side of Ozark uplift. See also Figure 18.
10. Pennsylvanian "Dutcher" sands of east-central Oklahoma.
11. "First Wilcox" sand of Seminole district. See also Figure 6.
12. Cromwell sand of east-central Oklahoma. See also Figure 5.
13. Woodbine sand of East Texas. See also Figure 7.
14. Lower Cretaceous and Jurassic wedge of southern Arkansas shown also in Figure 21.
15. Productive sands of Texas and Louisiana Gulf Coast shown in Figure 8.
16. Government Wells and other Tertiary sands of southwestern Texas.
17. Strawn sands of eastern Bend arch of north-central Texas. See also Figure 19.
18. "Sand Belt" province of West Texas shown in Figure 10.
19. Part of the Ordovician province of West Texas.
20. Reef limestone province of West Texas. See also Figure 9.
21. Panhandle province along north flank of Amarillo buried granite ridge.
22. East Coalinga-Kettleman province of San Joaquin Valley of California.
23. "East Side" province of San Joaquin Valley of California.
24. Midway-Sunset province of California shown also in Figure 11.
25. Santa Maria province of California.
26. Cutbank area of northwestern Montana.

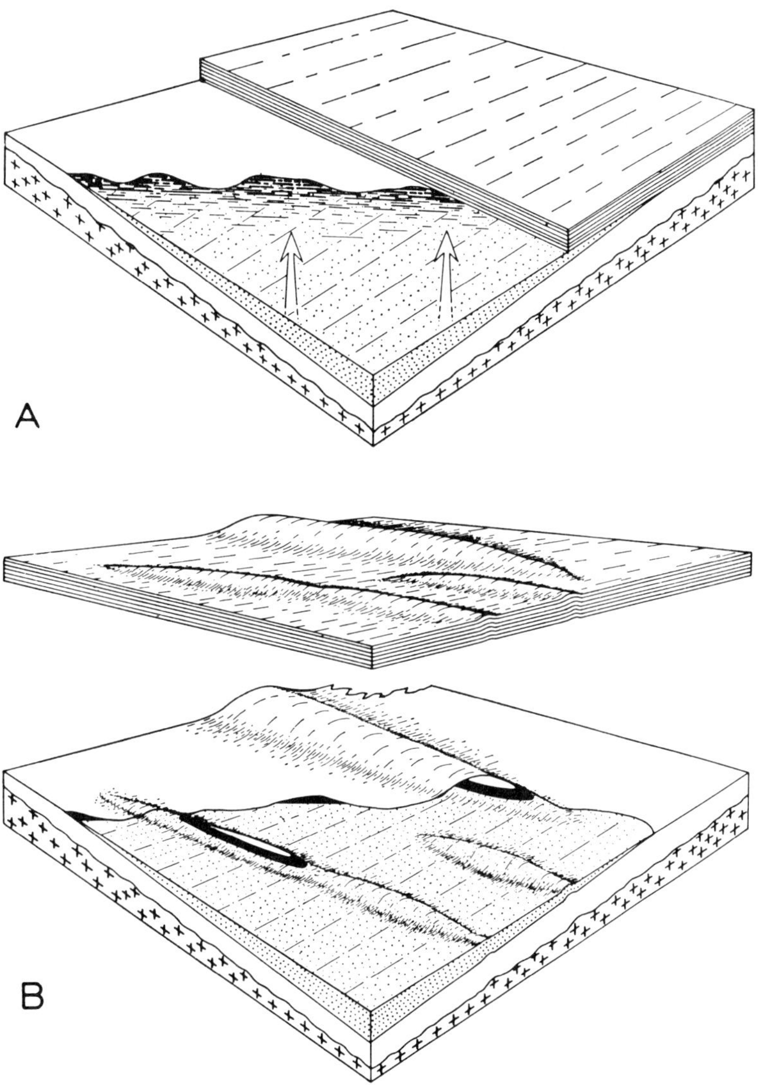

F_{IG}. 13.—Preferred explanation of association of oil and gas pools with updip wedge belts of porosity in reservoir rock. *A* shows overlap of wedge belt of porosity by impermeable cover and movement into wedge-edge area of whatever oil and gas are in reservoir rock under laws of anticlinal or updip accumulation. Subsequently, area is folded, as in *B*, and oil and gas in vicinity of edge become relocalized into traps, some structural and some irregularities in local porosity. Folds too far from edge remain dry or non-commercial since oil had by-passed these areas in its primary movement into wedge-belt area. See also A. I. Levorsen, "Studies in Paleogeology," *Bull. Amer. Assoc. Petrol. Geol.*, Vol. 17, No. 9 (September, 1933), p. 1118.

rock. Since wedge belts are commonly regional in extent the oil that has accumulated in them is on a regional scale, and, in the later folding and deformation in which local and lesser traps are formed, the primary regional accumulation is relocalized into many lesser pools and the area then becomes known as an oil province.

It is the explanation which best seems to account for the relation of many oil provinces to a wedge belt of porosity in the reservoir rock. It may, however, be combined with the other two explanations, to account fully for the relationship in some examples. The principle is illustrated in Figure 13.

As a test of the importance of the time of trap formation in the accumulation history of an oil field, two examples of a negative nature are given. They may be considered as a test "in reverse" of the principle of the close association of updip wedge belts of reservoir porosity with oil provinces.

(1) UPPER CRETACEOUS OF ROCKY MOUNTAIN REGION

Theoretically, the Dakota group of sandstones at the base of the Upper Cretaceous of the Rocky Mountain region should be one of the most prolific oil- and gas-producing formations of the United States. It consists of relatively thick sands covering large areas and underlain and overlain by thick shales, many of which are dark in color and should satisfy the most skeptical geologist as a good source for petroleum. Yet, with scores of large closed anticlines tested, the result is most disappointing and the Dakota group now stands near the bottom of favorable prospective producing formations in the minds of most geologists. Why is this?

It may be due to the absence of any primary oil formation within the sandstone or within the neighboring shales. This would be contrary to the apparently favorable environment, which is nearly everywhere present, of thick dark shales above and commonly below the Dakota horizon. Moreover, there are small pools and numerous showings of oil in the sand, which indicate that conditions were not totally unfavorable for the generation of oil in it.

Another explanation for the general lack of commercial production is in the geologic history and in the absence of the wedge belts of porosity which seem to be so common in more prolific provinces. The sequence is illustrated in Figure 14.

The upper diagram, *A*, shows an idealized section across the Rocky Mountain region at the end of Cretaceous time. The Dakota formation is in the form of a great geosyncline rising from a depth of 14,000 feet near the center of the basin to less than 4,000 feet around the edges. Any oil or gas in the Dakota formation at that time would be expected to move outward and up the dip in the direction of the arrows, in accordance with the anticlinal or structural accumulation of oil and gas. Later, during the Laramide revolution, large anticlines and domes were formed at many places in the region, as shown in *B*, and any oil then remaining in the Dakota sands would be expected to be concentrated in the tops of these structures. The Big Horns, the main ranges of central Colorado, and the Zuni area of New Mexico are typical of the folds formed during this time.

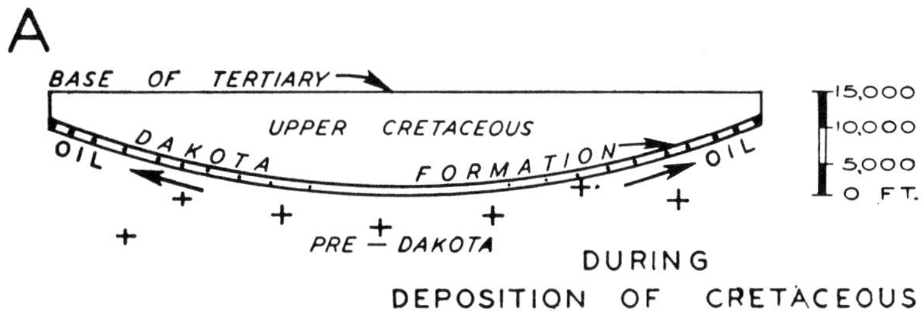

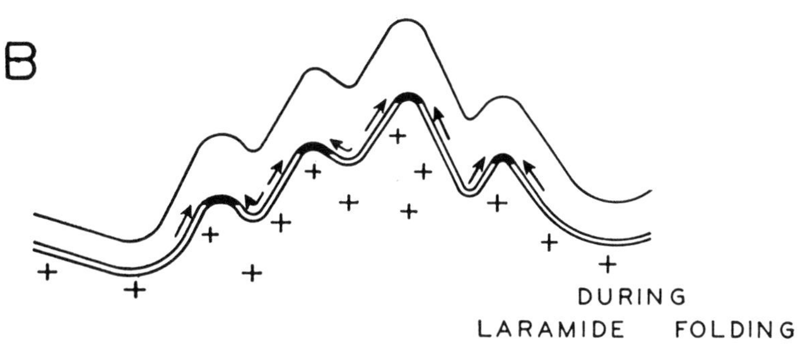

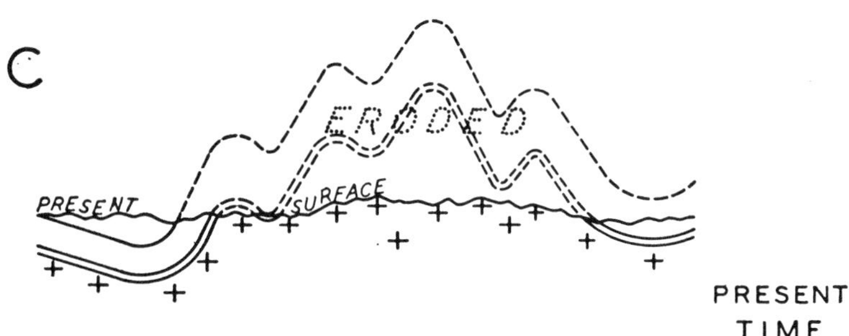

Fig. 14.—Idealized sections across Upper Cretaceous of Rocky Mountain region. *A* shows thinning of Cretaceous rocks from 14,000 feet near center ot basin to less than 4,000 feet along edges. Any oil in Dakota sands during end of Cretaceous time would be expected to move updip toward edges.

B shows condition during and after Laramide period of folding. Any oil remaining in Dakota sands would now be expected to accumulate in tops of large anticlines which were forming.

C shows present conditions, with oil fields eroded during post-Laramide erosion period which exposed granite cores in many areas of Rocky Mountains.

Thus, oil had such excellent opportunities to escape that very few remaining folds in Dakota sand are found to be commercially productive.

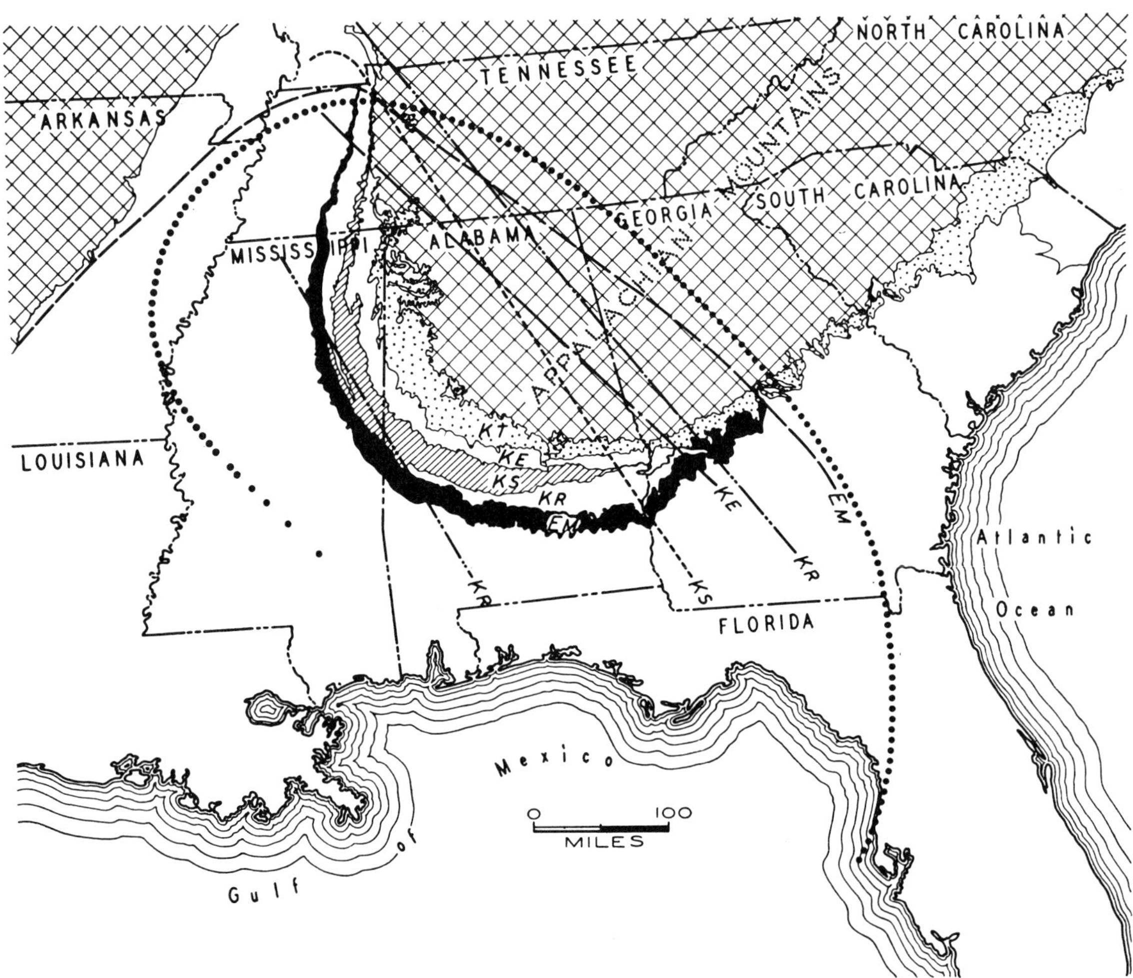

Fig. 15.—Southeastern states, showing present distribution of rocks at surface. EM is Midway formation (Tertiary) and cross-hatched area is basement or pre-Cretaceous complex. Intervening formations are Cretaceous rocks which wedge out north and east. Lines KE (Eutaw), KR(Ripley), KS (Selma), EM (Midway) connect points on northwest and on southeast of like facies. Dotted line marks approximate edge of Tuscaloosa sand.

Subsequently, as indicated in *C*, the high parts of the Rocky Mountain region were eroded away, and with the erosion went whatever oil and gas pools had accumulated during the preceding period. The favorable structures remaining are the lesser ones and relatively small compared with those that were removed. A reason for their generally poor oil showing may be that the region experienced two periods when any oil in the Dakota sands probably escaped, leaving only small remnants of the original oil and gas content. In this connection it is interesting to note that most of the production from the Rocky Mountain structures comes from sands such as the Frontier, Wall Creek, Muddy, Tensleep, and Sundance which wedge out and are regionally lenticular.

(2) UPPER CRETACEOUS OF SOUTHEASTERN UNITED STATES

Another example of the importance of updip wedge belts of porosity, but "in reverse," is in the thick sands of the Eutaw and Tuscaloosa formations in the region south and west of the plunging Appalachian Mountains in Mississippi, Alabama, and Florida. The sequence is illustrated in Figure 15 and Figures 16–A, 16–B, and 16–C.

In this area, the evidence from the areal maps is strongly suggestive of a facies change within the Cretaceous rocks from northeast to southwest. The straight lines in Figure 15 connect points along the outcrop of similar facies or stratigraphic relations. These lines connecting points of like stratigraphy strike northwest and southeast as do also the isopach contours on the total thickness of the Cretaceous rocks where buried in Mississippi and Alabama. The wedging-out of the Cretaceous, north and east along the outcrop is evidence of the conditions which prevailed on the northeast before the Appalachian arching occurred.

The sequence of events is indicated in Figure 16–A, which shows the extent of the Tuscaloosa-Eutaw sands at the time they were overlapped by the overlying Ripley and Midway formations. This edge is an updip wedge edge of porosity, probably largely a shoreline deposit although in part the result of truncation and overlap. Any oil and gas in these porous sands at that time would be expected to move toward the structurally higher edges as shown by the arrows.

Later the region was arched, as shown by the open arrows in Figure 16–B, in the area which is now the southern end of the Appalachians, in northern Louisiana, and in the Ocala, Florida, region. Oil and gas previously concentrated along the wedge belt would now be expected to move along the edge toward the regions of higher structure (hatched) as shown by the solid arrows.

A still later stage was the erosion of the uplifted, arched Appalachian Mountain region with the removal of any oil and gas that may have accumulated there during the arching period. The other areas of possible accumulation are unexplored with the exception of the Tinsley and Pickens pools of Mississippi, which lie within one of the areas of accumulation shown in Figure 16–B.

A sequence of events such as outlined may explain the absence of oil and gas in the eastern Mississippi and southern Alabama region in the structures where

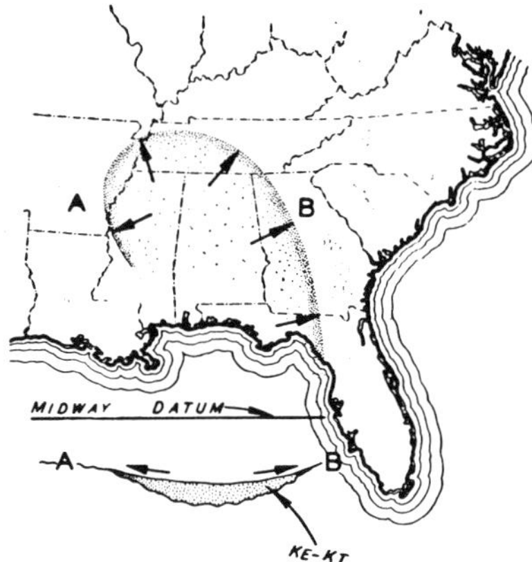

Fig. 16-A.—Map showing area (stippled) underlain by Tuscaloosa and Eutaw sands (Cretaceous). Edge on north and east is due to primary lateral gradation during deposition; edge on west is result of uplift, truncation, and overlap. Any oil in these porous formations, under law of anticlinal or updip accumulation, would be expected to move toward higher sides of basin and become lodged along belt of wedging porosity.

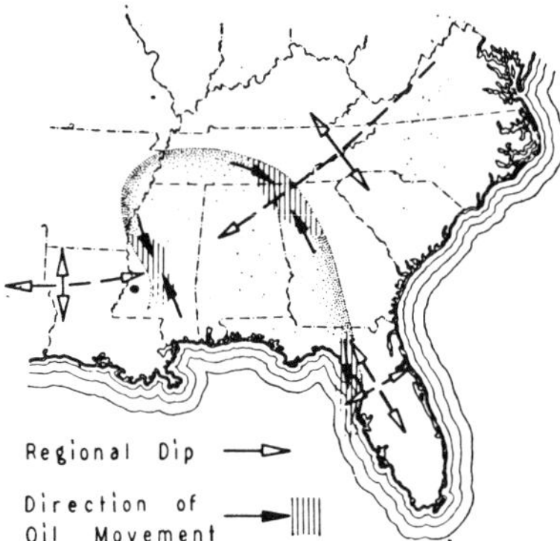

Fig. 16-B.—Later arching of Appalachian Mountain region, Ocala dome in Florida, and arching across northern Louisiana caused general relocalization of oil already accumulated along wedge belt into hatched areas.

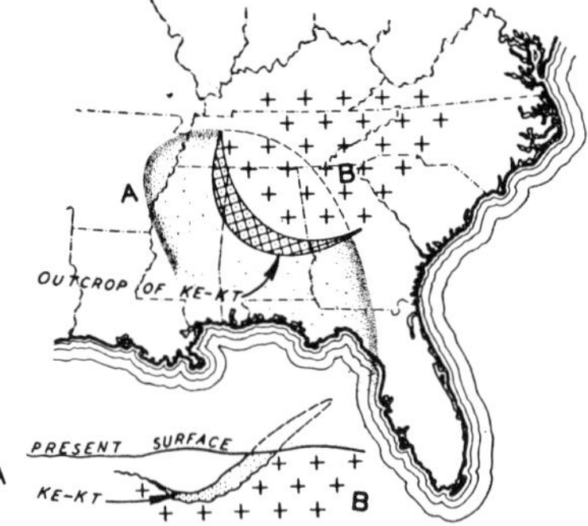

Fig. 16-C.—Erosion of Appalachian arch removed large area of possible accumulation shown in Figure 16-B, located in the vicinity of common state corner of Tennessee, Alabama, and Georgia. Production at Tinsley and Pickens is in region of primary accumulation shown in Figure 16-B.

the Tuscaloosa and Eutaw sand sections have been tested. Whatever oil and gas there ever was in these formations had a chance to escape or by-pass this area before the local folds were formed; thus, the dry structures may be explained by the poor timing of the local folding. Other reservoir rocks of this area may not necessarily be dry since their geologic history may be quite different.

APPLICATION

The application of the principle of wedge belts is simple. If you accept in your thinking an updip wedge edge of porosity in the reservoir rock, whatever the cause, as a fundamental prerequisite to the formation of many or most oil provinces, then you have *immediately* expanded your concept of the undiscovered petroleum reserves of the United States almost without limit. This is because there are thousands of miles of buried, updip, wedge belts of porosity in rocks which are known to produce oil and gas elsewhere, but are to-day only partly explored or completely unexplored. A wedge belt may be called the clue to a province whereas a local fold is the clue to a pool.

Figure 17 shows more than 10,000 miles of such wedge belts remaining to be explored. Fundamentally, they represent the same type of geology which has in the past produced half a billion barrels of oil in the dolomite wedge belt of the Trenton formation of Indiana and Ohio; the same kind of geology which has and will produce 4 billion or more barrels of oil in the 60-mile porosity wedge known as the East Texas field; or which has produced more than a billion barrels of oil along the wedge known as the Midway-Sunset district of California. The evidence for these unexplored wedges is almost entirely in the geologic literature and they are each well known to most geologists. In some places the evidence is largely the result of surface mapping, in others it is the result of a few test wells, and in still others it is the result of various combinations of well known surface and subsurface data. If the confidential data in the hands of the oil companies were available, it is probable that as many more miles of unexplored wedge belts of porosity would be known as are known to-day. If to that is added the new wedge belts of porosity which will most certainly be discovered in the exploration of those wedges now known, the ultimate number of miles which remain to be explored should be sufficient to satisfy the most pessimistic of geologists for many years.

LAYERS OF GEOLOGY

A second principle of geology which has a wide application to petroleum geology is the concept of successive layers of geology in the earth, each separated by an unconformity. They are present in most of the sedimentary regions of the United States and will probably be found to prevail the world over. The discovery of these layers of geology is generally the result of drilling, and many discoveries of oil have been contemporaneous with discoveries of totally unsuspected geology as well. In a few of the well developed oil areas, a rather complete understanding of these layers can be obtained; in most other regions, these layers will probably be

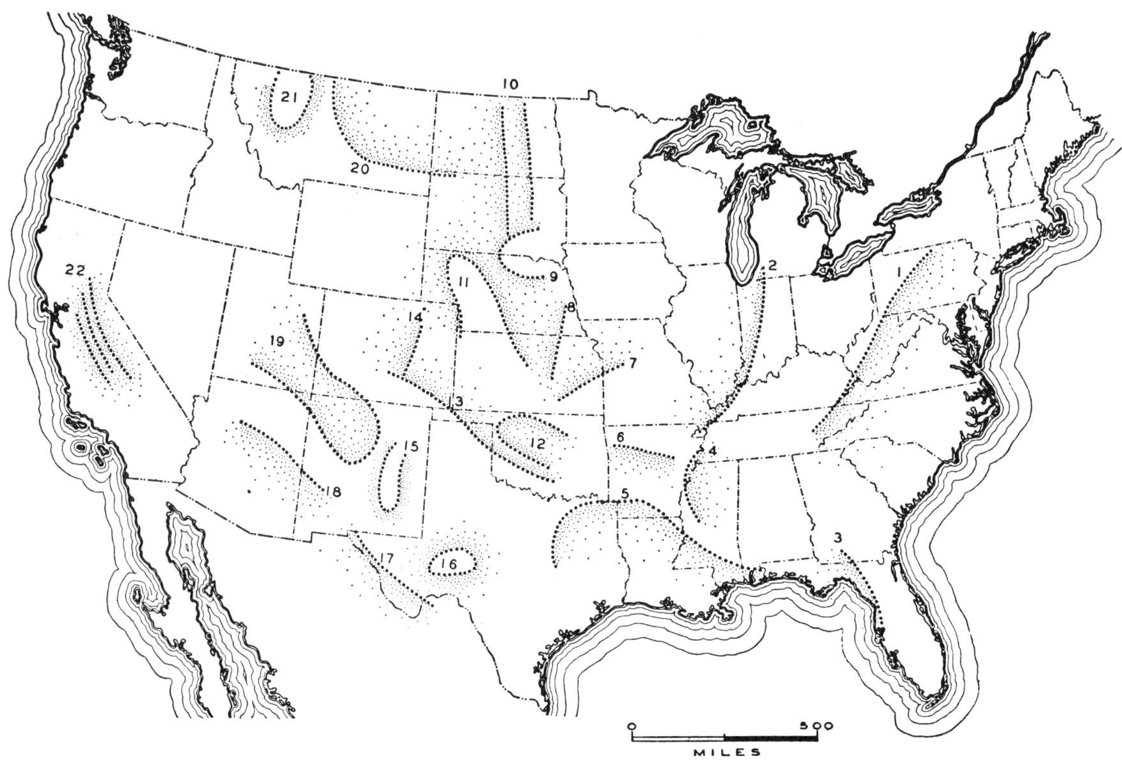

Fig. 17.—Outline map of United States showing more than 10,000 miles of well known essentially unexplored updip wedge belts of porosity. Confidential data in files of oil industry will probably double number and mileage of these belts and an additional like amount will undoubtedly be added during development of those now known. Prospective provinces shown are:

1. Updip westward wedging-out of early Paleozoic rocks which crop out in mountains on east.
2. Eastward wedging-out of St. Peter sand of middle Ordovician age.
3. Eastward wedging-out of thick sands of Eutaw and Tuscaloosa formations of Cretaceous age. See also Figure 16.
4. Overlapped, updip wedge of Eutaw and Tuscaloosa sands of Cretaceous age. See also Figure 16.
5. Wedge of rocks of Lower Cretaceous and Jurassic age which are producing in southern Arkansas. See also Figure 21.
6. Northward thinning-out of many Paleozoic formations.
7. Overlapped edge of upper and middle Ordovician formations on northwest flank of Ozark uplift.
8. Overlapped edges of entire pre-Pennsylvanian section along west flank of Nemaha granite ridge of Kansas and Nebraska.
9. Wedging-out of entire pre-Cretaceous section around Sioux uplift region of southeastern South Dakota.
10. Porous wedge belts of Jurassic to Ordovician age extending across Dakotas.
11. Wedging-out of all pre-Permian formations in western Nebraska as probable extension of Central Kansas uplift.
12. Wedging-out northwest and south of thick Pennsylvanian section of Anadarko basin of western Oklahoma.
13. Southward wedging-out of thick early Paleozoic rocks along north flank of Arbuckle-Amarillo uplift.
14. Eastward wedging-out of thick Pennsylvanian and Permian sands and arkoses along west flank of Sierra Grande arch of eastern Colorado.
15. Wedging-out of Pennsylvanian formations around Pedernal Hills uplift of eastern New Mexico.
16. Wedging-out of pre-Permian rocks around Fort Stockton "high" of West Texas.
17. Northeastward wedging-out of porous rocks of Lower Cretaceous age along Rio Grande in trans-Pecos area of West Texas.
18. Wedging-out of pre-Permian rocks along south side of Zuni-Fort Defiance-Grand Canyon area of New Mexico and Arizona.
19. Thick series of Pennsylvanian and older rocks which thin out north against Uncompahgre uplift of Colorado and south along Zuni-Fort Defiance-Grand Canyon arch of New Mexico and Arizona, and east around east side of San Juan basin in New Mexico.
20. South and west wedging-out of Big Snowy (Chester) series of upper Mississippian age.
21. Several sands in Cretaceous and Jurassic systems which thin out around Sweetgrass arch of northwestern Montana.
22. Numerous wedges of porosity in Eocene and Cretaceous rocks of central and northern San Joaquin Valley of California.

86

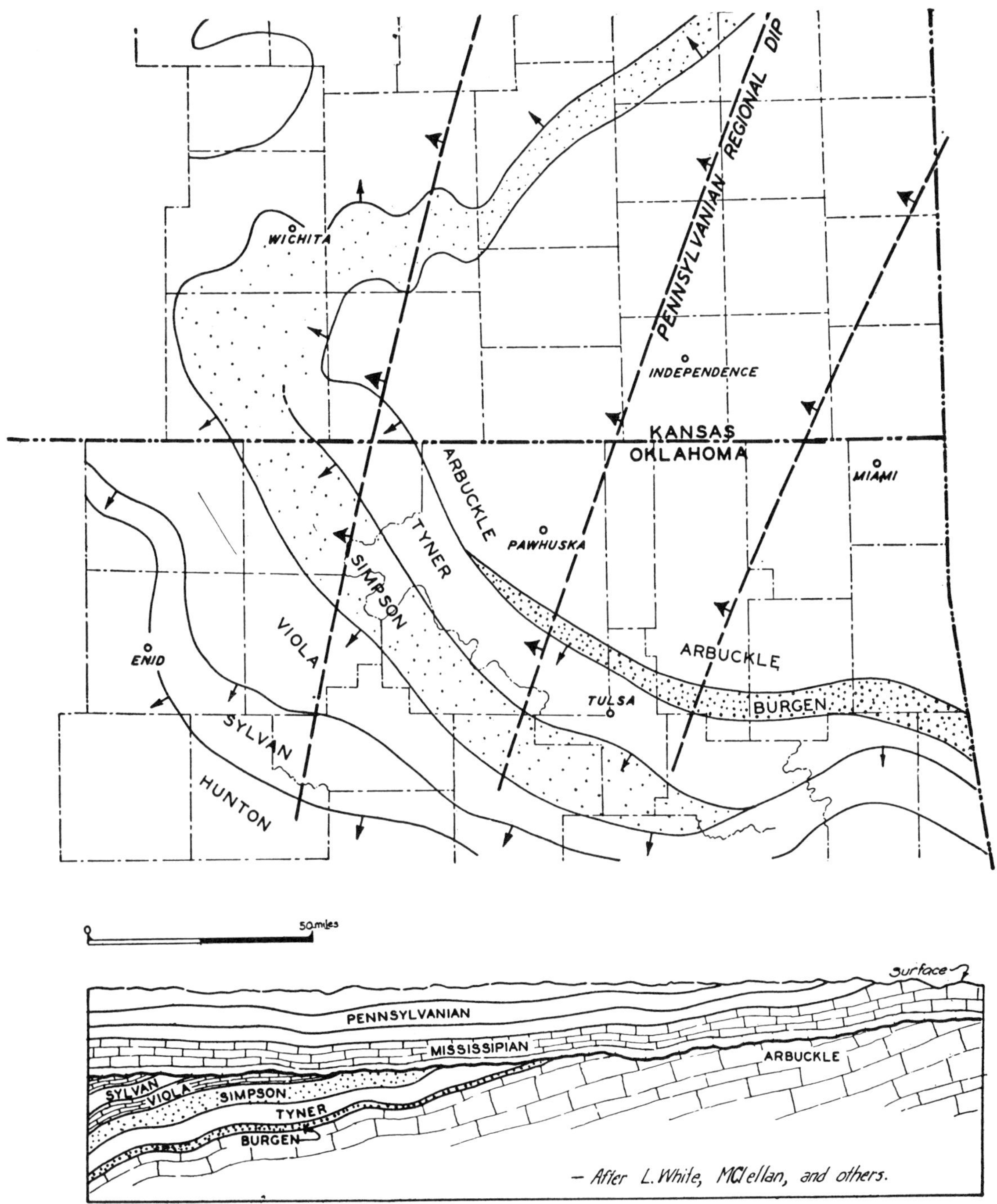

FIG. 18.—Southeastern Kansas and northeastern Oklahoma, showing two layers of geology. Upper layer, Pennsylvanian and Mississippian rocks, has regular north-northeast strike and west dip. Lower layer, Devonian and earlier rocks, is broad arch, pitching west, and forming west end of Ozark uplift. The two layers are separated by unconformity; general geology and petroleum geology are quite distinct in each layer.

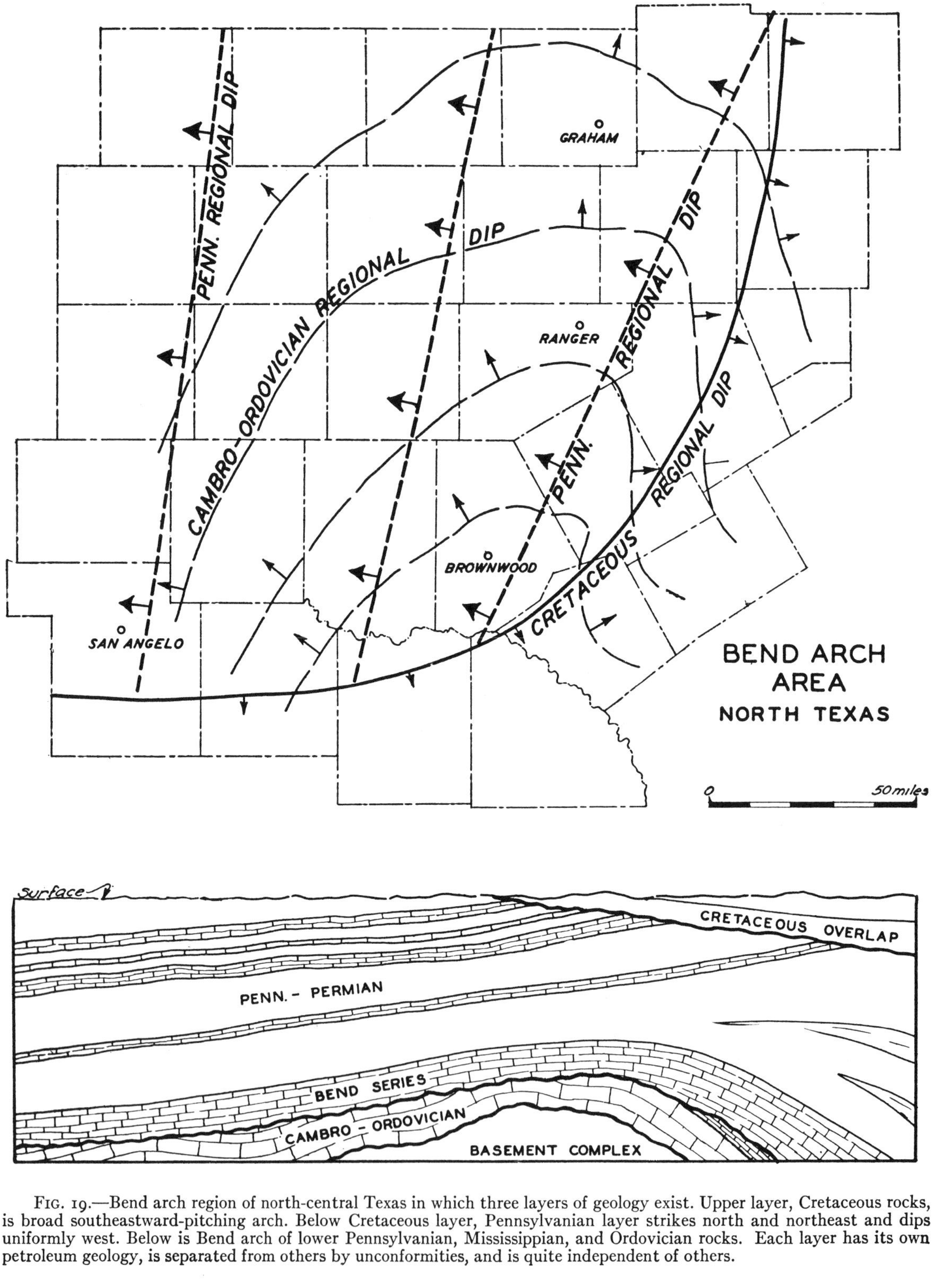

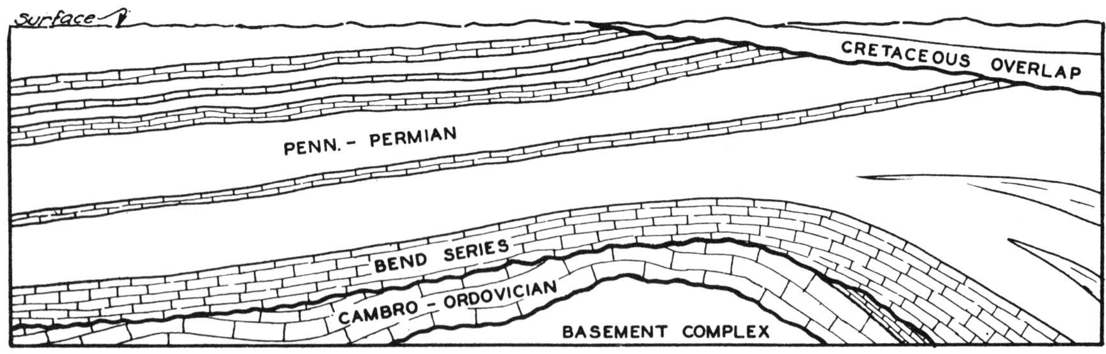

FIG. 19.—Bend arch region of north-central Texas in which three layers of geology exist. Upper layer, Cretaceous rocks, is broad southeastward-pitching arch. Below Cretaceous layer, Pennsylvanian layer strikes north and northeast and dips uniformly west. Below is Bend arch of lower Pennsylvanian, Mississippian, and Ordovician rocks. Each layer has its own petroleum geology, is separated from others by unconformities, and is quite independent of others.

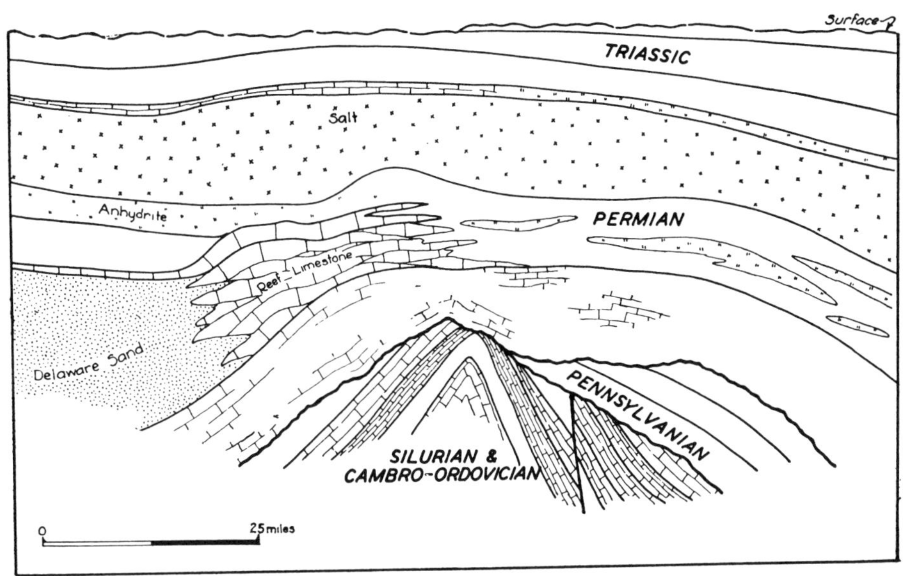

Fig. 20.—West Texas and southeastern New Mexico are underlain by several layers of geology. Upper layer of Triassic and Cretaceous rocks has uniform, low, southeastward dip. Permian, with its highly productive reef limestones grading into thick sandstones on one side and into anhydrites and shales on other, occurs next below. Underlying Permian is variable section of Pennsylvanian rocks ranging from thick dark shales to almost solid limestones. Next below is complicated layer of early Paleozoic (Devonian, Silurian, Ordovician, and Cambrian). Petroleum geology of each layer differs from each other layer; each layer is separated by unconformities; and each is quite independent of others.

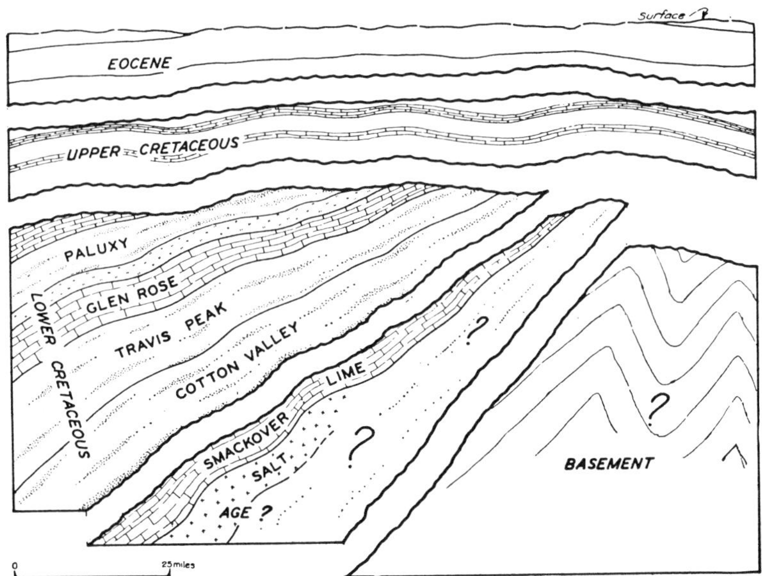

Fig. 21.—Northern Louisiana and southern Arkansas: another area in which numerous layers of geology are found superimposed, one on another. Layers are separated by unconformities; each has peculiarities of oil and gas geology; therefore each must be prospected differently from others, and each is independent of others. There are no clues in upper layers about what may be expected in next lower layer and each layer must be considered a unit.

discovered with additional drilling and at this time we can only guess at their extent and nature.

Typical of the better known areas where several layers of geology have been discovered are the four areas shown in Figures 18 to 21, inclusive. Each layer is separated from the other by an unconformity; each layer of geology is completely independent of other layers above and below; there is no clue in the upper layer of either the existence or the character of the next layer below; and each layer has its own oil and gas geology, completely independent from each of the other layers.

It is in this last characteristic that we particularly find application to the future discoveries of petroleum, for it means that no sedimentary area can be considered as explored until many test wells have been drilled through all of the layers of geology down to the basement rocks. As a matter of fact, every sedimentary region in the United States has unknown and unexplored geology waiting to be drilled. It is probable that there is a volume of undiscovered and therefore unknown and unexplored geology equal to that which is already explored, and it is inconceivable that these unexplored layers will not some day be translated into equally large volumes of oil and gas. The infinite complexity and at the same time the infinite opportunity in the petroleum geology of any locality becomes apparent when we remember that each layer has its own wedge belts of porosity, its own facies and lithologic changes, its own structural history, and its own oil and gas geology.

This concept as a part of one's thinking immediately gives hope and promise to countless square miles of territory which are at present condemned because of their monotonously uniform and uninteresting structure and stratigraphy when considered from the viewpoint of orthodox petroleum geology. As in a layer cake, where the sweetest layer may be deep in the cake, so with these layers of geology, the richest layer at any locality may be the second, or third, or fifth layer from the top, with no indication of its presence until the drill has crossed the overlying unconformity.

Discovery

A number of published surveys as well as numerous governmental hearings have shown the steadily declining rate of new oil discovery during recent years and 1942 appears to have one of the poorest records of all. This decline has been, according to Lahee,[5] in spite of an increase in the wildcat footage drilled and in spite of an increase in the total number of wildcat wells drilled. The new fields average smaller in size, and even if they are greater in number, the total amount of new oil discovered is annually becoming less and less. Obviously something is seriously wrong.

An attempt has been made to show that from two lines of geological thinking we have reason to believe there are still vast volumes of rocks in which geological conditions are known to exist which may reasonably be expected eventually to be-

[5] Frederic H. Lahee, "Wildcat Drilling in 1941 with Comments on Drilling Rate," *Bull. Amer. Assoc. Petrol. Geol.*, Vol. 26, No. 6 (June, 1942), pp. 969–82.

come translated into equally large petroleum reserves. The question of the failing rate of discovery is then not yet to be answered by saying it is due to a lack of oil to be discovered.

On the other hand, the oil industry has spent vast quantities of money in an effort to make discoveries; it has established large and expensive exploration departments; and it has apparently spared no expense or effort in utilizing to the fullest extent everything that it thought might contribute to the finding effort. Great as this finding effort has been, it is not enough to do the job. Some think that an increase in the price of oil, a subsidy to the wildcatter, or some other form of increased advantage to those who are charged with discovery will result in the necessary increase in reserves. While it seems obvious that a prerequisite to any successful discovery program on a national scale is a price which will provide enough incentive to take the necessary risks, our particular concern, as petroleum geologists, is to increase discovery and decrease risk under whatever economy prevails.

It is often said that what the oil industry really needs is a new exploration tool, just as the fisherman needs some new tackle. But, like the fisherman analogy, when it is remembered that the new fishing tackle is generally designed to catch the eye of the fisherman and not necessarily the eye of the fish, it gives cause to wonder whether or not some of the new gadgets or devices continually coming into the discovery picture may be more of the nature to catch the eye of the contour-minded executive rather than the elusive oil field which it is hoped to find. Some of the most successful fishermen are those who use only the old-fashioned hook and line with a little bait, and maybe we can learn from them.

Whether the economic framework within which we must operate is good or bad, we have an unused tool as old as geology itself. It is simply—more geology. Not only should such a procedure prove effective, but it is relatively inexpensive as compared with current exploration costs, and in addition, it has the distinct advantage of being something with which we are now familiar. We can start off with it now without waiting for some unknown device to be discovered.

Before exploring the matter further, it might be well to point out that more geology means either more geologists, better pay for geology as such with a consequent better geology, or both. The young geologist, when he starts out, equipped with his full quota of ambition, energy, enthusiasm, and a desire to do things, looks about him and observes that the big-pay jobs are not for geology as such but for office, administrative, and production jobs in which geology merely furnishes the backgound. The well sitter or pipe setter is held in higher esteem than the oil finder. Is it any wonder that he gradually turns his attention and energy to the business of the oil industry and lets geology lag? He finds there is a salary level, different in different companies, yet nevertheless present, above which he can not rise as a geologist dealing in rocks, samples, and field observations, but which is readily penetrated if he can do administrative or office geology dealing in maps, appraisals, reports, surveys, and data secured by others. The fault is not with the individual but with the system which gives so much more tangible recognition to the superficial and leaves the more fundamental factors to chance.

The correct identification of an unconformity, for example, and the correct interpretation of its meaning from well cuttings thousands of feet down in the ground, is not an administrative job; it can not be done by just anyone; and its identification may well be of greater direct importance to oil discovery than all of the other geological, geophysical, and administrative factors together. Yet almost any office job in the department is more desirable, pays better, and is more directly in line for further advancement.

When this Association began, more than 25 years ago, the objective of the few petroleum geologists then in existence was to find oil fields. Their full time was devoted to mapping and interpreting geological conditions with discovery as the objective. As time went on, and with the advent of geophysical methods, the rôle of discovery has gradually shifted to a large measure from the geologist as such, to the geophysicist, and the geologist is becoming more and more interested in the development of oil and gas which has already been discovered rather than in the discovery of new fields and new provinces. As a matter of fact, the modern oil-company geological organization has become a general service department. It serves the production department continually with geological advice on drilling wells, sitting on wells, advice on setting casing, advice on coring prospective sands, advice on well abandonments, and repeated advice as to whether the well is running high or low. It serves the land department with advice on rental payments, lease renewals, bonus payments, farm-outs, lease abandonments, and drilling deals. It serves the executive department with innumerable conferences and a continuous stream of reports, of statistical information, appraisals, competitive production records, and answers to the thousand and one questions that are always being asked. It serves the legal department as witnesses and developing evidence in lawsuits, and it serves the pipe-line department in advising on the prospects for discovery in the wildcat wells that are drilling near the company pipe lines. The geologist has in effect buried himself in routine service work.

All of this work is important to an oil company, and, in order to do it properly and efficiently, an increasing amount of administrative work becomes necessary. Governmental forms have to be filled out, voluminous records have to be kept and have to be instantly available, costs must be kept down, and all sorts of miscellaneous maps must be kept continually up-to-date. The daily conferences, reports, and administrative details, whether in the district offices or in the main office, require an ever increasing part of the total effort so that the thinking necessary to oil discovery gets secondary consideration or is turned over to the geophysicists. By his training, his ability to think in terms of earth variables, and his understanding of the problems involved, the geologist is particularly well fitted to do all of these tasks which have been assigned to him, and all of them are necessary and desirable, not only for the company but for the geologist as well. They give a stability to his employment and a constantly visible reason for his existence which is most comforting. But, they are not directly concerned with the thing which he can do best and which is of greatest importance to the petroleum industry, and that is the discovery of new oil fields.

In an attempt to analyze further the modern discovery process, a survey was made of the number of seismograph units as compared with the number of geologists in seventeen representative oil-company exploration departments. The results are shown in Figure 22. It is seen that the number of geologists per seismograph ranges from 3 to 9. If an average figure of $10,000 per month be used as the

RATIOS — GEOLOGISTS TO SEISMOGRAPHS

Company	Seismo-graphs	Geologists	Geologists per Seismograph
A	3	9	3.
B	2	8	4.
C	10	40	4.
D	16	48	4.
E	23	97	4.+
F	9	41	4.5
G	15	85	5.7
H	8	46	5.7

Company	Seismo-graphs	Geologists	Geologists per Seismograph
I	11	64	5.8
J	4	24	6.
K	16	98	6.+
L	4	30	7.5
M	5	38	7.6
N	8	66	8.2
O	6	50	8.3
P	6	50	8.3
Q	4	36	9.

	Seismo-graphs	Geologists	Geologists per Seismograph
TOTALS	150	830	
AVERAGES	9—	49—	5.5

Fig. 22.—Ratios of geologists to seismographs in exploration departments of 17 typical oil companies operating in United States outside of California.

cost of a seismograph and $500 per month be used as the average cost of a geologist, which includes office help, draftsmen, and filing clerks attached to the geological department, it is seen that the money spent on geologists ranges from 15 to 45 per cent of the amount spent on seismic surveying. The average is 30 per cent.

As has been pointed out, the modern geological department spends only a relatively small part of its time in searching for new oil fields and by far the major effort—probably fully 90 per cent—on routine connected with production and development problems in fields which have already been discovered, and on routine clerical and administrative work. The amount of money spent directly by the geological department in search for new oil and gas fields—in what might be called "discovery thinking"—is thus only an average of one-tenth of the 30 per cent, or 3 per cent of the amount spent by the geophysical department, when these two departments are considered separately.

Stated still differently, all of the money spent on seismic surveying, plus most of the small part of geology devoted to discovery, goes into a search for structures, in which it is *hoped* oil will be found. When one remembers that at least a third and probably half of the past production and present reserves are located primarily as a result of some stratigraphic variation or anomaly of sedimentation, and in addition, as has been shown, that even most regions where the individual oil pools are structural, the province as a whole is fundamentally related to major stratigraphic variations called wedge belts of porosity, it is seen that the big majority of all our past discoveries have in them a stratigraphic factor, either directly or indirectly responsible for the accumulation of oil. Since most of the 3 per cent of the strictly geological time and thought spent on discovery is concerned chiefly with structural studies, we might truthfully say that less than one per cent of our total discovery effort—both geological and geophysical—is devoted to the kind of geology which is known to be responsible, either wholly or in part, for most of our oil. It does not make good sense.

Nothing herein should be interpreted as meaning there should be less seismic surveying, but rather that there should be a greater proportion of geology put into the exploration effort. Not only is it proposed that a greater proportion of geology be employed but that this increase be devoted primarily to discovery rather than to development. There are many geological discovery tools rusting away in our kit for want of use. The need is for "creative" geology as compared with what may be called "routine" geology.

An opportunity must be provided for those men with the capacity and with the ability for doing creative geology to actually do geology, and in addition, to be able to spend a part of their time with their feet on a desk looking out of the window where they can generate ideas and where they can reconstruct in their mind the conditions and the environments of past geologic ages. What we need is more of the sort of geology that Philip King did in West Texas and New Mexico, or the kind that Marshall Kay is doing in the Appalachians, or that Ross Heaton has done in the Rocky Mountains, or the kind that was so ably done by Ralph Reed in California. There are enough discovery ideas and prospective provinces in reports such as they have published to justify any operator looking for new ideas to keep men continuously on this sort of work as a practical and economical oil-finding attack.

We have been inclined to lean too heavily on mechanical devices to carry on the discovery process for us and not enough on coördinated personal observation, individual interpretation and understanding. These mechanical crutches, if used too long or too intensively, gradually dull our ability to carry our discovery thinking through to a successful answer, and, as we become increasingly dependent on outside help, we become decreasingly able to do without it. The unused talent finally is lost.

There are several kinds of geology, the doing of which should result in more discovery, and a few of them are discussed in the part which follows.

WEDGE BELTS OF POROSITY AND LAYERS OF GEOLOGY

If you agree that wedge belts of porosity are commonly a fundamental prerequisite of an oil province and that successive layers of unknown geology underlie much of the sedimentary region of the United States, then you must reach the conclusion that the only known way of reducing these conditions to practical application is through a strictly geological approach. Such an approach means the preparation and study of facies maps, lithologic maps, isopach maps, paleogeographic maps, paleogeologic maps, quantitative and qualitative sedimentation studies, and detailed stratigraphic analyses of every porous sedimentary rock unit in the United States and eventually in the world. It means a resolving of the great mass of geophysical data into stratigraphy and sedimentology as well as into structural geology. The basis for this work is first, a restudy of all surface rocks most of which will have to be re-examined from a stratigraphic and sedimentational viewpoint as well as structurally; and second, a restudy of the detailed subsurface data now in the files and laboratories of the oil industry in the form of well cuttings, cores, electric logs, porosity and permeability measurements, water, oil and gas content, geophysical and geochemical data, and anything which will throw light on the location of a wedge edge of porosity or the crossing of an unconformity into new and unexplored geological conditions.

When you accept into your geological thinking the philosophy of wedge belts of porosity and layers of geology, you must at the same time agree that there remains many times as much fundamental commercial geology to be done as has been done in all of our past geological experience. Most of the information in the unexplored regions is and always will be fragmentary and inconclusive, and it is for this reason more than any other, that experienced, broadly trained, and speculative-minded men will have to be developed. Moreover, this kind of geology is not the type that can ordinarily be done as a casual task between conferences, but it is the kind that will require hours, months, and years of the most careful study, thoughtful analysis, and imagination without end.

LIMESTONE-REEF FIELDS

Petroleum geologists in West Texas and New Mexico, working with the limestones and dolomites of Permian age, have led the way to a better understanding of the rôle that limestone reefs play in the accumulation of oil and gas. While there is a voluminous geologic literature concerned with the problem as it is developed in the Permian basin, there is little or no work being done on the larger problem of reef production generally. Reef production is not confined by any means to the Permian basin. The southern fields, or "Golden Lane" of Mexico, for example, produce their oil from reef limestones in the middle Cretaceous. Undoubtedly much of the early Devonian production in southern Ontario is related to reef-like deposits, and many wells of thousands of barrels daily production were found at exceptionally shallow depths. Likewise, the limestones producing in Michigan probably have in many of them reef elements of which a better understanding would provide more discovery.

A few years ago, Laudon and Bowsher described the numerous reef deposits which crop out in the Mississippian limestones of southeastern New Mexico[6] and now we learn of a number of rich oil pools being found in similar but buried Mississippian limestone reefs in the North Texas region. Even more exceptional are the several pools producing from thick limestone reef material in the otherwise uniformly bedded shales and thin sands and limestones of the Pennsylvanian of North Texas.[7]

There are enough oil pools which are now known to produce from limestone reefs or from reef-like deposits to justify the conclusion that there are a great many more yet to be discovered. Every company concerned with the discovery might well have an expert on "reef" production—a person of broad geological experience, skilled in the interpretation of geological conditions from well cuttings, able to visualize subsurface conditions and geological environments, and one who would devote his entire working time over a period of years to this problem. The problem of reef production is fundamentally geological and the cost of such work in terms of the discovery possibilities is insignificant.

LENSING AND SHOESTRING SANDS

The amount of production from shoestring and lensing sands in the United States is probably much greater than is generally supposed. Individual pools are both large and small, there are many of them, and in the aggregate they account for a substantial part of the national supply. There are dozens of such pools in Pennsylvania, in West Virginia, in Ohio, Illinois, Indiana and Kentucky, in Kansas, in Oklahoma, and in Texas, and with very few exceptions they were discovered without benefit of geology, geophysics, or much, if any, scientific reasoning.

There was a time during the development of the shoestring and lensing sands of southeastern Kansas and northeastern Oklahoma when they attracted considerable scientific interest and articles descriptive of them were published by Rich, Cadman, Charles, and others. When the shoestring sands of Greenwood County, Kansas, were being first developed, Sidney Powers was much concerned over their origin and how they might be more scientifically prospected. He was particularly interested in getting different views and explanations of some of the possible origins and once said, "I hope before they are all developed that someone who has seen a sand bar along the ocean shore will become interested in studying them. As a matter of fact," he continued, "those who are doing all of the present

---

[6] L. R. Laudon and A. L. Bowsher, "Mississippian Formations of Sacramento Mountains, New Mexico," *Bull. Amer. Assoc. Petrol. Geol.*, Vol. 25, No. 12 (December, 1941), pp. 2107–60.

[7] J. K. Murphy, Paul E. M. Purcell, and H. E. Barton, "Seymour Pool, Baylor County, Texas," *Stratigraphic Type Oil Fields*, Amer. Assoc. Petrol. Geol. (1941), pp. 760–75.

philosophizing are 100 per cent landlubbers and I have a hunch that these are not stream channels but are buried ocean sand bars. I would be most happy to have someone who has even seen the ocean take a look at them."

Bass[8] went to the ocean and came out with his classic shoestring sand studies which answered for a long time many of the perplexing questions about these deposits. Since then, however, interest has lagged and apparently the oil industry is content to leave their future discovery to the random wildcatter as in the past.

Yet, here is a field of study which most certainly will yield results if attacked scientifically. Such attack would be purely geological in nature and probably largely in the field of detailed sedimentology. It would depend in a large measure on a thorough understanding of all the phenomena of sand deposition along varying sea coasts, on the ecology of such regions, and on the ability to identify sand forms accurately with few data. Is it not probable that detailed sedimentation studies will enable one skilled in lensing sands to tell whether the test well is on the lagoonal side of the sand bar or on the offshore side? And thereby accurately advise which direction to move in order to encounter more sand? Or whether the well is near the end or the sides of the sand deposit? Or that the sand encountered is a stray sand or a blanket sand? Sedimentationists are thinking in these terms and one has only to read the writings of Krumbein, Pettijohn, Krynine, Russell, and others, to see that they are thinking in terms far in advance of what we as commercial geologists are content to call satisfactory.

In terms of probable results, the cost of keeping experienced geologists on this type of exploration problem over a period of years in negligible. As in other fields of geology, however, it is a kind of work that calls for broad experience, imagination, keen interest in the geological problems involved, and the confidence of the management which will permit much drilling on reasons that might now be called unorthodox. Any oil company looking to the future might well put one or more geologists permanently on such work, men who will eventually become expert in the identification, prediction, and understanding of this much neglected type of production.

DETAILED SURFACE STRUCTURAL MAPPING

We hear it said that the day of surface mapping is over. Certainly it has declined in importance in the normal exploration programs to the point where it is often difficult to find geologists capable of doing even an ordinary detail job when such emergencies develop. It is probably true that the hey-day of reconnaissance surface mapping is past in the United States but it can be seriously questioned whether detailed plane-table mapping is not still an extremely effective discovery tool. There has been done actually only a relatively small amount of really careful, detailed surface plane-table mapping, particularly with the assistance of aerial photographs. Over and over again we hear of geophysical or subsurface

---

[8] N. W. Bass, "Origin of the Shoestring Sands of Greenwood and Butler Counties, Kansas," *Kansas Geol. Survey Bull. 23* (September, 1936).

prospects which it is later remembered or determined underlie surface anomalies, and we likewise hear of geophysical and subsurface failures which underlie readily workable yet perfectly normal surface structures. The value of detailed surface geological mapping is too often recognized after it is too late rather than as a primary discovery technique.

Two examples of recent discoveries resulting from surface mapping show that such methods are by no means ready to be put on the shelf. The first is the discovery of the East Watchorn pool in T. 23 N., R. 3 E., Pawnee County, Oklahoma, by the Alma Oil Company. Everett Carpenter, known to many of you as having practiced petroleum geology in the Mid-Continent region for the past 30 years, was in the employ of the Alma Company looking after some drilling in the old Watchorn pool. While driving back and forth to work he observed some abnormal surface dips and spent two or three days in working out a rather detailed reconnaissance map of the area of the disturbance.

On the basis of this work the Alma Oil Company decided to shoot the area and results of variable quality were obtained, probably due to the considerable alluvium in the area. Enough was indicated, which, when added to the surface mapping, caused the company to take a block of leases at a dollar per acre and drill a test well. The result was the discovery, in the "Wilcox" sand, of the East Watchorn pool, which it is interesting to note follows the surface mapping more closely than it does the seismic interpretation.

Two things stand out in this case history. First, if it had not been for the fact that Everett Carpenter, with his wealth of experience, was actually out in the field looking for surface anomalies, there would probably still be no pool known at this locality; and second, it is absurd to say that even in such an old and established area as northern Oklahoma the surface geology has been adequately mapped when a structure such as this can be discovered by reconnaissance methods.

The second example of discovery resulting from surface mapping is that of the Hawkins field in Wood County, northeastern Texas, as told to me by Morgan Davis. This area was mapped by E. A. Wendlandt, of the Humble Oil and Refining Company, in the summer of 1930. At that time not much significance was attached to the structure since it appeared relatively insignificant in comparison with other structures in the basin. However, in 1933 and 1934, the Long Lake and Cuyuga fields were opened under small surface structures and the Humble Company leased a small block of leases on the fold which Wendlandt had mapped. Results from seismograph surveys were poor and thought was given to dropping the block of leases or farming it out.

However, in 1937, at the insistence of Wendlandt that the area had possibilities, it was core-drilled under the direction of T. J. Burnett with the result that the company acquired a much larger area of leases.

Although the Humble Company did not actually make the discovery, their intention was to drill a test well and they were curing titles when Manziel completed his discovery well in December, 1940, at a location which later marked the edge of the pool. Their block of 7,000 acres of leases now has 361 wells and an oil reserve on the order of 400,000,000 barrels.

Here, too, the lesson to be learned is that surface detail work pays big dividends when done by one with experience and skill in this kind of geology. When done right it deserves top consideration in the search for oil fields.

Another lesson is that here is an enormous reserve under an apparently insignificant structural anomaly. It must have been classified far down on the Humble's list of prospects for many years. Yet, when unconformities and variable geological history are added to what appeared to be fourth- or fifth-rate prospects, they have a way of changing into something entirely unforeseen and unexpected.

These are only two examples of many that might be presented to show that there is still a place for surface mapping; that it leads to oil and gas discovery; and that it is not a kind of work to leave entirely to the young geologist; but that it can be most effectively done by men of experience—which also means such men must go where the outcrops are located. In surface geological mapping more attention than heretofore should be given to the sedimentary variations, facies changes, and stratigraphic studies which will bring it into more practical relation to subsurface studies, wedge belts of porosity, and problems related to overlaps, unconformities, and local and regional geologic history. Moreover, when it is remembered that eight or ten experienced plane-table parties can be kept continuously in the field for the equivalent cost of one seismograph crew, the cost in relation to oil discovery is insignificant.

COÖRDINATION OF DATA

In the language of the automobile manufacturer, every oil field may be called a "custom-built job." Superficially they may look alike, but if we lift the hood we find the insides vary from field to field. They differ in geologic history, in structural development, in sedimentary and stratigraphic relationships, in position with respect to the basins, in reservoir-rock conditions, and in the physics of the contained liquids and gases. The possible combinations of conditions are almost infinite in their variety. This leads to the conclusion that each oil field is different from every other field and that the next field to be discovered will have a combination of geological factors different from anything that has been seen before,—it will in fact be a new geological experience.

There is much in the science of geology besides structural geology which can be applied to the discovery of new oil and gas fields. Yet we persist in applying mass-production methods—in our case we might call it mass structure surveying as practically the sole discovery tool. True, geologic structure is the most obvious, the most readily determined, and therefore the most quickly exhausted of any of the factors which it is believed have a bearing on the occurrence of oil and gas, but it is by no means the whole answer. And even in our structural work, we find ourselves too often adhering to one of the several kinds of structural surveying— surface, subsurface, or geophysical—to the exclusion of the others. The discovery process is complex and we dare not leave it to any single method or technique. With the several dozen kinds of devices for obtaining detailed and accurate geologic data of all kinds, our problem of blending, coördinating, and interpreting all of this information, and putting it to work in discovery becomes more and more important.

The story of the discovery of the Pauls Valley field in Garvin County, Oklahoma, well illustrates the possibilities of coördinating variable data by one with broad geological experience. As far back as 1928, Ira Cram, then district geologist for the Pure Oil Company, recognized the fundamentally favorable regionally high structure of the area, and started working on it. In an effort to localize structural closure on this broad arch, surface detail mapping was done which resulted in some ideas, but generally the beds exposed were such that detail mapping was impossible over much of the area. Magnetic work failed to show anything of interest and, in 1933, seismic operations were commenced. The subsurface geolology was complex and the difficulties of intepretation were great, but during seismic reconnaissance, enough encouragment was obtained to justify the leasing of several anomalies, one of which later included the Pauls Valley pool.

Between 1934 and 1937, thirteen Ordovician dry holes were completed in the Pauls Valley district, none of which offered any definite promise of future discovery. Cram, however, re-examined all of the seismic data in the light of the detailed sample information revealed by these dry holes, coördinated the information, and as a result, reworked the geology of the entire area. From this restudy, the geologic history was worked out with fair accuracy, additional detailed seismic work performed, and relationships were established which had been indefinite or merely suggested.

Between 1937 and 1941, four more dry holes were drilled by other companies, all of them in the vicinity of what is now the Pauls Valley field, but none of them gave any indication of the presence of commercial oil. These later holes, however, did confirm the geological interpretation which had been developed as the result of the detailed geological and seismic work, and in 1942, the Pure Oil Company's Teter No. 1 was drilled in Sec. 31, T. 4 N., R. 1 E., Garvin County, Oklahoma, as the discovery well of the Pauls Valley field. Production was obtained in the Bromide sand of Ordovician age and there are now 14 flowing wells in the pool and the limits are not yet defined.

Many other companies held leases in the vicinity and many geologists at one time or another saw the possibilities of the area. However, the continuously discouraging results of 17 dry holes drilled during the 8 years between 1934 and 1942 dulled their enthusiasm, and the general opinion of the possibilities of the area at the time of the Pure discovery was far from favorable.

The lesson for all of us in this discovery is that if it had not been for the persistence of Cram, his technical experience and interpretive ability to coördinate detailed well-cutting data, surface structural maps, and geologic history with geophysical data, and the faith the Pure Oil Company placed in his judgment in drilling when nearly everyone else had given up hope, the Pauls Valley field would still be undiscovered and probably would remain unknown for many years to come. How many more similar areas are now lying in the files, more or less condemned, yet waiting to be reworked in the light of modern concepts, with modern tools, by men of experience and with the ability to coördinate geology, geophysics, and all the host of technical data now available, and with the imagination necessary to visualize geological history in terms of oil and gas accumulation? Nearly every company has prospects by the dozen such as this, just waiting for a fresh viewpoint, a restudy, or a more experienced attack to transform them from something of merely historical interest into something of potential future worth.

GEOLOGICAL RESEARCH

The time is rapidly approaching when some extensive fundamental research is called for in petroleum geology. Most of the research laboratories of the oil companies are devoted chiefly to studies of refined products, new uses for petroleum, development of geophysical instruments, and development of engineering and production principles and techniques. In these modern research laboratories, as well as in the exploration departments, we find the main efforts are directed chiefly toward the improved utilization of petroleum which is already discovered or produced rather than toward the more fundamental problem of discovering new supplies to replace those that are now being consumed.

If there is any doubt as to the ability of the industry to maintain a supply of petroléum through the years ahead—and we all realize there is such a doubt— then through some means or other, research on many of the fundamental questions concerning the occurrence of petroleum can not start too soon. Most of our oil provinces have been discovered contrary to the orthodox geological opinion prevailing at the time, which is rather good evidence that our fundamental thinking is not sound. Other industries, which use petroleum products, are developing new uses through research, but our thinking has been standing still on the problem of discovering the supplies these increased demands call for.

An industry that spends a quarter of a billion dollars or more each year on exploration can well afford to set aside a small percentage of this money in an attempt to make the effort more effective, if not to save the entire situation. This is in fact a public responsibility. It can be done as a broad industry-wide approach as is being done on a small scale by some members of the American Petroleum Institute in the study of the origin of petroleum, or it can be justified as a legitimate private-company expense if only for the simple reason that it is a practical and economical insurance against the day of insufficient supply.

Some of the subjects of research that can be studied profitably are suggested in the following section.

"What is it and where does it come from?" has been a question in the minds of many since the first oil, pitch, or asphalt was discovered in the beginnings of human history. At no time, however, has there been any demonstrable proof of the mode of its origin and, although there have always been many theories and explanations, even to-day we have nothing on which we can depend as satisfactory. As a matter of fact, the most recent thought on the question of the origin of petroleum indicates that it may well be a much more complex and involved process involving chemistry, bacteriology, physics, and geology, than anything we have even thought of heretofore. The more we know of the subject, the more difficult appears to be the answer.

Yet, in spite of the evident lack of information, the oil industry has either spent or withheld millions of dollars and great quantities of time and materials, on what was the currently prevalent thought on the subject. As long as discovery was adequate, the question of origin was of purely academic interest, and it did not make much, if any, vital difference to anyone what the ideas were on the source material of petroleum—whether it was the urine of whales as some early investigators thought, or the product of subterranean volcanism as was thought by others, or the organic material deposited in marine sediments according to more modern ideas.

The chief geologist of one of the major companies told me last summer that while attending college, and after considering the subject of source rocks, he wrote a thesis with the title, "Why It Is Impossible to Find Oil in West Texas," and it received much favorable comment from the faculty. That was the prevalent thought of the time and it acted as a stone wall beyond which the orthodox geologist was unable to see. The presence or absence of source rocks in West Texas was the chief fundamental criterion on which exploration out there was either done or not done. In the early twenties it was a real stroke of luck for the Gulf Company to have on its staff men like Kip Harper and Ben Belt, who had escaped in some manner from the mental fence concerning the lack of source rocks in West Texas, within which the average geologist had retreated. Thus, while one company with a liberal viewpoint on this question acquired many millions of barrels of oil, the others, largely because of a different viewpoint on this one question, were prevented from participation.

Now, as a matter of fact, no one at that time knew whether or not there were source rocks in West Texas for the simple reason that even to-day no one knows what a source rock is. The chief reasoning on which the prospecting or lack of prospecting was based was prejudice—pre-judging—and insofar as a source rock is concerned, it is still prejudice. Yet the argument comes up continually that an area is unfavorable because it does not contain a "source rock." "I like to see some black shales, or at least some dark shales," "There ought to be some fossils in all that limestone to indicate a source rock," "where is your source rock," or a similar expression of doubt is a common excuse for not drilling in new territory.

Is it not ordinary common sense to spend some money on trying to find out something definite about source rocks? Or at least to recognize applications to which lack of such knowledge applies? How many future productive areas will be condemned because some geologist thinks there are no source rocks? With the exception of the relatively small appropriations available to the A.P.I. research committee, in the work of Trask and the present project on the origin of oil, there has been only intermittent and casual interest in the subject of the source material of our oil fields. Yet the industry is paying in a big way for its lack of such knowledge and will continue to pay in a big way until something is done about it.

FRESH OR LOW-CONCENTRATION WATER

A few years ago if one were to listen to the average geological conversation in the offices or on the streets of Tulsa—and probably in many other towns in the northern Mid-Continent area—he would probably hear a heated argument on the possibilities of oil and gas in the Forest City basin of northwestern Missouri. This area was very much in the limelight at that time and attention was directed toward it largely because of its similarity with the Michigan basin and the Illinois basin, both of which had been proved productive. There was something different, however, in the Forest City basin and that was that the older formations were known to contain either fresh or low-concentration waters. What did it mean? Some said that it meant there would be no commercial oil fields—and so far at least they have not been disproved. Other companies thought that such a condition was not particularly detrimental and they waived this objection and sent geologists and lease men into the area and later drilled test wells there. In the aggregate, large sums of money were either spent or withheld by various companies largely dependent on the philosophy or attitude of their geologists toward this question of the influence of fresh water on oil accumulation.

As in the case of source rocks, the interesting fact is that no one has yet shown any demonstrable reason or observation that fresh or low-concentration waters have any significance either for or against the accumulation of oil and gas. In other words, we know nothing about it, yet this lack of knowledge in the form of prejudices—pre-judging—translates into untold amounts of money, time, and effort either withheld or spent. A small percentage of the money involved, if applied to research on the question of the function of such waters in oil accumulation, may or may not give the answer, but the problem certainly would be put into a form that the industry could evaluate more correctly than at present and it would not be dependent on half-baked theories, most of which "jelled" while in college.

When oil fields such as some of those in northern Sumatra, in Venezuela, and in our Rocky Mountains, are found to occur associated with fresh or nearly fresh waters, how many others might there be in areas that are now condemned for this reason? There are too many oil fields throughout the world in which oil is being produced along with or in close association with fresh or low-concentration waters to continue our present loose thinking on this subject.

SEDIMENTATION

Sedimentation, with all of its related studies, promises to be of increasing importance and one of the key discovery tools of the future. Some of its possible applications have been pointed out and we may well see the day when it rivals structural geology as a way of finding oil.

Yet, we must admit, that in spite of the magnificent work that has been done in the past by the little band of sedimentationists scattered through some of the college geology departments of the country, we really have made only a start on a workable understanding of the subject as it can be applied to petroleum geology. Every oil field in the world involves sediments of some kind, and those sediments, whatever their character, are more intimately associated with the oil and gas than any other factor. In addition, they either precede, or are contemporaneous with, the formation of the oil.

In our approach to discovery, however, sedimentation as a practical scientific device to find new oil fields or extend old ones is almost ignored. Probably one reason for this lack of appreciation by the industry generally is that the sedimentationists have not yet discovered a way of presenting their ideas in the form of contour maps with some kind of a closed-contour symbol indicating the favorable areas for drilling. Let us not forget that in spite of this oversight which may some day be corrected, these men have ideas that may profoundly change our prospecting technique.

Gradually coming into the picture are the quantitative methods exemplified by the work of Krumbein, Pettijohn, and Russell; the application to basic problems of oil movement and reservoir conditions as developed by Waldschmidt; the single-handed attack on many of our fundamental problems in petroleum geology that is being made by Krynine; and over it all the balanced thinking of Twenhofel. All of this pioneer work is being done by men in colleges and is without cost to the oil industry. Almost the only really fundamental work being done by the oil industry on sediments is with clays, and then only because of their application to drilling muds.

An inkling of what is ahead of us in sedimentation was well stated by Payne in a recent issue of our *Bulletin.*[9]

It is now coming to be realized that the future must bring a new form of stratigraphic research, middle-road research which will bridge the gap between the *old stratigraphy* and the *new stratigraphy* that is now unfolding. This middle-road policy will call for complete restudy of the stratigraphic record and for lithologic descriptions and environmental interpretations which recognize and evaluate in simple semi-quantitative manner the fundamental attributes, properties, and structures of sediments and the environmental factors and sedimentational processes which control its genesis.

. . . The new stratigraphy must include a breakdown of the stratigraphic record into lithologic and environmental types and their varieties and must place emphasis on *lithologic classification* as well as stratigraphic classification.

---

[9] Thomas G. Payne, "Stratigraphical Analysis and Environmental Reconstruction," *Bull. Amer. Assoc. Petrol. Geol.*, Vol. 26, No. 11 (November, 1942), pp. 1698 and 1699.

A real understanding of sediments would probably do more than any other single factor to improve the geological interpretations which are put on geophysical data and in like manner it would improve the speculative geology which is necessary to all subsurface studies. Would it not be wise for the oil industry, either collectively or individually, to set aside even an infinitesimal fraction of its discovery costs to gain a better understanding of one of the very few factors common to all oil fields? Especially since our present knowledge has merely touched the fringes of a phase of geology, the applications of which even now give indication of being critical to much of our future discovery thinking?

PERSONNEL

A very practical and natural question in times such as these is where to find the men to do the geological work which has been described. There is probably no complete answer until the war is over. There are, however, several partial solutions, which taken together may help to tide over the situation. They are these.

First: "First things first." When the problem of first importance becomes the discovery of new oil fields and not the development of old ones, there will be many geologists now doing routine or "hack" geology in local areas who can be advanced to the more important creative job of discovery in new areas. They can be put into the field or they can be trained to interpret, in terms of discovery, some of the volume of data, the collection and filing of which is now their chief concern.

Second: Many is the consulting geologist with a background of broad experience and with the technical ability to take on problems such as have been described. The discovery showing which they as a group continue to make with the small amounts of information, equipment, and money they have available strongly suggests that with more data to work with they would have commensurately greater results. Probably one reason for their success is that they know they can not make a living if their objective is merely to collect and file information; as a consequence, they learn to think in terms of discovery and direct their supreme effort toward that end.

Third: With the falling-off in college enrollments in geology departments there are many instructors, assistants, and professors who will be available for commercial engagements. There are literally thousands of special geological reports of the type previously suggested, waiting to be made and which they can make equally as well as, or even more intelligently than, the strictly commercial geologist.

CONCLUSION

In conclusion, what has been said resolves into the simple statement that the oil industry is by-passing an important discovery tool. It has attempted to mechanize, routinize, and standardize what DeGolyer has so aptly termed the "art of discovery"—an inherently individualized and infinitely variable process—and the job is not being done. The proposal is to return to geology as such, but on a much higher scientific level than heretofore. As a discovery tool, there is enough geology now rusting away in the sheds of our filing cabinets and in the background of our thinking to cultivate a really substantial and continuous crop of new oil fields. We have, in the form of more geological observation, more geological reasoning, and more geological understanding, discovery tools that will take hold as our present methods let go, and they are fully able to carry, not only the geologist but the oil industry itself, to new levels of achievement. Moreover, the United States is the great proving ground for exploration ideas and the problems we have here and the solutions we get become the guide to discovery throughout the world. If we are only encouraged and permitted to exercise our latent abilities we can do much toward lifting the clouds of doubt and uncertainty concerning our oil supplies which have settled round about us. This means that all of us must think discovery if we are to realize it.

AAPG Bulletin, v. 34, no. 7 (July 1950), p. 1607–1611.

PLEA FOR LOOSE THINKING[1]

E. DeGOLYER[2]
Dallas, Texas

"I design not to confute or convince you, but only to present and submit my thoughts to your consideration and judgement."—Sir Isaac Newton.

Past awards of this medal have been acknowledged upon occasion with a statement of something of the recipient's way of thinking about our chief concern, the art of prospecting for oil, or about the science of geology upon which it is based. This is an agreeable procedure which I shall follow in the hope that it will crystallize into custom.

History has long been my favorite study. It is not strange that this should

[1] Response on receiving the Sidney Powers Memorial Medal, 35th annual meeting of the Association, Chicago, April 25, 1950.

[2] President of the Association in 1925. Honorary member, 1944.

be so for a geologist. During the past year particular attention has been given to the history of early science. Most history is political, concerned with king-lists and the rise and fall of empires. It tells us, with any acceptable degree of precision, only what men did. What they thought is but imperfectly recorded. One of the beauties of the history of science is that it is essentially a history of ideas; of thinking.

The sixteenth and seventeenth centuries, when science began to emerge from the older and broader field of philosophy and to achieve independent status, are particularly important. This was the period when printed books began to be common. This was the period when the first telescopes and microscopes were made and used. It was the period of Copernicus, Galileo, and Kepler; of Descartes, Newton, and Leibnitz; of Vesalius, Harvey, and Hooke. Geology had not yet begun to be a science but this was the time of publication of one of the earliest books dealing with fossils, mountain-climbing Conrad Gesner's *De Rerum Fossilium*, 1565, and of one of the earliest descriptions of sedimentary processes, Nicolaus Steno's *De Solido intra Solidum*, 1669. It was a period when men's minds were more than ordinarily active. It was the Rennaissance.

The history of early science is the history of discovery, as indeed is the history of all science. One can not read about discoveries for long without becoming curious about how they are made. We want to know and understand the scientific method. At this point the doctors disagree. Let us consider a review and analysis of the methods of initiating scientific inquiry recently made by Professor F. S. C. Northrop of Yale.[3] Sir Francis Bacon would have us collect data, refrain as long as possible from hypothesis; proceed inductively. Descartes prescribes deductive procedure from what is indubitable; reasoning from the hard core of solid truth. Cohen and Nagel go for hypotheses almost immediately and test by trial and error. John Dewey emphasizes analysis and definition of the problem and proceeds promptly thereafter to hypothesis.

It is not proposed to compare these methods of attack. You will analyze your own thinking and decide for yourself which of the procedures seem to be most fruitful. It is noteworthy that the moderns proceed so quickly to hypothesis. It is difficult to believe that one does not begin to formulate or at least search for a hypothetical solution almost as soon as he realizes that he has a problem.

These students of scientific method are not agreed on the manner of initiating research but there is one point upon which they as well as other students of method are in full agreement: their skepticism with regard to traditional belief. Descartes states the position clearly in his first rule of method, "Never to accept anything as true when I did not recognize it clearly to be so."

And so, having indicated that the problem is as old as science itself, I come to my thesis: we should re-examine those problems for which we believe we have

---

[3] *The Logic of the Sciences and the Humanities*, Macmillan Company, New York (1947).

found satisfactory solutions in order to determine whether some other solution may not be feasible and whether or not our solution is unique. Particularly is it urged that we divest ourselves of prejudice and reconsider problems which are not solved but are regarded as practically solved by what Chamberlin called the "ruling theory." Such problems as the origin and migration of oil are of this type. If we achieve no more than the simple classification into the categories of "proved" or "unproved" of the theories upon which our art and science are based, we will have taken a long step forward of very material value.

The Drake well was less than a year old when the anticlinal theory was proposed. Oil, gas, and water, gravitationally stratified, occur in domes and inverted troughs. Our understanding of oil pools then was a scant knowledge of a few fields in the Appalachian region and eastern Canada. There, anticlinal occurrence was the exception rather than the rule.

Our theorists were not theoretical enough or they would have recognized sooner that the dome or inverted trough performed a function that could be the result of conditions other than anticlinal structure. As it was, after more than a generation of attempts to force Cinderella's anticlinal slipper onto every new oil field, we had to amend our theory. The anticlinal gave way to the structural theory. Finally, with full recognition of the controlling importance of function as the least common denominator, we have come to the trap theory.

If our first oil pools had been studied by physicists, the function of the trap rather than its presumed anticlinal nature might have been emphasized. The geologist then should have forseen the many conditions which might perform such function before the drill forced his hand. We have made the error of stressing a single solution for a problem which has no unique solution.

Many of our judgments of the value of individual prospects are based upon assumptions that certain theories are true. The theories may or may not be true and the assumptions are thus true or false. The history of the art of prospecting is strewn with the wrecks of what were once "ruling theories," many of which were accepted because they seemed to be so reasonable. It was widely held during the first half-century of the industry, for example, that regions of the occurrence of profuse and numerous oil and gas seepages were unfavorable for the occurrence of substantial oil or gas pools in the subsurface. It was obvious that the seepages were the result of leakage and, geological time being what it is, it appeared to be eminently reasonable that any important pool would have leaked away. This was the official viewpoint in Mexico as lately as the early years of this century. Almost a corollary of this proposition was the belief that a faulted structure was not a favorable prospect because the oil and gas would have leaked out along the fault plane.

J. Peter Lesley, the most experienced oil geologist of early Pennsylvania, soon discovered or accepted the discovery of another, that gas occurs in solution in oil in the subsoil. He went further however and held that gas, oil, and water were intersoluble, an error from which he did not recede during his long lifetime. Dumble held at one time that oil would not be found in the Gulf Coast in formations older than the Miocene. Josiah Dwight Whitney, State geologist for California and an outstanding economic geologist of his day, held that flowing oil wells were not likely to be found in California and, if they should be found, that they were likely to be in the northern rather than the southern part of the state.

Many of us can remember, and it was not too long ago, when drilling contracts for wildcat wells in Oklahoma and Kansas provided for abandonment when the Oswego lime was reached. Even younger men will remember when it was not considered probable that commercial oil would be found in eastern Oklahoma south of a certain imaginary line, a line which lies north of the prolific fields of the Seminole plateau.

"Yes, yes," you reply, "but that was a long time ago and we have learned a lot since then." True enough, but it is not so long ago that the highly important fields of the Cuyama Valley were discovered in an area which had been turned down by geologists probably as often as any other. During a recent California trip an earnest effort was made to learn why this was so. The most direct and commonest answer was that Cuyama lies west of the San Andreas rift and it had been commonly held that no important fields would be found west of the rift. Just why this should have been believed is not apparent since the fields of both the Los Angeles basin and the Coastal districts lie west of the rift. Likewise, a generation of competent geologists, American, English, Swiss, and Mexican, mapped and remapped the subsurface of the Golden Lane in Mexico before one of them was able to look at the problem with fresh vision and clear enough eyes to locate the Alazan pool. Neither of these fields was discovered with new techniques. They were discovered with new viewpoints.

A new viewpoint, many new viewpoints are needed. It is a matter of thinking about the same group of facts differently from the manner in which we have been accustomed to think. This is a difficult thing to do. Fresh ways of looking at a problem are important in other industries and sciences, however, and must be in ours. In a recent advertisement an officer of the company which developed the polaroid camera states, "We had to side step the conventional and banish a hundred years of 'habit thinking' in our industry to even *envision* a camera that takes a picture, develops and prints it in less than a minute." An eminent historian, in his study of the origins of modern science, points out that revolutionary changes in the strategic fields of both astronomy and physics came about, not through the development of the telescope and microscope, important as they have been, but through changes in the minds of pioneer scientists; changes in their manner of thinking.

Most of us probably hold many geological prejudices; beliefs that we must admit are not fully proved as yet. So common is this habit of thought that one is surprised to find oneself with an open mind on problems which have been under examination and discussion for a long time. On origin of oil, for example, have we proved or disproved our commonly held theories? I confess deep ignorance as to what source rock may be. I am catholic-minded enough, however, to accept the "ruling theory," but only as a working hypothesis. On long or short distance migration of oil, the case is not proved and it is doubtful whether it is likely to be proved. These may be the exceptions, however, which prove the rule.

In conclusion may I emphasize the possibility, as it seems to me, that as much oil may be found in the future with new viewpoints as with new techniques. The most important tool of which any geologist is possessed is his mind, the one tool which is sharpened and not dulled by use.

Volume 29   Number 7

BULLETIN
of the
AMERICAN ASSOCIATION OF PETROLEUM GEOLOGISTS

JULY, 1945

RESOURCES AND RESOURCEFULNESS[1]

IRA H. CRAM[2]

Chicago, Illinois

INTRODUCTION

Standing at the very head of a long list of difficulties which consistently confront the oil industry is the problem of discovering adequate quantities of crude oil and gas. To date the industry, by developing an unexcelled resourcefulness, has been eminently successful in solving this major problem. The members of this Association—geologists, geophysicists, geochemists, teachers, petroleum engineers, scouts, production men, executives, operators, and others—have made a signal contribution to the industry's success. Since the organization of the Association 28 years ago, almost 26 billion barrels of crude oil have been won from the earth, and a little over 49 trillion cubic feet of natural gas have been produced and marketed in the United States. In addition, the estimated proved reserve of crude oil has been enlarged by approximately 14 billion barrels, and the estimated proved reserve of natural gas has been increased between 90 and 100 trillion cubic feet.

The accomplishment of finding 40 billion barrels of crude oil and between 140 and 150 trillion cubic feet of gas in 28 years is a record of which the industry can be justly proud. But that accomplishment is now history; what of the future? Two questions immediately present themselves. How large are our resources of oil and gas? Whatever they may be, are we going to continue to be resourceful enough to develop them economically? Nobody knows or has any way of knowing even the approximately correct answers to these questions. Partial answers to both questions ventured by students are numerous and varied. I do not propose to offer even partial answers, but merely to analyze certain aspects of the questions in a way that may prove to be helpful.

[1] Presidential address, thirtieth annual meeting of the Association, Tulsa, March 27, 1945. Published by permission of the chief geologist of The Pure Oil Company.

[2] Assistant to the chief geologist of The Pure Oil Company. The writer expresses his appreciation to A. Rodger Denison, A. I. Levorsen, Carl C. Addison, Theron Wasson, C. Plummer, W. E. Wrather, and Carey Croneis, all of whom read the manuscript and offered helpful suggestions.

RESOURCES

The nation's resources of oil and gas consist of the undiscovered reserve and the discovered reserve. The latter is subdivided into the proved reserve and the reserve already discovered but not recoverable at present prices and with present production methods.

DISCOVERED RESERVE

The total area in the United States covered by oil and gas fields is approximately 26,000 square miles. Approximately one-third of this area is productive of oil. From these fields approximately $29\frac{3}{4}$ billion barrels of crude oil and almost 61 trillion cubic feet of gas have been produced, and the proved reserve is estimated at 20 billion barrels of oil and 114 trillion cubic feet of gas. The reserve of crude oil not recoverable at present prices and with present production methods may well be 25 or more billion barrels. We have discovered not only the commonly estimated 50 billion barrels of oil; we may well have discovered over 50 per cent more. This presently unrecoverable reserve is a resource of great importance, insufficiently stressed by many observers. How to *recover* an ever-increasing percentage of the oil we *discover* presents a challenge to the industry's resourcefulness in general, and to the resourcefulness of geologists and engineers in particular.

UNDISCOVERED RESERVE

The undiscovered reserve of oil and gas lies beneath certain oil and gas fields that have not been thoroughly tested to basement and beneath an area of approximately 1,600,000 square miles outside the producing fields. Scarcely a sizeable part of this prospective territory has escaped attention; yet enormous areas have hardly been scratched by the drill. These largely undrilled areas make up the horizontal frontier. The vertical frontier lies below the formations normally tested and below the depths customarily reached in drilling.

Faith in the existence of a large number of oil and gas pools in these frontiers as well as in the more densely drilled areas is justified by the general geology and history of discovery. In the horizontal frontier of thick, warped, changing sediments, traps capable of holding oil and gas can not fail to exist. Some of these traps will be detected in advance of drilling by geological investigations; others will be found by geologists using the geologic data revealed by drilling. There are undiscovered traps and known traps that have not been drilled to sufficient depths in the more densely drilled areas between the horizontal frontier and the producing fields.

In considering the prospects of the more densely drilled areas, and of both the horizontal and vertical frontiers, the amount of condemning drilling is a most pertinent factor. Only structurally or stratigraphically well-located dry holes that have been drilled through all known producing formations or through the entire sedimentary column or to 15,000 feet can be considered as truly significant. Only those drilled through the sedimentary column may be judged as truly condemning.

Without attempting to present a complete discussion of deep drilling, certain significant facts should be pointed out. In a prospective area of approximately 150,000 square miles, most of which is in the Gulf Coast states, the sedimentary column is thicker than 15,000 feet, and in this large area only 5 wells have been drilled to 15,000 feet or deeper. In the area in which the sedimentary column is thinner than 15,000 feet, test wells to basement are rare indeed, except where the column is thinner than 5,000 feet. Only approximately 325 wells have reached basement between the depths of 5,000 and 15,000 feet in an area of approximately 850,000 square miles, and approximately half of these are located in a few California producing fields two of which produce from the basement. Even if all the dry holes reaching basement below 5,000 feet were well located structurally or stratigraphically, the amount of truly condemning testing is small. The number of test wells reaching basement above 5,000 feet in an area of approximately 600,000 square miles is inadequate.

How much of the oil and gas cached in the enormous horizontal and vertical prospective territories will eventually be found is anybody's guess. Until well-located dry holes have been drilled through the sedimentary column or to the limits of drilling equipment on a high percentage of the geologic anomalies that could be traps and on producing traps, there will be no nearly conclusive evidence that the end of discovery is in sight. However large the undiscovered reserve may be, it is of no practical importance unless it is turned into the proved reserve. The conversion of undiscovered reserve to proved reserve has never been easy, except in retrospect, and it can only become less easy. We who are charged with the duty of finding and developing oil and gas fields must first find and develop a resourcefulness adequate to the task.

RESOURCEFULNESS

Introduction.—The nation's resources of oil and gas are developed through the efforts of an army of men and industries cooperating in an atmosphere of free enterprise and private ownership. Various phases of the complicated procedure are dominated by men with special training and knowledge. The all-important initial phase—exploration—is naturally dominated by geologists who are specialists in the composition and structure of that part of the shell of the earth in which the deposits of oil and gas lie hidden. Resourceful geologists have applied their science to the solution of a great variety of problems of exploration and exploitation, so that today geology permeates the producing branch of the industry, and geologists occupy positions ranging in rank from trainee to company president or owner. Their horizon is virtually unlimited, and the successful capture of our hidden reserves of oil and gas depends in large part upon their performance. The resourcefulness of geologists is, therefore, an important component of the industry's resourcefulness, and, as such, it is worthy of special consideration.

                    IRA H. CRAM

During the past 25 years the precision of geologic investigations has been increased by the use of new geologic tools—plane table and alidade, core drill, magnetometer, torsion balance, refraction seismograph, reflection seismograph, gravimeter, various logging devices, *et cetera*. The invention of additional new instruments is not unlikely. But the development of a new geologic tool does not necessarily depend upon the invention of a new instrument. The torsion balance, for example, was not invented specifically for use as a geologic tool. Resourceful geologists made it a geologic tool by using it to determine geologic anomalies.

The geologic data available on a given area may be scant or voluminous, but in the aggregate the amount of information on the geology of the United States jamming the libraries and files of surveys, companies, and individuals is staggering; and the rate at which factual material is furnished by geological and geophysical surveys and wells is no less staggering. Geologists have by no means exhausted the possibilities of making better use of this information in unraveling the geology. Mastery of the various methods of obtaining geologic data is necessary to the understanding of the facts and to the intelligent use of the methods. Referring particularly to the information acquired by geophysical means, geologists can add materially to the interpretations made largely by non-geologists by taking the trouble to learn the principles of the various geophysical methods, and, more important, by working with the data and realizing that they are, in fact, geologic data. Failure to work into the geologic picture this additional geological information is akin to ignoring certain well logs or outcrops in interpreting the geology of an area. If geologists are hesitant to delve into reflection seismograph work, for instance, they need only recall that they use well logs daily and that a reflection seismograph record, like an electric log, is merely a well log expressed in physical, rather than geologic, terms. As a matter of fact, geologists have long used geophysical instruments—the compass, the dip needle, the alidade, the binocular microscope, and the petrographic microscope. Geologists expand their understanding of the composition and structure of a fragment of rock by using the petrographic microscope. By using geophysical instruments they expand their understanding of the composition and structure of what is merely a larger fragment of the earth.

It is realized that not all geologists now have the opportunity of delving into all branches of geoscience, and, further, that there must be specialists in geoscience. Such specialists, many of whom have not been formally trained in geology, have done and will continue to do a most creditable job of acquiring and interpreting new data on the earth. They are indispensable. They can, however, improve their acquaintanceship with the science of geology just as every geologist can strengthen his acquaintanceship with the other geosciences. As geologists and other scientists attain a better understanding of each other's fields, the results of their combined efforts cannot fail to improve. Independent interpretations of each line of investigation by specialists in each field are important but inadequate. It is definitely necessary to develop an increasing number of geologists capable of interpreting, correlating, and reconciling all earth data in the preparation of a single interpretation of the geology of a given area.

CONTROLLED IMAGINATION

Every geologist knows that information sufficient to solve any sizeable geologic problem is seldom, if ever, available. There are gaps between the geologic facts that can be spanned only by the imagination. Failure to bridge these gaps means failure to draw conclusions, however tentative they may be. And practically all decisions in exploration, ranging from the decision to explore a remote area to the decision to abandon the last well in an exhausted field, are based upon geological conclusions of a tentative nature. These tentative conclusions, or geologic ideas, grease the gears of the exploratory machine. They frequently turn out to be wrong, but without them little is started, little is accomplished.

To bridge the gaps between geologic data by leaps of an imagination under control is to be resourceful. Controlled imagination is not guesswork. It is imagination tempered by experience, knowledge and sound reasoning. Like a muscle it grows through enthusiastic exercise. Imagination plays its greatest role in the early stages of exploration. The choice of areas to work with seismographs or other tools capable of obtaining details of the geology, is ordinarily based upon few facts and much imagination. The geologist whose imagination assists in leading the right tool to the right area has contributed immeasurably to the discovery of oil or gas. The most competent reflection seismograph party can find no prospects if it is always assigned to unpromising areas by those in charge. In the progress of detailed investigations the geologist who detects in the incomplete work a suggestion of an important structural anomaly in an inaccessible area, and proves its existence by going to the trouble and expense of exploring the area in detail, has used his imagination to great advantage. The geologist who daringly places the right geologic interpretation upon a mass of data that the unimaginative are afraid to interpret has also demonstrated the potency of an imagination under control.

It would indeed be a calamity if geologists failed to develop further their imaginations by adopting the false premise that the machines will do their thinking for them. A new machine may make it unnecessary to imagine any longer what certain conditions may be, but each new machine provides geologists with new data upon which to ply their imaginations. The machines merely guide imagination into new channels. Thus, by using the reflection seismograph, geologists may no longer have to imagine what the subsurface may be beneath a surface nosing, but the problem of using the seismograph to the best advantage and of interpreting correctly the seismic data puts as great or greater burden upon their imaginations. Likewise, in the days before electric logging when faulting was supposed to be extremely rare in the Gulf Coast, geologists did not have to work their imaginations on the solution of complicated fault systems.

Through experience in drawing geologic conclusions from a miscellaneous assortment of data, every geologist has developed to some degree the necessary powers of controlled imagination. Petroleum geologists as a group have matured to the point that their imagination, when under control, is in effect a geologic tool. In investigating the geology of the world for economic purposes, this mental tool must play a part increasing in importance as weaker and weaker geologic evidence must be employed as the basis for interpretations.

CONTROLLED IMAGINATION IN APPLIED GEOLOGY

It need not be pointed out to a group of petroleum geologists that in this highly competitive oil business a constant stream of economic decisions, as well as geologic decisions, must be made based upon whatever geology may be available. Just as the undiscovered reserve is useless unless converted into the discovered reserve, geology is useless to the industry unless it is translated into economics. The petroleum geologist's job is not done when the geologic investigations have been completed; it has just begun. In translating geology into economics, no machine will assist geologists to read into the geology the practical application of it. Geologists have only their experience, knowledge, theories, reasoning powers, and imagination to guide them. In other words, their only tool is the head.

The choice of areas to investigate and the choice of geologic anomalies to drill are two of the most serious economic problems in the search for oil and gas fields. In attempting to solve these problems, geologists draw heavily upon their knowledge of the geology of known oil and gas accumulations. It helps considerably to possess worldwide knowledge. It is not far-fetched to state that a geologist can find more oil in Michigan, for instance, if he understands the geology of the oil deposits of a remote area like Iran. This knowledge of the past, however, is inadequate. Something must be added, and this something springs from the imagination. It takes imagination to apply the one fundamental principle governing the occurrence of oil and gas deposits, namely, that important deposits of oil and gas occur only in traps in permeable rocks in sedimentary sections that are mainly marine in origin. What are traps? They are any kind of geologic anomaly that can hold oil and gas. One cannot forecast accurately the kind of structural anomaly it takes to hold oil and gas in a given area, cannot forecast accurately the underground stratigraphic and structural changes that may form or assist to form a trap, and cannot forecast precisely the probable producing formations. The next great oil or gas field may well be stratigraphically and structurally quite unlike any presently known field. The geologic setting of the next petroliferous province may be most surprising.

The unimaginative appraisal of the economic significance of geology in the light of the past—orthodox economic appraisal—leads to mental condemnation of those geologic anomalies that are unlike those upon which nearby or remote oil and gas fields are located. The libraries and files of surveys, companies and individuals are full of geologic data indicating a number of good prospects mentally condemned as of this date. Any geologist at any time may see one or more fields or new petroliferous provinces in these libraries or files by the not altogether simple process of tearing himself away from orthodox economic appraisal of the geology and adopting a fresh viewpoint. The more imaginative geologist can achieve this fresh viewpoint; the less imaginative achieve it, if at all, with difficulty.

Most geologists can draw upon their experience and prepare a rather imposing list of oil and gas fields located in areas where oil and gas were not supposed to exist, or located on anomalies known to many but mentally condemned by all but a few or all but one geologist. In attempting to analyze such a list, I reach the conclusion that the geologists responsible for the discoveries had controlled imagination, and pushed the use of it, while their competitors were complacently hiding behind the screen of orthodox economic appraisal, styling themselves practical men. These competitors thought they knew, among other things, that a certain number of feet of closure was necessary, that certain formations could not produce, that a county line or river separated the good and bad areas, that a certain area might produce gas but certainly not oil, and that the rims of certain basins might be good but certainly not the bottoms. In short, they knew where the oil was not, and built mental walls around these supposedly undesirable areas in order to contain their efforts and money in the supposedly desirable areas. The oil finders, recognizing the impractical procedure of applying a little knowledge unimaginatively to the solution of so vast a problem, merely adopted the truly practical procedure of drilling those geologic anomalies which stood the test of their own interpretation of the economic significance of the geology. The oil finders, directing their mental efforts to developing reasons for the presence of oil and gas fields, were positive thinkers. The non-oil finders, directing their mental efforts to developing reasons for the absence of oil and gas fields, were negative thinkers.

SALESMANSHIP

The oil finders do something else. They sell themselves on their interpretations and in turn sell their ideas to others capable of completing the discovery process. The loss of a good idea through poor salesmanship may postpone indefinitely the discovery of an important oil or gas field or new producing province. Salesmanship is the all-important follow-through.

The selling of geologic conclusions of the more imaginative variety admittedly calls for strong salesmanship. The potential buyers are frequently steeped in orthodoxy and inclined to place little trust in projects that depend heavily upon the human equation. Moreover, since geologic conclusions are tentative and subject to rapid change, geologists commonly find themselves in the position of having to unsell the potential buyer on an idea previously sold to him. The job of unselling is as big as the job of selling. Neither job can be done well if geologists have not translated their geology into economics, the language of the potential buyer. The geologist who understands thoroughly the economic implications of his geology has gone a long way toward selling his ideas, imaginative though they may be.

There is definite need of selling more wildcats in the wide-open spaces where geologic evidence may be thin but where the much-needed new provinces may be waiting to be tapped. There is need for deeper drilling in producing provinces. More effective salesmanship on the drilling of more test wells in the search for stratigraphic traps or in the exploration of complicated structures is a necessity. The industry and geologists should be more widely sold on the idea that the drill is a geologic tool, a tool for gathering geologic information as well as a production tool. The burden of developing and selling an intelligent drilling campaign of the proportions needed rests upon geologists.

CONCLUSION

Resources and resourcefulness. We have them both in quality and quantity. They are equally important as national assets. However blessed our land may be with deposits of oil and gas, they must be captured before they are, in a practical sense, resources. Petroleum geologists lead the way in the initial phase of capture —exploration—and are playing an ever increasingly important role in the succeeding phase—exploitation. It is imperative, therefore, that their resourcefulness, although but one element of the industry's resourcefulness, be maintained at the highest possible levels.

The problems of pure and applied geology will continue to depend heavily upon the imagination for solution after the development of many new fact-finding devices, including those which may detect oil and gas accumulations directly. Hence petroleum geologists have no alternative but to develop further, and to employ fearlessly, controlled imagination in interpreting and applying their science. Furthermore, there is little choice but to be alert to the development of new geologic tools and concepts, to push to the limit the use of old and new tools and concepts, to acquire and apply worldwide knowledge and experience, to think positively rather than negatively, to fuse into all interpretations every scrap of earth data however obtained, and to evaluate thoroughly the economic implications of the geology. By increasing the resolving power of geology through the continuous improvement of geological resourcefulness, a greater and greater percentage of the nation's undiscovered reserve of oil and gas will be captured for the benefit of man.

Reprinted by permission of the Tulsa Geological
Society from *Tulsa Geological Society Digest*, v. 26
(1958), p. 84-101.

OIL IS FOUND WITH IDEAS[1]

by

Parke A. Dickey[2]

Next year the oil industry will celebrate the hundreth anniversary of its birth at Titusville, Pennsylvania. Some industries have had humble beginnings, but not the oil industry. Its birth was spectacular, and its growth since that time has been marked by intermittent, extravagant booms, or waves of gushers which have often resulted in an acute oversupply. These periods have usually been followed by periods of low discovery rate during which it looked as if we were going to run out of oil.

A review of the history of petroleum technology shows that each of these bursts of increased production was caused by a new idea. Sometimes the idea was mechanical — for example, new drilling methods like the rotary made it possible to drill on the Gulf Coast, and heavier rigs made it possible to drill for deeper horizons in old areas. Sometimes it is technological like the discovery of waterflooding or hydraulic fracturing. Often it is geographical —oil is discovered for the first time in a new area. Most often it is geological — a new theory on where oil can be found impels men with faith and vision to raise money and spend it wildcatting.

We usually find oil in new places with old ideas. Sometimes, also, we find oil in an old place with a new idea, but we seldom find much oil in an old place with an old idea. Several times in the past we have thought we were running out of oil whereas actually we were only running out of ideas. This principle can be illustrated most vividly by reviewing the story of the principal discoveries in each producing area of the world.

Pennsylvania

The oil business started with an idea—the mechanical one of drilling for oil. In Ontario, West Virginia, and Kentucky oil was produced in a small way from hand-dug wells. The promoters of the Pennsylvania Rock Oil Company in 1855 got Prof. Benjamin Silliman, a chemist and geologist at Yale University, to analyze a sample of the oil from Titusville. His analysis showed its great industrial potential for illuminating and lubricating. Apparently the founders of the company, Jonathan G. Eveleth and George Bissell, intended to dig and enlarge the oil springs at Titusville. Bissell accidentally stopped one hot summer day in a drugstore in New York [2,3] where he saw an advertisement of Kier's "Rock Oil" "discovered in boring for salt water near the Bank of the Allegheny River about four hundred feet below the earth's surface." Drilling was the key to getting oil in commercial quantities, and the reason why the Drake well started the industry off when the other oil wells did not.

Bissel hired a friend, Edwin F. Drake, who was a conductor on the New Haven Railroad, and sent him to Titusville to try drilling for oil. He went to Salina, Pennsylvania, and persuaded a driller and blacksmith experienced in drilling for salt to come to Titusville. He had trouble getting through the soft gravel, and bought some cast iron pipe which he drove 32 feet to bedrock. He drilled to 69-1/2 feet, and shut down over a week end. Sunday afternoon the driller, "Uncle Billy" Smith, noticed that oil had filled the well to a few

---

[1] President's address

[2] Jersey Production Research Company, Tulsa, Oklahoma

feet from the surface. The well produced about 10 barrels a day by bailing. Later when it was tubed and pumped from the bottom it produced about 25 barrels per day. He had hit a shallow stray sand that produces nowhere else in the area. This sand apparently is open to the gravels of the buried valley of Oil Creek at Titusville where there was a lake during glacial times.

The Third Sand is productive 400 feet beneath the Drake well and richly productive a little way to the north. It was not long before wells were drilled to it, and they produced tens and hundreds of barrels a day. Land was quickly leased for miles all up and down Oil Creek, and even up and down the Allegheny River. By the end of the year 1859, 74 producing wells had been drilled. One of the first big flowing wells was drilled at Rouseville in April 1861. It caught fire and 19 people were burned to death. Tremendous excitement prevailed and people flocked to the "Oil Regions" of Pennsylvania. It was not long before the limited market for oil was flooded and the price dropped. The creek valleys were pretty well drilled up and production started to fall off. This set another pattern that has been followed ever since. New discoveries are followed by overproduction and falling prices. The situation has always cleared up quickly—not by a reduction in the supply but by an increase in the demand.

The effect of new prospecting methods can be clearly shown by following the daily oil production of Pennsylvania, which is plotted in Figure 1 on semilogarithmic paper. This paper has the advantage that equal percent increases show as equal rises on the curve. It has the additional advantage that a constant percent decrease per year plots as a straight-line decline. When no new oil is being found, previously discovered fields show declines that often are approximately straight lines on semilogarithmic paper. As a result of this fact, if the rate of the decline is known, a rough idea of the amount of oil discovered by each new development can be obtained. Each new area or each new idea on where or how oil may be found is followed by a burst of discovery, each greater than the last. In a surprisingly short time the field is drilled up—the method is exploited, and production declines until a new idea or area is discovered. In Pennsylvania a pattern was set that has since been repeated in other states. We find oil by new methods in new areas.

Oil finders of that day, like those of today, looked for oil in places that resembled those in which it had previously been found. They therefore first confined their wildcatting to the narrow deep valleys that dissect the high and forested Allegheny Plateau. They drilled up and down Oil Creek, and found much oil down stream from Titusville although none above. They drilled down the Allegheny River from Oil City with much success, finding the Reno and Franklin fields. Up the Allegheny they found little oil until they got to Tidioute, where the Third Sand is coarse and pebbly and only 150 feet deep. They had less success in 1863 and 1864, and the Civil War was in progress. This first burst of drilling in five years discovered about 60 million barrels of high gravity, low sulfur oil.

It was not until about 1864 that a venturesome operator, probably contrary to the best advice of the experts, drilled a well on the Old Ocean farm on a hilltop west of Oil Creek. He hit a rich streak of Third Sand, and started a new burst of development which is clearly shown in the curve. This excitement caused some famous booms. Pithole was a shortlived field in which the wells were drilled so closely and cased so poorly that the sand was virtually exhausted and watered out in a year. But many other fields were found under the hilltops.

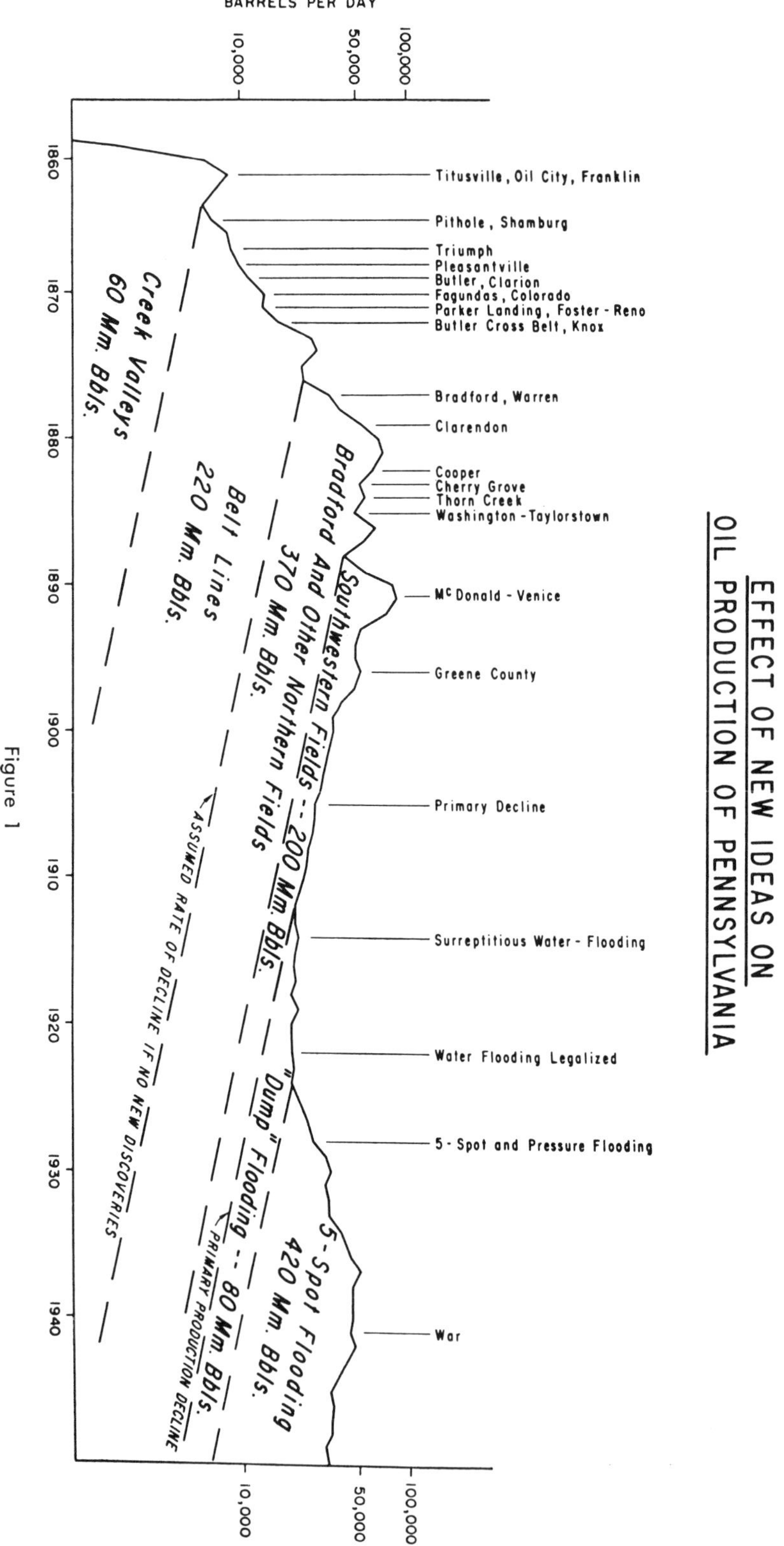

Figure 1

As drilling proceeded, it became apparent that the oil fields lined up in long, narrow, belts and trends. This must have been noted by many operators, but the most astute and literate was C. D. Angell [4,5] who owned a lease at Scrubgrass in the Allegheny River valley, which is here very narrow. He noticed a similar small spot of production up the river at Foster. He was not a geologist, but was well educated and a former school board member. Studying the well records, he found that the sand thicknesses and intervals were similar at Scrubgrass and Foster, so he decided that they were both on the same belt. Having faith in his idea, he had engineers run a line connecting these pools through the plateau east of the river, and leased the land on either side of this line. His idea proved correct, and he was wonderfully successful, becoming very wealthy. The linearity of oil trends is very well marked in this area [6] (Fig. 2).

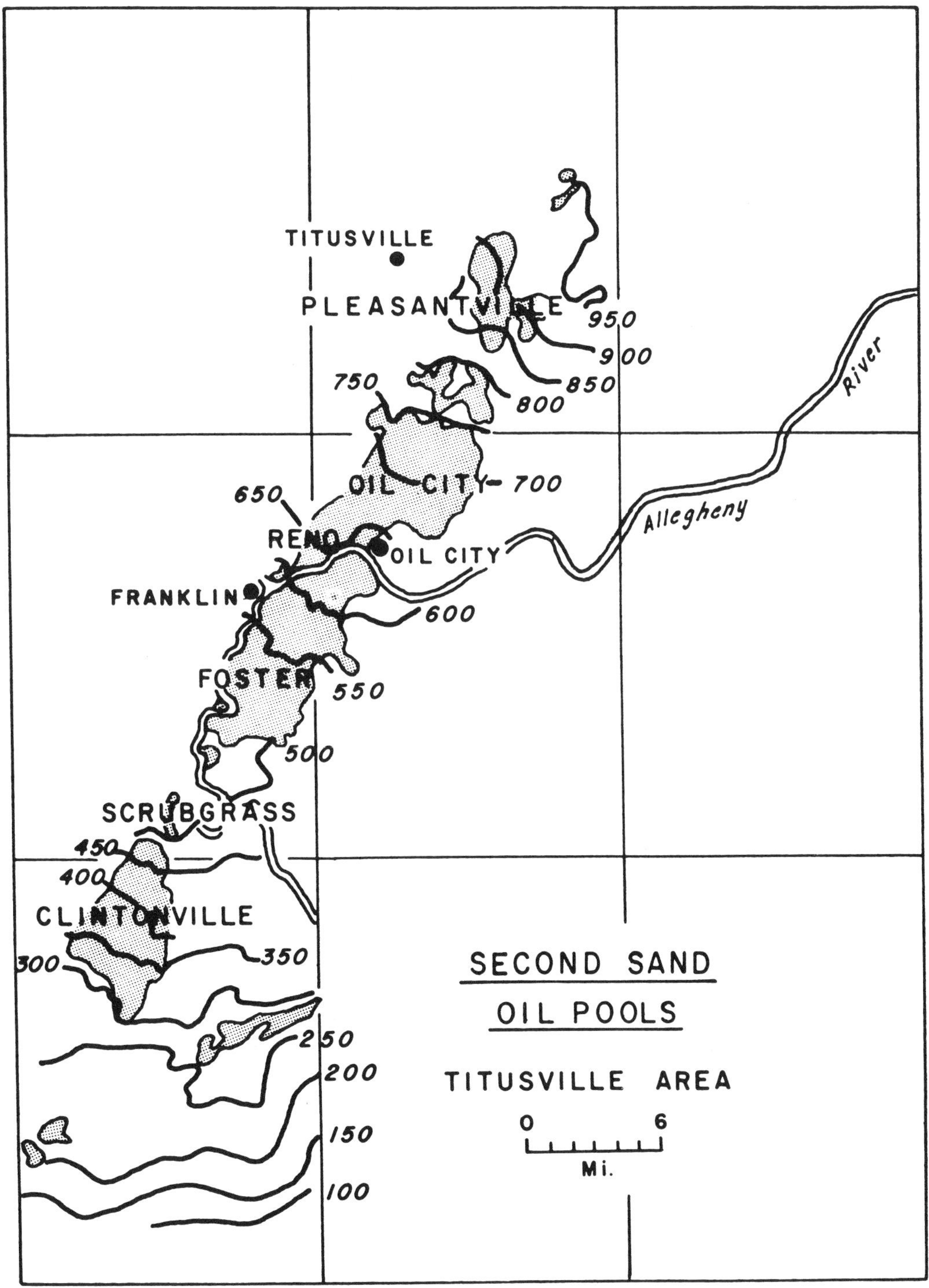

Figure 2

One of the most remarkable aspects of oil discovery is the speed with which all the oil found with a new idea is usually developed. Within 10 or 15 years from the very beginning of the industry, in a rugged, forested area, with no well developed technology or modern transportation, they found almost all the oil there was. In spite of 100 years of wildcatting, *almost none has been found since.* This story has been repeated in most but not all new oil areas as will be shown later.

The Second Geological Survey of Pennsylvania was organized in 1874 under the direction of J. P. Lesley[7,] who had been an assistant of Henry D. Rogers during the first Survey in the 1850's. Lesley was a man of unusual brilliance, ability and charm, and a fine geologist. He assembled an able group of assistants and gave them much freedom to do their own work and publish their own reports promptly. Lesley himself had shown the importance to geology of contoured topographic maps and had introduced contours to represent geologic structure. Any geologist who ever worked in Pennsylvania has been impressed by the accuracy and completeness of the reports of this Survey, and, in fact, they are still in many respects superior to more recent work.

The geologist assigned to the oil regions was John Franklin Carll, originally trained as a civil engineer, who had already been some years in the oil regions. He at once proceeded to examine samples and run levels between wells. At the end of the first year's work he published a progress report. In this is a structure map of the area in which the principal oil fields occur[8]. The fields themselves are not shown on this map, but they are shown on a larger map to the same scale folded in the same report. The linear character of the oil pools was by now well established and Carll refers to them quite naturally as "belts" and "leads."

In the meantime oil had been discovered in western Ontario at localities that were recognized by T. Sterry Hunt, a geologist of the Canadian Survey, as anticlinal[9]. E. B. Andrews, of Marietta, Ohio, followed the prominent and very sharp Burning Springs Anticline in western West Virginia, and noted the occurrence of oil seeps and some producing oil wells along it. He reasoned that oil occurred in fissures[10], and that the sharp folding favored their formation, while the flat-lying rocks to the west lacked fissures and therefore were barren. The Ontario fields have had some economic importance, but those along the Burning Springs anticline have never been of much account. In 1874 most of the oil in the world was coming from Venango and Warren counties, Pennsylvania, where there were no anticlines, but only a gentle regional dip to the southwest. Lesley, in a note appended to a cross-section of the Warren area, in the 1874 report on the Oil Regions[8] wrote:

"This section will serve another useful purpose. It must impress the reader with a conviction of the undisturbed condition of all this part of Pennsylvania. Yet the most productive oil fields of Venango, Clarion, Armstrong, and Butler counties are equally undisturbed. The popular notion that petroleum wells are dependent upon anticlinals, faults, or other disturbances, is a pure fancy of the imagination. And it is very desirable to get rid of it. Even geologists of standing and reputation are still more or less affected by it. . . . . The district of greatest oil production in Pennsylvania is precisely the district where there never has been any disturbances whatever."

It has been suggested that the geologists of the Pennsylvania Survey were prejudiced by Lesley against the anticlinal theory. However, these geologists had collected thousands of well logs and had drawn many maps which clearly showed that the theory did not hold in what was then the principal oil-producing area of the world. This fact was clearly recognized by DeGolyer in 1939 in a lecture at Princeton University, but it seems to have been lost sight of since then.

The next report on oil and gas of the Second Survey was by Henry E. Wrigley in 1875[12]. Wrigley says that the truth of the oil belt theory "lay in the fact, that the general course of the grand current which deposited the sand rock was in the direction named" (p. 6). Farther on, (p. 42), he says that "what we are searching for today is the location of the vent holes by which this oil reaches the surface of the earth, whether such vent hole consists of an open sand-rock or sponge as in Pennsylvania, or of an anticlinal or system of broken rocks as in West Virginia."

The structural theory was revived and applied to gas by I. C. White[13], and later by White and others to oil in West Virginia. There are big closed structures in West Virginia which affect both oil and gas accumulations. Even here, however, most oil pools are strongly affected by lensing of sands, so the theory was not very useful in prospecting for oil, although it was notably successful during the big expansion of the natural gas industry which occurred at Pittsburg starting about 1882. The Lima-Indiana fields, which were the next great set of oil fields after Pennsylvania, were not on well-defined anticlines, although Orton[14] pointed out five years after their discovery that the gas, oil, and water lay at definite structural elevations. The anticlinal theory remained unused by oil men, not because they were conservative, but because it did not apply to the principal oil fields of the time, and obviously could not be used to find oil in these areas.

Carll later wrote a more comprehensive report on the "Geology of the Oil Regions"[15] which may be regarded as the foundation of both petroleum geology and engineering. It has received some recognition [16,17] but not as much as it deserves. Carll saw the need for keeping careful records and examining samples, he explained porosity and permeability, he explained clearly and correctly the role of gas and water in expelling oil from the reservoir, and described secondary recovery by waterflooding. It is a remarkably clear and careful analysis of the principles of petroleum geology and reservoir behavior. Little of importance in this field was added until the reports of the Bureau of Mines began to appear about 1915, and the concept of relative permeability in the 1930's.

Returning to the consideration of the production curve of Pennsylvania, Figure 1, oil production increased pretty steadily until 1881. Most of this increase was due to the gradual development of the Bradford field. This pool is by far the largest in Pennsylvania, and it was developed slowly between 1870 and 1880. When it was drilled up, the inevitable decline set in. Only two important fields remained to be discovered, Washington-Taylorstown and McDonald-Venice, both south of Pittsburg, over 200 miles away. Since 1891 there have been no discoveries in Pennsylvania of any consequence, in spite of much wildcatting, some of it quite deep. The reason is geological—there are no good reservoir rocks between the Devonian in which the oil and gas occur, and the Ordovician which is deeper than 10,000 feet and is still mostly beyond reach of the drill.

From 1899 to 1912 production declined in a remarkably smooth manner, dropping from 85,000 to 21,000 barrels per day. At this point there is a distinct break in the curve. Most of the older fields had been flooded by water from leaky casings or improperly plugged wells soon after their discovery and water was generally considered very harmful in its effects. Laws against admitting water to the oil sands were passed in the 1880's. For some reason, possibly because of its more orderly development and slower decline, this watering-out had not taken place at Bradford. The permeability in this pool averages only 10 millidarcies, and the water moves very slowly through the sand, giving an effective drive. It was found that if the casing were ripped or pulled so that the groundwater (which is abundant in this area of heavy rainfall and deep forest) could enter the sand, the oil production of neighboring wells was increased. No one would admit to doing this, however, until 1922 when the practice was legalized. Shortly after this, water began to be introduced under pressure from pumps, and about 1925 the 5-spot pattern became common. In the 1930's air and gas drives were applied to the old Venango fields with some success, but the production of the rest of the state mostly continued to decline. Waterflooding at Bradford resulted in giving new life to the oil industry of Pennsylvania from 1920 to 1954, after which the field was pretty well drilled up and a new steep decline set in.

Ohio

After Pennsylvania the next important producing state was Ohio. Oil was produced in a small way in southeast Ohio near Marietta from 1860, but production from the state did not become important until 1885 with the discovery of the Lima-Indiana field at Findlay. The story of the development of this field is unique and interesting [18,19].

Gas had been known to occur in water wells in the vicinity of Findlay, Ohio, since 1836. Following the big development of gas for industrial purposes around Pittsburg in 1884, a company was formed at Findlay, a driller procured from Bradford, and gas was obtained. The fourth well drilled also produced some oil. The thirteenth well produced 12 million cubic feet of gas per day. The city bought the property and established a municipally owned gas system. It then offered free fuel for five years to manufacturers who would move to the city. This naturally caused quite a boom, and the population increased from 4,500 in 1884 to 25,000 in 1889 and land sold at exorbitant prices.

Other cities in the area could not be out-done in this manner and they also started municipal gas plants. The city trustees, unacquainted with the technology of the gas industry, wasted much money and enormous quantities of gas, none of which was metered. In a few cases the city made money and in some the bond issues were paid off, but at other places large sums were lost. By 1900 the supply of gas was virtually exhausted everywhere.

This rapid drilling for gas turned up some oil which was first regarded as a nuisance for it interfered with the gas. It was low gravity and sulfurous and of little use to refiners equipped to refine high-grade Pennsylvania oil. An acceptable fuel oil could be made but the kerosene smelled badly, and the crude was called "pole-cat" oil. The price dropped from 40 cents per barrel in 1886 to 15 cents in 1888. Against the opposition of his associates, John D. Rockefeller actively entered the field. The Standard Oil Trust in 1885 built an experimental refinery, organized the Buckeye Pipe Lines, and later the Ohio Oil Company to produce Lima oil. This was Standard's first major venture into the producing end of the business[20].

Standard purchased all the crude offered until in 1888 it held more than ten million barrels in storage. They turned with no success to several chemists for help in finding a way to make saleable kerosene. The problem was finally solved by Herman Frasch, a German pharmacist of Philadelphia. He had previously developed a process for refining paraffin wax which had been purchased by Standard. Frasch had moved to Canada where the Trenton formation also produces sour crude. In 1887 he applied on his own for a patent on a method of removing the sulfur, which consisted of distilling the oil in in the presence of copper oxide. The copper reacts with the sulfur and forms copper sulfide, leaving the oil sweet and odorless. Standard bought Frasch's process and employed him to set up a laboratory in Cleveland. To assist him they employed a young chemist named William Burton who had just graduated from Johns Hopkins. Between them they improved the copper oxide process and developed new types of stills. Burton later invented the cracking of petroleum and finally became president of the Standard Oil Company of Indiana. Frasch later invented the hot water method for producing sulfur from salt domes. The production of Ohio increased very rapidly and overtook that of Pennsylvania in 1895.

The extensive wildcatting, based entirely on random drilling, sustained the production at around 50,000 barrels per day from 1890 to 1905 at which time a decline set in. Deeper drilling 20 years later in the Clinton sand pools of central Ohio offset the decline for a while, but since 1932 the state has produced only about 10,000 barrels per day. Secondary recovery methods have been tried in many places but have not yet proved very successful.

Kansas

After Ohio, Kansas was the next oil-producing state. Small oil seeps had been known in eastern Kansas and Oklahoma since its earliest settlement. Attempts to produce oil were made near Paola before the Civil War[21], but the unsettled condition of the frontier during the war put a stop to activities. The first commercial production began at Neodesha in 1892. Guffey and Galey, a Pittsburg firm who had been successful in southeastern Pennsylvania and West Virginia, took over the venture after it was successful. These partners drilled all over eastern Kansas and into Oklahoma. They had been spectacularly successful in Pennsylvania, and also later at Spindletop in Texas. Their principal prospecting technique is not clear but apparently they drilled mainly in the vicinity of oil seeps. They were very successful in Kansas but lacked a market for the oil. The help of the Standard Oil Company was enlisted, whose subsidiary, the Forest Oil Company, started a refinery and later laid a pipe line to Whiting, Indiana.

Oklahoma

Wildcatters drilling the shallow, stratigraphic shoestrings and lenses of sand in the Cherokee shale worked their way southward and crossed the line into Oklahoma (then Indian Territory) about 1900. The oil production curve for Oklahoma, constructed in the same way as that of Pennsylvania, is shown

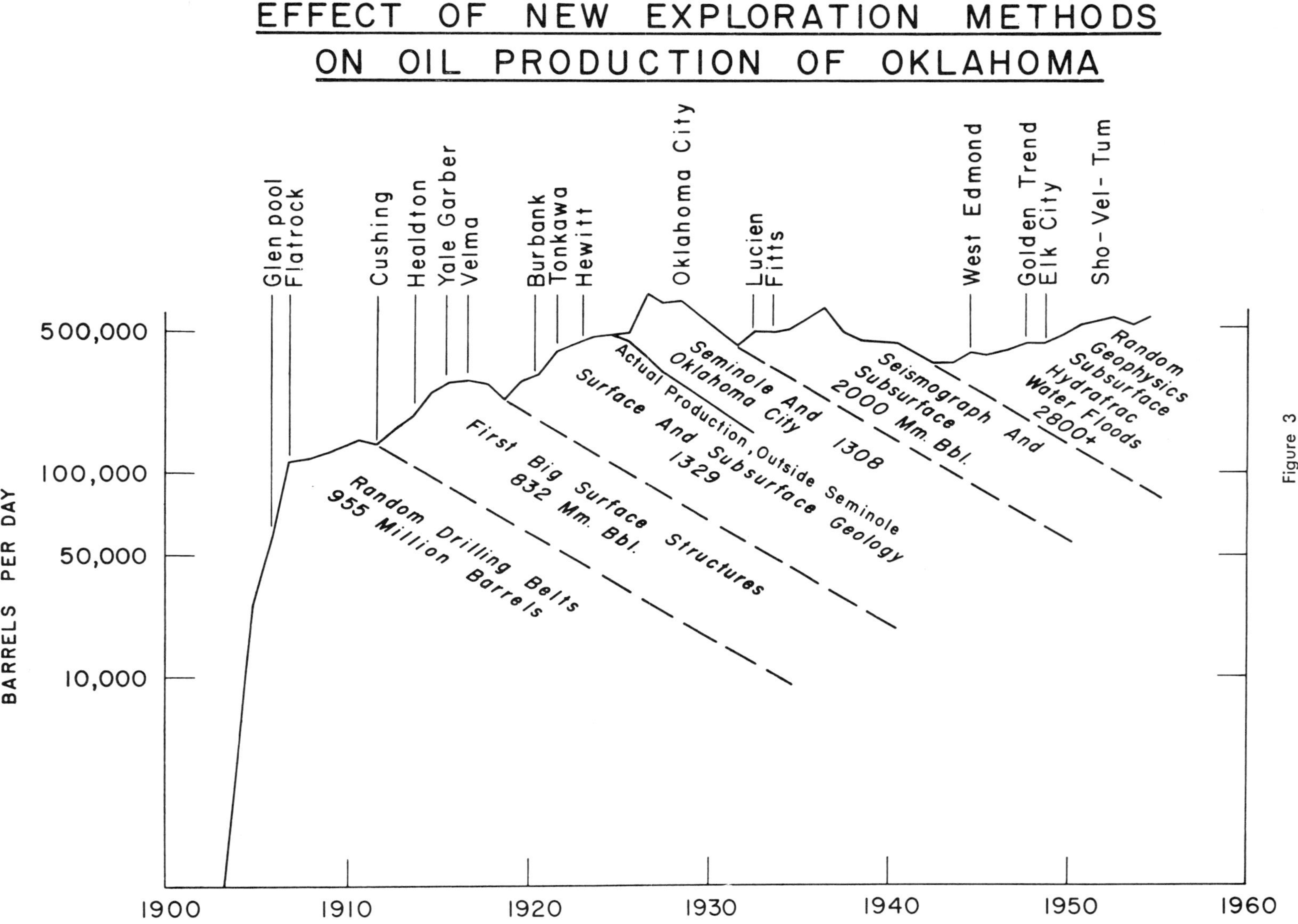

Figure 3

in Figure 3. The Bartlesville, Nowata, Delaware, and Alluwe fields were discovered by random drilling and by following the same southwest trends the
drillers had learned in Pennsylvania. The really big discoveries came with
Bird Creek, Flat Rock, and finally Glenn Pool in 1905, with a peak production

of 117,000 barrels per day in June 1907. These pools caused a line of supply stores to locate along the newly-laid tracks of the Frisco Railroad. The town took its name from a Creek Indian village nearby and became the city of Tulsa.

The first geologist to engage in commercial oil work in Indian Territory was H.B. Goodrich, who went to Ardmore in 1903 for the Santa Fe Railroad. He located an oil field near some asphalt seepages. The U. S. Geological Survey did some work starting with J. A. Taff about 1900, and the State Geological Survey was founded in 1908 by Charles N. Gould. Several oil geologists from the east also did consulting work about this time, but most of the big discoveries were still of stratigraphic type pools based on seeps or random drilling, and only a few oil companies employed geologists[22].

The fact that anticlines are favorable places to find oil finally became spectacularly evident with the discovery of the Cushing field in March 1912. Cushing is a prominent north-south structure with three separate domes. It can be seen on the surface and was mapped by W. S. Vandruff and his son in March 1911 for the Hill Oil and Gas Company[23]. The discovery well was drilled by Shaffer and Smathers in March 1912, without the aid of geology.[*] Development was rapid during 1913 and 1914, but the pool was so big that the peak rate was not reached until May 1915, when it produced 305,000 barrels per day. In May 1914, the Oklahoma Geological Survey published a structure contour map of the surface geology of the Cushing field by Frank Buttram[25] and others. It showed that the subsurface contours and the line separating the oil from the water paralleled the surface structure contours, and it made a big impression on many producing companies.

The discovery of the largest field yet found in Oklahoma, and probably in the world, on a well-known structure easily mappable from the surface, established geology in the oil business, and furnished the idea on which most new discoveries were based for the next 15 years. Many of the first geologists in Oklahoma are still alive and active in 1958; among them are Frank Buttram, E. W. McCrary, F. Julius Fohs, James H. Gardner, and Dorsey Hager. Among the companies who reaped the greatest oil-finding success from geology were Guffey and Gillespie and the Gypsy Oil Company (both later taken over by Gulf), Marland Oil Company (now Continental), Empire Gas and Fuel (now Cities Service). At this date it is difficult to ascertain exactly the extent to which geology was responsible for the new discoveries. Healdton is a great pool on a Pennsylvanian structure pretty well concealed by the Permian beds on the surface. However, Garber, Scholem Alechem, and Velma are all big surface structures. Even though many discoveries were still made by random drilling, geology came to play a part in the thinking and plans of most companies.

By 1917 the surface geology of northeastern Oklahoma had been pretty well investigated and there were no more big structures like Cushing and Garber left. For the next three years little oil was found although this was a period of inflation and sharply rising demand. The price of oil was less than $1 per barrel in 1914 and rose steadily and rapidly to $3.50 in 1920. The shortage gave rise to a series of gloomy predictions that we could run out of oil. These statements are still frequently quoted and held up to ridicule. According to Sydney Powers, the geologists had already forgotten about the oil discovered by random drilling where there were no anticlines, and decided that

---

2 But with the aid of a smart landman. Thomas Slick, one of the partners is reported [24] to have hired all the livery rigs and locked them in their stables, and sent all the notaries off on vacations.

when the supply of anticlines gave out, the supply of oil would too[22]. This same mistake has been made since by geologists in other places.

But a new burst was already in the making. When the Government allotted the lands of the Osage Indians at so many acres per person, the mineral rights were reserved to the tribe. The whole reservation (now Osage County) was first leased in 1896 to Henry Foster for gas. Not much oil was discovered, partly because it was not actively sought for, and partly because of the lack of competition. In 1916 this lease expired and was not again renewed, and the Federal Government instituted the policy of holding periodic open auctions for leases. In order to evaluate this enormous area of potential oil lands (1,570,000 acres) the U. S. Geological Survey undertook a detailed study[26]. The geologists assigned to this task were a distinguished group and almost every one later made a name for himself in the oil industry. Among them were K. C. Heald, L. H. White, K. F. Mather, M. I. Goldman, Sydney Powers, O. B. Hopkins, F. R. Clark, and others. They instituted detailed plane-tabling on the surface and careful subsurface studies utilizing sample examination. These geologists and their new and detailed techniques spread throughout the state and their work resulted in a new wave of discoveries based on, or at least strongly influenced by, geological thinking of a somewhat more profound and intensive nature than the search for simple surface anticlines. Small noses mappable on the surface often turned out to be closed anticlines at depth. Another and larger burst of discoveries took place. Blackwell (1917, 11 million), Hewitt (1919, 200 million), Burbank (1920, 470 million), Tonkawa (1921, 129 million), Stroud, Knox, Papoose, Cromwell, and Wewoka (1923), Chandler and Davenport in 1924, and many smaller fields can all be credited to surface and subsurface geology. This way culminated in the enormous discoveries of the Seminole area in 1926 and 1927, and Oklahoma City in 1928 and 1929. Robert H. Dott wrote a paper in 1934[27] which pointed out that continual new discoveries are required to maintain production. While the discoveries at Seminole and Oklahoma City pushed up the state's production, the older fields declined rapidly, as shown in the Oklahoma production curve, Figure 3.

The Seminole area produces from the Ordovician Simpson or "Wilcox" sands, which are arched into a series of gentle domes. The structures can often be recognized in the unconformably overlying Pennsylvanian. Some of them were worked out by surface and subsurface geological studies, and some were discovered by random drilling. The Wilcox sands are continuously permeable and all are in a common aquifer with a strong water drive. Wells were large and sustained their production. Seminole became the biggest of the long series of booms. By July 1927 production from the Seminole area reached 527,000 barrels per day.

Oklahoma City was discovered in 1928 by the Indian Territory Illuminating Oil Company by drilling on an anticline exposed at the surface. This field also produces from the Simpson sand truncated over the crest of the anticline by the pre-Pennsylvania unconformity. It is here permeable only in the neighborhood of the field, possibly because permeability was developed near the unconformity by weathering. There was no active water drive as at Seminole, but production was sustained by gravity drainage, and it was here that this recovery mechanism was first clearly shown to be important in oil production. The wells were very deep for the time, and stimulated the development of high-pressure fittings and deep pumping equipment.

But the geologists were too much carried away by their enthusiasm for structure. For example, they took credit for discovering Burbank, which actually should be credited to random drilling. Before its discovery, many dry holes had been drilled on surface structures in the western part of Osage County. Apparently there were no productive sands in the Pennsylvanian, and the operators became quite discouraged. In May 1920, the Marland Oil Company drilled the discovery well of Burbank on a small dome mapped on the surface. In September of the same year The Carter Oil Company drilled another discovery on another small anticline 3-1/2 miles away to the southeast. L. M. Neumann and George Dorsey of Carter had noted sand at about the Bartlesville horizon in the record of an old well at Burbank and then mapped the nearby small closed structure. Both companies bought their acreage at the same sale. As a matter of fact, either could have discovered the pool just as well by drilling in the intervening syncline, and the biggest wells of all were several miles from the highest point structurally[28] and 100 to 150 feet lower. The sand body is a large shore-line type sand and its location could not then, or now, be predicted from surface geology. It was the apparent impossibility of predicting the location of lenticular sands from the surface that made geologists over-emphasize the role of structure, because it could be found from the surface. The advent of geophysics further emphasized the role of structure so that geologists almost forgot about stratigraphic traps.

In 1929 Oklahoma produced 698,641 barrels a day, a peak that has not since been equalled. The development of Oklahoma City in 1930 and 1931 coincided with the depression and the flood of oil from East Texas. Exploration was curtailed and production declined from 761,027 in 1927 to a low of 419,847 in 1932. It is well known and often emphasized that a period of low prices restricts exploration activity and decreases the rate of discovery. During this same period the price fell from $1.30 to 75 cents per barrel. But the decrease in rate of production may have been technological as well as economic, resulting from the exhaustion of surface geology. Note that in contrast production rose from 238,000 barrels per day in 1919 to 760,000 in 1927, while the price fell steadily during the same period from $2.75 to $1.30. This period of falling prices coincided with a fantastically rapid increase in production. The reason, of course, is that the rapid discovery affected the price, instead of vice versa. What finds oil is new ideas as much as high prices.

The seismograph, developed in Texas in the late 1920's where it was very successful in finding Gulf Coast salt domes, was introduced into Oklahoma in the early 1930's. In 1934 Dott was able to say "within the past year practically all major operators have gone 'seismic,' with the result that a large number of highs have been leased." The seismograph did not have the great success in Oklahoma that it had in Texas. The larger structures all show on the surface, and had previously been found. Many of the smaller structures discovered with the seismograph were dry because the sands, being so often lenticular, were not present. Besides, large areas of the state give poor reflections. Lucien was the biggest pool discovered with the seismograph during this period, with an ultimate reserve of about 50 million barrels. The period from 1932 to 1937 showed a modest increase due mainly to Lucien and Fitts; the latter a surface and subsurface geological discovery, assisted by some good luck. From 1937 to 1944 was a period of steady decline.

A new series of discoveries began in 1944 and has continued to the present. The main cause has apparently been improvements in the use of the seismograph, partly technical but mainly better interpretation of the data in the light of subsurface geology. Random drilling and accidental discoveries are still important, however.

West Edmond lies just north of Oklahoma City on the same structure, but oil occurs in the Devonian Hunton lime at the unconformity instead of in the Wilcox. The structure was, of course, well known and had been under lease many times, but no geologist knew the Hunton would be saturated at this point. Nobody thought highly of it because the Hunton is not a very prolific formation in Oklahoma. The pool was found in 1944 by Ace Gutowsky, a promoter who is said to have located the well with an out-and-out doodlebug of the swinging-ball type. This pool of 120 million barrels was developed in an orderly fashion, and was the first of a new set of discoveries.

The Golden Trend with over 300 million barrels is a series of Pennsylvanian sands in the Deese formation which are above the same unconformity but pinch out against it. The first discovery was Southwest Antioch in 1948, on a false seismic "high." When The Carter Oil Company, who had previously shot the area, restudied their records, they found no high but clear evidence of the pinchout of the Deese formation around the south flank of an old uplift. The company leased for miles along the belt where the formation apparently pinched out, and much of it proved productive. This is one of the first cases where a company both intentionally and successfully used the reflection seismograph to find stratigraphic traps. It might be regarded as a scientific development of C. D. Angell's "belt" idea, 80 years later.

Elk City was discovered by Shell in 1947 with the seismograph. It is a low dome in an area that elsewhere is complex structurally. Since 1947 production has continued to increase as a result of small discoveries, some of them based on reinterpretation of structure from old well records, some seismic, and some of the result of hydrafracing oil sands that were previously noncommercial.

Secondary recovery by waterflooding was started in the early 1930's in the shallow old northeast Oklahoma fields, almost entirely by Pennslyvania operators. It was very successful in certain areas and less so in others. The major companies showed little interest in this technique, and as recently as the 1940's some of them sold or traded their holdings in old pools such as Burbank, Glenn, Cushing, etc. It was not until the 1950's that waterflooding began to constitute a substantial fraction of the production. Most of the new geological ideas described above were quick to be exploited and their potentialities in a given area were quickly exhausted. The story of waterflooding is in marked contrast. The phenomenon was observed in the 1860's or 1870's, clearly and accurately described in 1885, but it was not until 1910 that it was exploited at all, and not until 1925 that it was extensively used in Pennsylvania. It is extremely hard to understand why it was not used on a large scale in Illinois and Oklahoma until 25 years later.

Illinois

The most remarkable example of the principle that oil is found with ideas is exhibited by the production curve of the State of Illinois (Fig. 4). Oil was found along the LaSalle anticlinal belt at Westfield in 1904. The boom spread rapidly, and the peak of drilling was reached in 1907, when 4,988 wells were completed[29]. Peak production was reached the following year. By 1911 most of the major pools in the LaSalle uplift area were found. Large reserves only

EFFECT OF NEW IDEAS ON
OIL PRODUCTION OF ILLINOIS

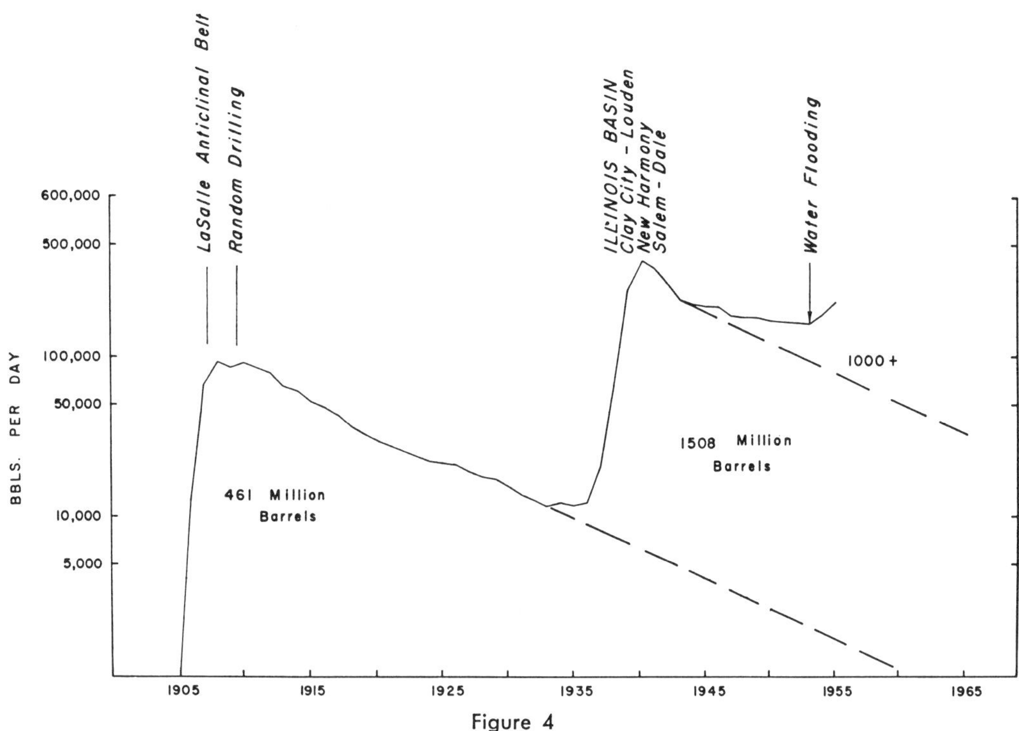

Figure 4

a few miles to the west and only a few thousand feet deeper lay undiscovered for 20 years. The curve shows a smooth decline from 1911 to 1933.

Bell [30,31,32] ascribes a re-awakening of interest in the Illinois Basin to the discovery of the Mt. Pleasant field on an anticlinal fold in the deepest part of the Michigan Basin in 1928. A paper read by him in 1931[30] pointed out the significance of this discovery to the Illinois Basin, and the Illinois Geological Survey consistently maintained that there were important oil possibilities in the Illinois Basin. However, it was not until the late 1930's that seismograph surveys located apparently favorable structures. In 1937 Patoka and Clay City were discovered. An active leasing and drilling campaign followed and many large pools were discovered in the next few years. Production in 1940 reached a peak of 400,000 barrels per day. Continued active prospecting for stratigraphic traps maintained production pretty well until 1953. Waterflooding of the Patoka pool started about 1940 and of the Siggins pool in 1942. It was slow in catching on, however, and although secondary production became appreciable in 1946, it was not until 1953 that waterflooding amounted to a considerable fraction of the total production. Production in 1954 and 1955 increased rapidly as the large pools, Loudon and Salem, were placed under flood.

The 20-year cessation of oil discoveries in Illinois is almost unique. Oil discovery in other states has come in cycles, each dominated by a new method or a new area. These followed each other with a pause of only a few years. In Illinois the hiatus between cycles lasted for 20 years during which almost no oil was discovered. It is interesting to inquire into the reasons for this period of stagnation.

Many shallow holes had been drilled in the basin area and it was supposed that they had condemned it. Ver Wiebe[33] said in 1930. "During the years from 1889 to date so many scattered wells have been drilled in the Eastern Interior Coal Basin province that no hope remains for any large pools." The oil sands that produced on the LaSalle anticlinal belt were found to contain salt water, and it was assumed that they, and the deeper Chester sands also, would contain salt water all across the basin. This idea was explained and fortified by the anticlinal theory which became widespread in the oil industry about this same time. It was also believed that the carbon-ratio theory propounded by David White[34] made the deep basin an unfavorable area in which to prospect. The scientific theories which were fashionable at the time had a part in preventing discoveries.

One company had obtained control of most of the oil production of the oil fields. The enterprising oil men who had developed them moved on to greener pastures in Oklahoma and Texas. Among these were Mike Benedum and Bill Skelly. In 1926 only one company employed a geologist on their regular staff, although other operators may have employed consulting geologists from time to time[35]. There was the industry, money, equipment, enterprise, and highly favorable territory—all that was missing was ideas. The new oil was found by geologists from Oklahoma and Texas with ideas from outside. The company that controlled most of the older production was very successful in many other places in the U. S., but in Illinois it never succeeded in finding much new oil even after the other companies had shown the way.

Conclusion

Each of the big waves of discovery can be credited to the influence of a new idea. The industry began with the idea of drilling bore holes for oil instead of digging pits by hand. New oil field discoveries swept over Pennsylvania when it was found that oil pools lined up in northeast-trending belts. Ohio's oil production came as a byproduct of natural gas development and also depended on new chemical research on how to refine sulfur-bearing crude. The anticlinal theory had its first important effects in Oklahoma and the Rocky Mountains about 1915. The torsion balance found one crop of salt domes, the refraction seismograph another crop, and the reflection seismograph still another. Secondary recovery by waterflooding shows in the production history of several states. Hydraulic fracturing is another new development that has resulted in its own crop of oil.

Once a new discovery is made, excitement results in a large amount of drilling based on the idea that led to the first discovery, so that the new idea is soon exhausted. This pattern was set early. Within 10 years after the Drake well almost every oil pool of any consequence in the area was discovered. Within 10 years from the time the anticlinal theory was generally accepted most of the big anticlines of the U. S. were drilled. When this happened in the 1920's, the geologists began to fear that we were running out of oil when the trouble really was that they were out of ideas. Within 10 years these same geologists were drowning in a flood of new oil from stratigraphic traps.

All the salt domes on the Gulf Coast discoverable with the early refraction seismograph equipment were found within a few years of its introduction. More recently, after Leduc was discovered in 1947 the operators in Alberta found almost all the good Devonian reefs in the area in three or four years. During this time they showed little interest in Cretaceous sands, although it was long known that they were productive. It was not until 1950 that there was a new wave of drilling for Cretaceous sand discoveries in the same general area that had been drilled over for reefs. The history of the industry shows that every few years there has to be a new idea, for the old ones are quickly exploited and exhausted.

High prices for crude oil tend to increase the discovery rate and low prices tend to discourage it by decreasing the number of wildcat wells drilled yearly. However, there is by no means a one-to-one correlation between rising prices and rising discovery rate. This is because the law of supply and demand works both ways. A series of new ideas and big discoveries results in an oversupply of crude oil that depresses the price. This pattern was set early also. In 1861 the production of 6,000 barrels per day so exceeded the limited demand that the price fell to 10 cents per barrel and the producers got together and decided to restrict production. In this they were unsuccessful— the wildcatters kept on drilling. What happened was that the demand continued to increase—so that by 1864 there was a shortage again and the price was $10 per barrel. This sequence of events has repeated itself every few years since. It is lucky we never were able to restrict production in order to keep the price up. Ten dollars per barrel seemed like prosperity to the oil men at Titusville, but suppose they had been successful in keeping the price that high with respect to other costs and that crude sold for $60 and $75 per barrel now. We probably would be selling thousands of barrels of oil for lubricating oil and kerosene instead of millions of barrels for motor and heating fuel.

The main—in fact almost the only—value of geologists to the industry lies in their familiarity with how oil has been found in the past. Based on this experience, they try to find oil now. But history shows clearly and repeatedly that past experience is exhausted quickly; frequently in 10 years or less, which is before a new method can be generally accepted and incorporated in textbooks. Sometimes while geologists were probing around with old methods, finding smaller and smaller pools, wildcatters and doodle-buggers went out and started a new burst of discovery. It often happend that the geological interpretation which led to the wildcat location was wrong but the oil was there anyway. Instead of claiming a false credit, geologists should learn their lesson and revise just not their maps, but also their thinking.

This past experience is often quoted so as to make it ludicrous. Geologists are alleged to have offered to drink all the oil to be found under the red-beds of Oklahoma, or the Gulf Coast of Texas. Leaders of the industry as asked where, in their opinion, the oil in the future will be found, have nothing but their past experience to guide them, so they say it will be found in the future where it has been found in the past. An outstanding example is quoted by J. V. Howell[1].

"Not only are the localities named above (Pennsylvania, West Virginia, Ohio, Colorado and California) the chief petroleum producing districts in the United States, but the indications are that, with the possible exception of Wyoming, they will continue so to be . . . The Illinois field is an exceedingly small one, with but little promise for the future, while the Kansas and Texas fields will, at best, produce only a few thousand barrels each year of a high grade lubricating oil. However, there have been so many surprises in petroleum that these statements must be regarded as only setting forth the present indications."

The only certainty in making forecasts is that the oil of the future will not be discovered where it has been found in the past by methods that have been successful in the past.

Geologists consider themselves experts in oil finding by virtue of their knowledge of how oil has been found in the past. It is proper that they should do so, but with great humility and care to think straight. Many times in the history of oil-finding a knowledge of how oil had previously been found blinded geologists so that they could not see new oil lying within their grasp. We must separate the true inductive facts which can be weighed and measured from speculation and hypotheses, discover ways of getting new facts, and learn new principles on which to base sound deductive decisions on oil-finding programs.

BIBLIOGRAPHY

(1) U.S.G.S. Mineral Resources, 1889-90.

(2) Giddens, Paul H.: "The Road from Hanover," Dartmouth College (1953).

(3) Giddens, Paul H.: *The Birth of the Oil Industry*, Macmillan (1938).

(4) *History of Venango County, Pennsylvania*, (1879), p. 300.

(5) *Derrick's Handbook of Petroleum*, Oil City, Pennsylvania, (1898).

(6) Sherrill, R. E., Dickey, P. A., and Matteson, L. S.: "Types of Stratigraphic Oil Pools in Venango Sands of Northwestern Pennsylvania," *Stratigraphic Type Oil Fields*, Am. Assoc. Pet. Geologists, (1941), pp. 507-538.

(7) Ames, Mary Lesley: *Life and Letters of Peter and Susan Lesley*, Putnam, (1909).

(8) Carll, John F.: *Report of Progress in the Venango County District*, 2nd Geol. Surv. of Pennsylvania, Harrisburg, (1874).

(9) Hunt, T. Sterry: "Notes on the History of Petroleum or Rock Oil," *Canadian Naturalist*, (1861), vol. 6, p. 241-55.

(10) Andrews, E. B.: "Rock Oil, Its Geological Relations and Distribution," *American Journal of Science*, (July, 1861), Second Series, vol. XXXII, No. 94.

(11) De Golyer, E.: "Development of the Art of Prospecting," The Guild of Brackett Lecturers, Princeton, N. J., (1940).

(12) Wrigley, Henry E.: "Special Report on the Petroleum of Pennsylvania," Report J., 2nd Geol. Surv. of Pennsylvania, Harrisburg, (1875).

(13) White, I. C.: "The Geology of Natural Gas," *Science*, (1885), vol. 5.

(14) Orton, Edward: "First Annual Report, Geol. Surv. of Ohio (Third Organization)," (1890), p. 92, 107.

(15) Carl, John F.: "Second Geological Survey of Pennsylvania," Harrisburg, (1885), vol. III.

(16) Howell, J. V.: "The Geology of the Oil Regions of Warren, Venango, Clarion, and Butler Counties, Pennsylvania by John F. Carll," Bull. Am. Assoc Pet. Geologists, (Sept., 1948), vol. 32, No. 9, p. 1829-1831.

(17) Lytle, William S.: "John F. Carll: Pioneer Petroleum Geologist and Engineer," *Geotimes,* (March, 1957), vol. 1, No. 9, p. 8.

(18) Orton, Edward: "Petroleum and Natural Gas," Geol. Surv. of Ohio, (Columbus, 1888), vol. VI.

(19) Bownocker, John Adams: "Occurrence and Exploitation of Petroleum and Natural Gas in Ohio," Geol. Surv. of Ohio, (Columbus, 1903), Fourth Series, Bull. No. 1.

(20) Giddens, Paul: *Standard Oil Company (Indiana),* Appleton-Century-Crofts, (1955), p. 3-5.

(21) Rister, Carl Coke: *Oil! Titan of the Southwest,* U. of Oklahoma, (1949), p. 27.

(22) Powers, Sydney: "Petroleum Geology in Oklahoma," Okla. Geol. Surv., Bull. (1928), No. 40, p. 1-20.

(23) Beal, C. H.: "Geologic Structure in the Cushing Oil and Gas Field, Oklahoma," U.S.G.S. Bull. (1917), No. 658.

(24) Tait, S. W.: "The Wildcatters," Princeton U. Press, (1946).

(25) Buttram, Frank: "The Cushing Oil and Gas Field, Oklahoma," Okla. Geol. Surv. Bull., (1914), No. 18.

(26) Heald, K. C., et al.: "Structure of Oil and Gas Resources of the Osage Reservation, Oklahoma," U.S.G.S. Bull., (1922), No. 686.

(27) Dott, Robert H.: "Discoveries Required to Maintain Status of Production in Oklahoma," *Oil Weekly,* (April 30, May 7, and May 14, 1934).

(28) Beckwith, H. T.: "Osage County," Okla. Geol. Surv. Bull. 40, (1930), vol. III, p. 248-50.

(29) Arnold and Kemnitzer: *Petroleum in the United States and Possessions,* Harper, New York, (1931).

(30) Bell, Alfred H.: "Role of Fundamental Geologic Principles in the Opening of the Illinois Basin," *Econ. Geol.* (1941), vol. 36, No. 8, p. 774-785.

(31) ──────, Witherspoon, Paul, and Hautau: "Oil and Gas in the Illinois and Michigan Basins of the United States," (   ), XX International Geological Congress Symposium on Oil and Gas Fields, vol. III, N. America.

(32) Bell, Alfred H.: "The Relation of Geology to the Development of the the Petroleum Industry of Illinois," XX Trans. Ill. State Acad. Sci., (1931), vol. 23, No. 3, March.

(33) Ver Wiebe, Walter A.: "Oil Fields in the U. S.," McGraw Hill, New York, (1930), p. 133.

(34) White, David A.: "Some Relations in Origin Between Coal and Petroleum," Jour. Wash. Acad. Sci., (1915), vol. 5, p. 189-212.

(35) Bell, A. H., Letter to author, January 8, 1958.

Reprinted by permission of McGraw-Hill Publishing Co.
from Graham B. Moody, ed., *Petroleum Exploration
Handbook* (1961), p. 1-1 to 1-6.

Chapter 1

PHILOSOPHY OF EXPLORATION

By B. W. Beebe

*President, Exploration, Inc. Petroleum
Exploration Consultant. Formerly Vice-
president of Production and Director of
Keating Drilling Company. A Past Vice-
president of AAPG*
Boulder, Colo.

The petroleum geologist is the keystone of the arch of the petroleum industry.
Industry, commerce, and our civilization depend on adequate supplies of cheap crude
oil and gas. The petroleum geologist, with the cooperation of others, supplies this
demand. Petroleum geology can actually be divided into several classifications, the
most important of which is the exploration geologist—the prospector, the pioneer,
the man in whose mind an oil or gas field must be visualized before a discovery can be
made. No other professional individual has such a unique and peculiar position, the
blending of precise scientific and engineering observations with judgment, experience,
philosophy, and intuition. Thus we speak of the art of petroleum prospecting.

The petroleum industry furnishes most of the energy required for our rapidly growing and expanding, highly technical civilization, a civilization the foundations of
which are based on the availability of cheap, efficient sources of energy. These
sources are for the most part the fossil fuels, primarily oil and gas. A little more than
100 years ago, 90 per cent of our energy sources was furnished by human and animal
muscle; today, 98 per cent of our energy comes from inanimate sources, and more than
two-thirds of this total is supplied by oil and gas. Demand for these products,
furthermore, is increasing at an unbelievably rapid rate. Few of us ever stop to
realize how completely dependent we are in every phase of our everyday living on oil
and gas and the multitude of products derived from them. The acceptance of
geology by the petroleum industry and its rise to its current preeminent position were
products of sheer necessity. The pressure of demand was such that the haphazard
methods of exploring for oil and gas were no longer adequate.

The application of the earth sciences to the problem of exploration for hydrocarbons
barely covers a half century, although many of the basic principles governing the
search for oil have been recognized for well over 100 years. Two famous pioneer
geologists, H. D. Rogers and W. B. Rogers, noted as early as 1840 that the greatest
amounts of oil came from warm saline springs along the crests of anticlines. Shortly

1–1

141

thereafter, Sir William Logan, director of the Canadian Geological Survey, in a report published in 1844, recognized the relationship of oil accumulation and anticlinal structure. The first clearly printed statement of the anticlinal theory was made in 1861 by T. Sterry Hunt, geologist and chemist of the Canadian Survey. Through these developments and the subsequent work of geologists E. W. Evans, E. B. Andrews, and Alexander Winchell, most of the basic principles of the occurrence and accumulation of hydrocarbons were recognized by 1870 (Howell, 1934, p. 10).* However, little attempt was made to apply these principles practically. It remained for Israel C. White, justly recognized as the father of petroleum geology, to win recognition of the application of geology by the oil industry. White resigned from the Pennsylvania Survey in 1883 and became a consulting geologist. In 1885, he published a short paper in *Science* which, together with his demonstrations by concrete example of the positive value of geology, finally won acceptance of its application to exploration problems.

His early successes, however, did not result in immediate response, and he struggled for years, not only with oil men, but with fellow members of the profession. He was ably supported by Edward Orton with his monumental studies of oil and gas fields in northwestern Ohio. By 1900, the U.S. Geological Survey was studying oil-producing areas in Ohio and Pennsylvania, and it was generally recognized that oil, under certain circumstances, could be found in synclines. Neither White nor his predecessors believed or stated that all anticlines could be productive or that in any particular area an anticline was the best place to look for deposits of petroleum. It is unfortunate that the term "anticlinal theory" was applied to the results of White's studies. He recognized that anticlines were only one of many factors involved in searching for commercial deposits of oil and gas. All this pioneer work, however, gave a firm foundation to the application of geology to problems of exploration.

E. T. Dumble was employed as a geologist for Southern Pacific Railroad in Texas in 1897, W. W. Orcutt established the geological department for Union Oil Company the following year, and the Santa Fe Railroad employed H. B. Goodrich as a geologist in the Ardmore Basin area of southern Oklahoma in 1903. Ralph Arnold, W. W. Orcutt, Joseph A. Taff, G. C. Gester, and others were recommending drilling of anticlines mapped on the surface in California early in the twentieth century. Moody (1954, p. 1905, Fig. 11) shows that the peak of oil discovery resulting from drilling exposed surface anticlinal structures was reached in California in 1910.

It remained, however, for the increased demand for petroleum brought about by the development of the automobile in the early twentieth century to place geology on a sound footing in industry. Perhaps modern petroleum geology dates from about 1913, when Gypsy Oil Company, now Gulf Oil Corporation, and Empire Oil and Gas Company, now Cities Service Oil Company, established formal geological departments. In 1915, oil was discovered at Eldorado, Kansas, on a large dome which had been mapped geologically, and there was no doubt that geology had won acceptance by the oil industry. By 1917, Empire had a large extensive department with the first subsurface section. The structural period of prospecting had commenced, and the anticlinal theory became so thoroughly ingrained in the thinking of geologists and the petroleum industry that our progress still suffers from such restrictive application. The next 25 years marked the golden age of structural prospecting, and numerous methods and techniques were developed to aid the search for favorable structure. These included surface mapping, core drilling, magnetic and gravimetric prospecting, subsurface mapping, the refraction and the reflection seismograph. But by the late 1930s, it became apparent to thinking geologists that there was a finite number of closed anticlinal structures, that each one found reduced those remaining, and that

* References appear at the end of the chapter.

each discovery made the next one more elusive.   Obviously the larger, more obvious and prolific structures were found first, and ultimately the law of diminishing returns was reflected in the exploration effort.   It became more and more expensive to find less oil and gas.   There was a plethora of methods and a paucity of large, prolific structures.

By 1940 this situation had become quite apparent.   Many geologists were discovering that oil could occur in traps other than structural, and that these stratigraphic traps could be found only by careful, detailed, analytical geologic studies.   The same type of struggle faced by White and his supporters has again been raging in the industry and the profession for the past 20 years.   Although the industry and the profession pay lip service to the stratigraphic trap concept, little real progress and understanding, with accompanying development of methods and techniques to discover such traps, have been made.   Our ability to furnish our domestic economy with the required hydrocarbons is directly dependent on our ability to develop and apply principles, methods, and techniques which will result in the discovery of stratigraphic oil and gas reservoirs.

There is no magic in the term "petroleum exploration geologist."   The petroleum exploration geologist must first be a well- and broadly trained geologist skilled in the application of many branches of the science to the problem of finding oil and gas.   To borrow a term from medicine, a true exploration geologist is a general practitioner, applying the principles of structural geology, sedimentation, stratigraphy, paleontology, mineralogy, petrology, geomorphology, and historical geology to the problem of finding deposits of oil and gas.   The finding of hydrocarbons is statistical in character and is based on the principle of continuous exposure to situations and circumstances under which hydrocarbons are known to occur.   Oil and gas are not found by flashes of genius but are a product of rigorous observation and tenacious, dogged, often dreary intensive work and study.   Like most other scientific enterprises, the background of a discovery consists of 1 per cent inspiration and 99 per cent perspiration.   Levorsen (1954, p. 640) has stated so aptly: "The drilling of dry holes, holes that find only barren traps, or no traps at all, is a calculated risk of petroleum exploration."

The essential ingredients in any successful exploration program are capital, time, and technical skill.   The exploration geologist furnishes the most important of these ingredients: the technical skill, blended with an optimistic philosophy, perseverance, judgment, intuition, and perhaps a bit of mysticism.   His function is to employ properly the other two ingredients, capital and time.   His greatest enemies are dogmatism and prejudice.   It is most difficult with a vast quantity of miscellaneous data to separate actual facts from what appear to be facts, and there is a fundamental difference.   It is often difficult to classify and analyze properly, then to arrive at a clear, logical conclusion.   Unless the geologist is unusually objective and careful in his thought processes, constantly employing the scientific method of reasoning, he will succeed only in rearranging his prejudices.   The history of exploration geology is replete with examples of large, potentially productive areas which failed to yield to conventional methods of exploration and were condemned by the industry, only to afford production when attacked later by an operator not bound by prejudice and misconception.   One cannot help recognizing that, despite the conquest of the oil industry by geologists, many important oil fields have been found without technical advice in areas condemned by the profound manifestos of prominent members of the profession.   No geologist is infallible in his judgment.

Oil fields and oil provinces are found by positive, constructive thinking.   No oil or gas was ever found by condemning a prospect or an area.   A geologist may be correct in his recommendation most of the time by condemning prospects, but the end result is the liquidation of his company's assets through the pipeline.

The fundamental thesis of the exploration geologist is that oil is abundant in favorable areas and that any area of thick marine sediments is a potential producing province. Failure to find oil in such areas through an exploration program is usually the fault of the program; oil is not necessarily absent. Timing is also an important ingredient. Favorable areas often experience several cycles of exploration, until finally enough information has been gathered to reveal the flaws in the earlier exploration programs. A discovery results, and a new productive province has been found. In many cases, reasons for failure are so elementary that it seems inconceivable that so much time and effort have been required. It is fundamental that the more precise and positive the data and the more varied the indications of a favorable prospect, the better the prospect. Discovery of a new field which opens a new province is the ultimate achievement in the career of a petroleum geologist. The reaching of this goal involves a coordinated research: gathering, classifying, analyzing, and applying reason and experience to a great diversity of data.

What then are the essential characteristics of this paragon of scientific virtue, the successful exploration geologist? First and foremost, he must be a dedicated geologist, an individual to whom geology is not a means of making a living but a way of life, a man to whom there is nothing more interesting and fascinating than geology. He constantly is amazed that anything could be so intriguing and stimulating and yet be financially rewarding. The exploration geologist is the elite member of the geological profession, the pioneer, bold, adventuresome, courageous, ever ready to explore new vistas.

The same attributes that characterized our pioneer ancestors must be his. He must be an optimist, resourceful, curious, inventive, tenacious, with an abundance of initiative. Above all, he must have vivid but controlled imagination, remembering that the oil field must first be visualized in the mind of the prospector. Coupled with these qualities, he must have a prodigious memory and the ability to recognize the importance of data which at first glance might seem trivial. Then he must have the ability to recognize and appraise clearly changing situations and the mental adaptability to alter an opinion, no matter how dear to his heart. The importance of salesmanship cannot be overlooked, because an exploration geologist is continuously selling himself and his ideas. This means that he must be able to express himself fluently, clearly, and concisely both orally and in writing. Finally, he must have stamina, the drive which results in the maximum output per unit of time and the ability to spend uncounted hours in intensive concentration. Finding hydrocarbons is not an 8-hour-per-day, 40-hour-per-week job. The effort has nothing to do with retirement plans, vacations, fringe benefits. Security for the exploration geologist lies in the opportunity to take calculated risks. No other profession is so fraught with hazard, with complete disaster lurking in every venture, but no other profession offers the opportunity for such tangible and intangible rewards for successful pioneering.

There are approximately 15,000 petroleum geologists employed by industry, yet only a small percentage of these are real exploration geologists. Some are in exploitation, administration, evaluation, and related fields. There are also the few truly scientific specialists in various types of research. But unfortunately far too many are the gatherers of information, the artisans, the "geological clerks." All these categories support the exploration geologist in his efforts. Exploration geology is a highly individualistic and creative enterprise, not adaptable to the electronic calculator or the assembly line.

The exploration geologist nears the end of an era of prospecting and, with ever-increasing demand for the product of his efforts, must advance boldly into a new era. Exploration, for the most part, since its acceptance by industry, has been based on the search for structure, primarily anticlinal. The watchword has been: "Find a closed structure!" Most of the instruments, methods, and techniques which have been developed and successfully used have been designed to find abnormal structure in the surface and subsurface. The search for structure has been the practice despite the recognition by many thoughtful geologists that there are many important oil fields in which structure plays little or no part in the accumulation, that there are many other factors equally or more important. There are, for instance, unconformities, facies changes, depositional traps; these are the really determining factors controlling oil and gas accumulation. The anticlinal theory on which the industry has based its exploratory operations for so many years does not state that a local closed anticlinal structure is the best place to find deposits of hydrocarbons, and yet this is the assumption which has prevailed. Anticlines are the easiest type of trap to find, and, more than this, are the easiest to depict and sell to management. Many years ago an essay was written by the late E. DeGolyer (1923) entitled "The Seductive Influence of the Closed Contour." To quote Wallace E. Pratt (1942, p. 25): "Geologists appear to be infatuated with anticlines. The lure of anticlines has blinded them to the real controls in the occurrence of oil pools." A. I. Levorsen (1936, p. 523) summed up the situation most succinctly as early as 1936: "We have all, therefore, drifted along from one structure-finding method to another until we are now apparently approaching the real danger of an oversupply of methods and an undersupply of anticlines."

Most of the obvious, easily found structures within the continental United States have been drilled. The current high cost of finding and capturing a barrel of oil and the small reserve per productive discovery are ample evidence that our entire philosophy of exploration must undergo a complete transformation. The day of the geologist searching for oil with a structural or isopachous map is ending. The exploration geologist of the future must pay far more attention to sedimentation, facies changes, geological history, tectonic control of sedimentation, and environmental conditions and must learn to depict stratigraphic traps as he has so superbly learned to illustrate anticlinal structure. There are many undrilled anticlinal structures in areas outside the United States, but there are also many prolific oil fields to be found within this country. Drilling for oil in this country has given more geological information about North America than is known of any other area. There are still many potentially productive, relatively unexplored areas beckoning the pioneer. It remains only for them to be attacked by geologists with vision, courage, perseverance, and initiative. Those who blend the finest in geology with the knowledge that the end result must be converted to profits in terms of dollars will assure exploration geology its preeminent position in the industry.

REFERENCES

Ball, Max W.: "This Fascinating Oil Business," The Bobbs-Merrill Company, Inc., Indianapolis, 1940.

Ball, Max W. (ed.): Possible Future Oil Provinces of North America, *Bull. AAPG*, vol. 35, pp. 141–485, February, 1951.

Beebe, B. W.: The Development of Petroleum Geology, in "A History of the Geological Sciences," unpublished symposium, California Institute of Technology, Division of Geological Sciences, 1940.

Beebe, B. W.: "A Philosophy of Exploration," unpublished address, 1956.

Beebe, B. W.: "Geology Now Unemployed, Future Unlimited," unpublished address, 1958.

Buckley, Stuart E. (ed.): "Petroleum Conservation," chap. 3, pp. 77–83, American Institute of Mining and Metallurgical Engineers, 1951.

Clapp, Frederick G.: Fundamental Criteria for Oil Occurrence, *Bull. AAPG*, vol. 11, pp. 683–703, July, 1927.

Clifford, Jr., O. C.: An Oil Finder Looks at His Profession and Scales It Down to Size, *Oil and Gas J.*, vol. 56, pp. 156–160, May 26, 1958.

Cram, Ira H.: Definitions of Geology and Subsurface Geology, *Bull. AAPG*, vol. 29, p. 470, April, 1945.

Cram, Ira H.: Resources and Resourcefulness, *Bull. AAPG*, vol. 29, pp. 857–871, July, 1945.

Cram, Ira H.: Geology Is Useful, *Bull. AAPG*, vol. 32, pp. 1–10, January, 1948.

DeGolyer, E.: The Seductive Influence of the Closed Contour, *Econ. Geol.*, vol. 23, pp. 681–682, 1923.

DeGolyer, E.: Future Position of Petroleum Geology in the Oil Industry, *Bull. AAPG*, vol. 24, pp. 1389–1399, August, 1940.

DeGolyer, E.: "The Development of the Art of Prospecting," Princeton University Press, Princeton, N.J., 1940.

Heald, K. C.: Essentials for Oil Pools, in "Elements of the Petroleum Industry," pp. 26–62, Seeley W. Mudd Series, American Institute of Mining and Metallurgical Engineers, 1940.

Heroy, W. B.: Petroleum Geology, in "Geology 1888–1938," pp. 511–548, Geological Society of America, 1941.

Howell, J. V.: Historical Development of Structural Theory of Accumulation of Oil and Gas, in "Problems of Petroleum Geology," pp. 1–23, *AAPG*, 1934.

Lahee, Frederick H.: Degrees of Success in Wildcat Drilling, *Bull. AAPG*, vol. 35, p. 138, January, 1951.

Levorsen, A. I.: Stratigraphic versus Structural Accumulation, *Bull. AAPG*, vol. 20, pp. 521–530, May, 1936.

Levorsen, A. I.: "Geology of Petroleum," chap. 1, pp. 3–12; chap. 14, pp. 606–635; chap. 15, pp. 636–651; W. H. Freeman and Co., San Francisco, 1954.

Moody, Graham B.: Oil in California, *Bull. AAPG*, vol. 38, pp. 1888–1910, September, 1954.

Noble, Earl B.: Geological Masks and Prejudices, *Bull. AAPG*, vol. 31, pp. 1109–1117, July, 1947.

Pratt, Wallace E.: "Oil in the Earth," University of Kansas Press, Lawrence, Kans., 1942.

Price, Paul H.: Evolution of Geologic Thought in Prospecting for Oil and Natural Gas, *Bull. AAPG*, vol. 31, pp. 673–697, April, 1947.

Sloan, Raymond D.: "The Future of the Exploration Geologist, Can He Meet the Challenge," unpublished address, 1957.

Weeks, L. G.: Factors of Sedimentary Basin Development That Control Oil Occurrence, *Bull. AAPG*, vol. 36, pp. 2071–2124, November, 1952.

Reprinted by permission of Matthew Bender & Co., Inc.
from *Exploration and Economics of the Petroleum
Industry*, v. 7 (1969), p. 19-29.

2

Are Sophisticated Exploration Methods the Answer?

EDUARDO J. GUZMAN
*Head of Exploration Technology, Mexican Petroleum Institute,
Mexico City, D.F., Mexico*

Introduction

A title expressed with a question mark is more often than not used with an ironical implication, and frequently is followed by a series of dogmatic affirmations with the intent to prove or disprove a certain thesis that draws the author's own personal and preconceived conclusions.

It is hoped, however, that the youth, aggressiveness, and forceful character which are still typical—although gradually dwindling—of the oil industry, particularly in its field of exploration, will allow for some controversial ideas without the danger of my being unduly accused and tried for high sacrilege before a computerized court of technological inquisition.

At a time when it has become fashionable to advocate a trend toward increasingly sophisticated exploration methods, this plea for a thoughtful consideration of the basic, sometimes forgotten and despised fundamentals, may easily be considered as a retrograde attitude.

19

It therefore should be pointed out that my criticism is not against the advancement of science or the betterment of technology per se, but against the indiscriminate application of increasingly costly and complex methods and approaches to the solution of petroleum exploration problems, particularly in those areas outside the United States.

Most of what will be said has been said before, and probably said better, but in the majority of cases these discussions are kept within the restricted circle of specialists in exploration and do not reach the upper levels of management, where decisions and budget appropriations are made.

Geology and Geophysics

Both geologists and geophysicists, whose job and goal has been to find and produce more oil at the lowest possible cost, recognize their failure to coordinate efficiently their respective efforts. Both groups continuously talk of the need for integration, which should first take place among themselves in order later to have management seriously weigh their problems. Yet, the gulf that has opened between these two groups is, unfortunately, widening to the detriment of the oil industry. The reason, although not the excuse, lies in the approach that each group has developed over the years towards the problems of exploration.

While the geologist, in his effort to go beyond the conventional method of looking purely for structure, has concentrated on trying to understand better the deposition of sediments and their relation to hydrocarbon generation, migration, and accumulation, the geophysicist has devoted his major effort to making his methods of mapping structures more precise and quantitative.

The first relies more on natural observation and the development of hypotheses, while the second has trended more toward the mathematical and physical approach with the aid of modern electronics. Because the geologist has not succeeded in conveying his ideas or findings to the geophysicist, perhaps the latter gets more and more wrapped up in his methods and technology, and frequently loses interest in the geological factors responsible for the presence of oil in nature.

Although there are many respectable exceptions, the whole division starts with a problem of the philosophy and the background of the men engaged in both disciplines; it is a natural result of the modern human trend toward extreme specialization. In the meantime, the oil industry suffers the consequences.

Recent Exploration Trends

The critical situation that is rapidly growing in the exploratory efforts and results within the United States' domestic industry has made this lack of coordination either greater, or more apparent.

The United States oil industry is rapidly searching for more oil to bolster its swiftly declining reserves. This almost desperate race against time to maintain its present economic preponderance has spread to the rest of the world, particularly to the developing countries, as a dangerous psychosis.

The petroleum industry in the United States is channeling a large percentage of its economic resources into offshore prospects, at operating costs that are still questionably profitable. Many other countries with large undeveloped sedimentary basins on land are initiating offshore exploratory campaigns with an attitude of "keeping up with the Joneses." This attitude is causing them to attack these expensive programs with their own limited resources, or to insist that the foreign companies working in their land undertake such enterprises.

Onshore, the exploration problems of the United States call for the most precise instrumentation in order to detect the subtle traps still undrilled within the oil producing areas. This, of course, implies a higher cost of exploration, which has been in many cases exported to countries with large unproved or undeveloped areas where the additional refinement does not yet warrant the extra expenditure.

There are many examples in the world of these premature adventurous offshore campaigns involving the use of costly geophysical methods where less expensive exploratory approaches still could yield considerable success in the discovery of new reserves. This worldwide trend is not new in the history of exploration. It has happened repeatedly even within the United States, where every new method or tool has tended to displace all other previous ones, and usually at higher operating costs. As an example, a number of fields have been discovered as a result of core drilling or seismic work, which later have been mapped as perfectly clear surface structures.

Marine or offshore exploration, particularly involving seismic work, is easier, faster, and cheaper than almost any method on land, but we tend to forget that offshore drilling and production are several times more expensive.

In strict justice, however, it must be admitted that there are also many cases in which offshore exploration, with the use of more refined methods, has been totally justified. In most of these cases, we observe, however, that success has been preceded by sound geological reasoning.

One such example, I am proud to point out, is the extension of the Golden Lane Cretaceous reef trend (to which reference is made later) off the coast of Eastern Mexico.

The North Sea and Australia are two further examples, where costly exploratory campaigns have been justified by the results. In both cases, the geological background knowledge was a decisive factor. In the North Sea, the detailed geology of the surrounding countries was extrapolated adequately, leading to the discovery of the vast gas reserves which are now changing the petroleum economics of Europe. The establishment of production in the Mesozoic-Tertiary basins in the Bass Strait of Southeastern Australia was brought about by the correct deduction of American and Australian geologists, whose interpretations were the clue factors for this significant discovery.

Alaska's recent North Arctic Slope discovery, the numerous discoveries in Libya, and now in Egypt, are additional examples of the successful use of our modern exploration technology. Yet, there are some recent disappointing examples, such as the Santa Barbara Channel, off the coast of California, where the geological problems were either misinterpreted or incompletely understood. Modern geophysical technology has not failed: The structures are present, but the stratigraphy has not proven to be as favorable for oil accumulation as originally expected.

Success in Exploration

Criticism is, unfortunately, frequently negative. Nevertheless, it must be pointed out that success in petroleum exploration, or for that matter in any kind of exploration, be it for minerals or water, is not the sole reflection of the quality and reliability of the information, but to a much larger extent, it is the result of a skillful and correct interpretation. A correct interpretation can be obtained only when the problem is well understood by the interpreters.

The history of petroleum exploration is, of course, full of examples of large and small oil fields that have been discovered by mere chance, and also of many wells located on half-technical reasoning, which have resulted in commercial production. Among these reasonings, we may remember the famous "creekology" technique, which was applied in Pennsylvania following the Drake discovery well. Also, several fields were discovered in the Gulf Coast after Spindletop by drilling on topographic mounds similar to the famous "Beaumont Hill." But few of the prospectors of those days understood the geological significance of each method and why it sometimes worked. Exploratory results would have been the same, no matter how carefully or accurately the streams had been traversed, or the topographic mounds surveyed and measured. Yet, both landscape features on which the methods were based are, as we now know, a reflection of the geologic structure.

But it was not until oil was found by drilling the uplands between the creeks that a reasonable understanding of the part played in the accumulation of oil by the anticlines was attained. In the Gulf Coast, the topographic mounds were found to have been caused by salt diapirs, but it has taken several decades to comprehend the mechanics of oil accumulation in the many adequate traps produced by such geological features.

In Mexico, part of the rich Golden Lane Fields were found early in the century through the presence of seeps and by drilling on trend. In order to discover the offshore extension of this enormous atoll-like reef, to which the oil owes its presence, the conditions first had to be understood.

In its discovery within the past ten years, the betterment and advancement of geophysical and drilling technology, no doubt, has played a decisive role, but the previous idea of its existence had been predicted by geological reasoning. It is quite possible that this offshore trend of the Golden Lane would have been found sooner or later, but the cost of exploration and drilling would have been several times larger.

It is much easier to find something when you know what you are looking for, and this is where sound geological reasoning always should play its part in petroleum exploration.

The Exaggeration of Sophisticated Methods

Our technology has undoubtedly advanced from the crude geologic mapping with a compass to the spectacular development in geophysical methods that we witness today. But most of this advancement has not been brought about necessarily by a better understanding of the geological phenomena, but mainly by the fantastic and unprecedented progress in the science of electronics and communications.

New exploratory methods have permitted us to go from the obvious textbook anticlines, reflected on the surface, to the search for the deeper, more complex, and even water- or ice-covered geologic features that comprise our exploratory prospects of today. However, our knowledge of the geological conditions, the way they came into being, and our capacity for predicting their existence have not advanced at the same pace. In many ways, we are still looking for the same conditions that were sought for at the beginning of this century.

We have managed to make our geophysical instruments lighter, smaller, and more accurate, and our recording processes more reliable, faster, and, perhaps, even cheaper when one considers the volume of information and quality of the data obtained. But do we know how to interpret all this precise, accurate, and reliable information? Do we know any more of what it means in terms of finding new hidden accumulations of oil than we did fifty years ago? Are we not in a way making detailed traverses of the creeks and mounds?

We talk and write at considerable length about stratigraphic traps, facies changes, and reefs, and their origins, and we have all sorts of esoteric nomenclatures to define and classify geosynclines, sediments, and depositional environments, but how much of this knowledge is actually applied when we are "looking for oil"? I do not claim that geological knowledge has not advanced considerably, but a gap has opened between the knowledgeable geologist and the petroleum industry's exploration efforts. It seems as if the oil industry has a remorseful feeling that it owes something to geology, but at the same time does not know how to use it. Companies create, operate, and support small and large research laboratories and projects; yet they seldom apply their findings in geology. We see all around us how exploration departments are overwhelmed by, and how exploratory campaigns frequently are dominated by, computation programming, digital processing, mathematical modeling, and automatic plotting. At the same time, the few men who have any geological knowledge of the area being explored, or who have a training to acquire it, frequently are cast aside as useless and impractical dreamers.

Optimizing and computing information without a thorough knowledge of what the outcome should mean may result in a conclusion like that of a major modern research department that "proved" recently that nine men and one woman can produce a baby in one month.

Resorting to a comparison that I frequently find useful between diagnostic medicine and exploration, we must remember that X-rays, electrocardiograms, and encephalograms have never substituted for the interpretation and diagnosis of the medical doctor. They are undoubtedly of great help and value to him, but only when adequately understood and used. This understanding, I must emphasize, is necessary for the correct use of the so-called sophisticated exploration methods. They are a tool and an expensive one at that, and must be used discriminatingly and wisely.

The electronic computer will do only what we tell it to do, and in many instances, its application has been greatly overrated, yet, it still is an objective, extremely useful instrument that permits us to handle enormous amounts of information and calculations which would be impossible to manipulate otherwise. Its use, therefore, should not be restricted to geophysics in the exploration field, but should be applied further to store, classify, retrieve, and work out geological information that is now practically lost in the files and well cards of our exploration departments. Again, we need the integration between the mathematician's and the naturalist's mind.

It is recognized that the number of geologists who are using the powerful tools of our new electronic era is gradually increasing. Large volumes of well data have been collected in the past few years by several groups in the United States, which now make them available to industry in a form suitable for computer storage, retrieval, and handling. Programs for the application of this information are rapidly multiplying and diversifying and, when adequately used, they represent great time and effort savers. Unfortunately, the information fed into the computers is frequently unreliable, and lacks consistency in the original geological criteria. Although the programmer may throw out or correct some of these discrepancies, the outcome still should be carefully analyzed and controlled by a trained interpreter. A final logical interpretation cannot be left to the computer alone, no matter how mathematically perfect and elaborate the program may be.

This fact generally is accepted by most experienced geologists and geophysicists, but, unfortunately, much of the field of electronic computing is not directly in the hands of men who have the background of the natural phenomena they intend to resolve, and more emphasis is frequently put into the mathematical procedure than into the significance of the outcome.

The Human Factor

Just as much as sophistication in our exploration methods is not enough to find oil, integration of geology and geophysics in itself is not enough either. Teamwork in our efforts has always been emphasized. Apart from its obvious necessity, it has a deep demagogic value, which makes everybody feel good. However, every team needs a coach to tell the players what, when, and how to do, and to make the ultimate decisions. A good or bad coach can drive the same group of men to victory or defeat in any of life's endeavours. This has been remarkably true in the oil finding business, and there are scores of examples where the ingenuity, leadership, optimism, and foresight of one man have been decisive in the success of his company. This truth is not always recognized because of selfishness and the short span of man's life. In oil exploration, where results have a lag of about five years or more, we frequently witness the paradoxical situation in which one man sows, only to see someone else do the harvesting.

It is true that oil was more abundant and easier to find fifty years ago, but it is also true that oil companies were managed by aggressive, self-confident men, who lived and felt every well and prospect they drilled, and the ones who had what it takes came up on top. All of this may sound like romantic nonsense, but if we examine the history of our industry and the men behind it, we will discover that the oil business is extremely sensitive to the doings, enthusiasm, and personality of individuals. The men who have created and helped consolidate the petroleum industry in the United States, and half the world around, through their vast contributions in the exploratory field, have all had these personal characteristics, which convert them into the giants of our oil business.

I wish I could individually name them all, but in fear of making serious omissions, let us just think of the few who have reached the high stature of presiding over or receiving other honors from the two largest professional societies of oil explorationists: the American Association of Petroleum Geologists and the Society of Exploration Geophysicists. I repeat, these men are not the only worthy ones, but, for want of a more complete list, let me express, together with today's petroleum industry, our most hearty respect to these and all the other exceptional men who have discovered, and who continue to discover, the oil and gas on which our whole civilization is founded. None of these men has ever been afraid to use new methods and approaches in their efforts to find new oil. In fact, many of the methods were discovered or originated by them. But all of these men have been extremely conscious of the limitations and the effective uses of all techniques.

If, in parts of this brief article, I have seemed dogmatically to attack some of the extremes to which our modern technology has pushed us, the attack has been only to emphasize the need to maintain, in our exploratory effort, a proper balance between the mind and the machine.

Bibliography

Annual Petroleum Development Papers. AAPG Bulletin, Vols. 45-52. 1960-1967.

Barrow, "Future of Petroleum Geologists." AAPG Bulletin, Vol. 51, pp. 2192-2196. 1967.

Beebe, "Will Exploration Follow the Dodo Bird or Will Management See the Light?" Oil and Gas Journal. May 25, 1964.

Cordell, "Future Explorationist." Oil and Gas Journal. August 7, 14, and 21, 1967.

De Golyer, "The Seductive Influence of the Closed Contour." Econ. Geology, Vol. 23. 1923.

Dillon, "Expanding Role of the Computer in Geology." AAPG Bulletin, Vol. 51, pp. 1185-1201. 1967.

Forgotson & Iglehart, "Current Uses of Computers by Exploration Geologists." AAPG Bulletin, Vol. 51, pp. 1201-1224. 1967.

Guzmán, "Exploration by Petroleos Mexicanos." Fifty-Fifth Meetting of the Princeton University Conference. 1963.

—"Reef Type Stratigraphic Traps in Mexico." *The Proceedings of the Seventh World Petroleum Congress.* Elsevier Publishing Co. Ltd., England. 1967.

Halbouty, "Heritage of the Petroleum Geologist." AAPG Bulletin, Vol. 51, pp. 1179-1184. 1967.

Hubbert, "Degree of Advancement of Petroleum Exploration in the United States." AAPG Bulletin, Vol. 51, pp. 2207-2227. 1967.

Kent, "Progress of Exploration in North Sea." AAPG Bulletin, Vol. 51, pp. 731-741. 1967.

Levorsen, "Stratigraphic Versus Structural Accumulation." AAPG Bulletin, Vol. 20, pp. 521-530. 1936.

—"Big Geology for Big Needs." AAPG Bulletin, Vol. 48, pp. 141-156. 1964.

—"The Obscure and Subtle Trap." AAPG Bulletin, Vol. 50, pp. 2058-2067. 1966.

—*Geology of Petroleum.* W. H. Freeman & Co., San Francisco & London. 1967.

Panel Discussion. "Accentuate the Positive, and Exploration Will Rally." Oil and Gas Journal. May 25, 1964.

Smith, "Relationships Between Geologists and Geophysicists." AAPG Bulletin, Vol. 51, pp. 2189-2191. 1967.

Weeks & Hopkins, "Geology and Exploration of Three Bass Strait Basins, Australia." AAPG Bulletin, Vol. 51, pp. 724-760. 1967.

The geologist's perspective:

An interview with Robert J. Weimer

Reprinted by permission of the Society of Exploration Geophysicists from *Geophysics: The Leading Edge of Exploration*, v. 6, no. 4 (April 1987), p. 32-34.

(Editor's Note: Dr. Robert J. Weimer is an internationally respected authority on stratigraphy. He spent the early years of his career with Union Oil Company and then taught for 27 years at the Colorado School of Mines. He has held emeritus status at CSM since 1983. Weimer is an SEG member (and delivered the Society's distinguished lecture Influence of basement tectonics on depositional systems and seismic stratigraphy *in 1979), but is primarily identified professionally as a geologist. He is an honorary member of the American Association of Petroleum Geologists and a recipient of that organization's highest award, the Sidney Powers Medal. His identification with petroleum geology is the reason that* TLE *sought Weimer's views and opinions on a variety of issues. The Editorial Board felt that, because of the widespread restructuring caused by current economic conditions, the perspective from another segment of the exploration community would be of broad interest within the SEG membership. The interview was conducted by* TLE *Managing Editor Dean Clark.)*

What is your view of the current economic problems in exploration?

We have, of course, been through downturns before but not with the variables and uncertainties we have now. In the United States we need a return of price stability because that will have a great influence on explorationists and investors. We had price stability for a long time because it was more or less controlled by the large international oil companies and domestic regulatory systems; but they lost control and now the price of oil is a function of international economic and political factors and tax laws — things that are not now heavily influenced, if at all, by the energy industry. In that kind of scenario, when these basic factors are under somebody else's control, investors are always going to be leery and without risk capital, the industry's days are numbered.

At the moment, all indications that I've heard are that an import tax could not get through Congress. I think the industry needs to emphasize energy security and build a case for the importance of that. The biggest energy problem facing the United States is the shortage of domestic liquid hydrocarbons that supply the transportation sector. Our economy runs on transportation. Take that out of the equation and a lot of things stop, including the military.

Exploration has another problem with investors, however, that is not directly connected to the economics of the moment. It is connected with the exploration boom of a few years ago. Then everybody was hiring lots of geologists and geophysicists. The demand was too great for the educational system then in place. Many schools that weren't adequately staffed graduated many people who needed additional education before they were qualified to work. But lots of them did go to work prematurely and didn't perform very well. That in turn destroyed the confidence of many investors in their scientific and technical personnel. I think that is a consequence that could have a long-term impact.

Is the price of geophysical work too high, considering the low price of oil at the moment? If so, how can it be brought down?

I've always been concerned about that. The challenge for the US industry is to reduce the cost of finding and producing new domestic oil to compete in the international market place. It's expensive to collect geophysical data. Explorationists have always felt that way. But when it's used properly, there's a payoff and in many areas you can't operate without it. But in some other areas, I question that you get dollar value for the price you're paying. Exploration is not a monolithic operation. There are areas where new methods have had tremendous success but there are others where they are applied only because "that's the way we do things now."

I haven't seen the exploration success ratio change in some areas in 35 years even though we've gone through different exploration cycles. Of course, the ratio might have gone to zero without the new technology but that's not the whole story. When I started out, we recommended drilling without geophysics. Now most people won't drill without geophysics. Is that because geophysics is really critical? Or is it because it's a requirement that the boss has to see it to validate a prospect?

Our goal should be an integrated approach to reduce finding costs. We have to look at the value of each element. But

cross-sections in the last 15 years. But a lot of geophysicists don't know the extent to which geologic information has also changed in the same period. There have been tremendous advances in petroleum geology in the last few years and they have to be integrated with the new geophysics. You can't use geology of 1970 vintage with 1986 geophysics and expect to come up with what you need. The reverse is also true; it's generally unproductive to use 1986 geology with 1970 geophysics, although exceptions do exist.

What you have to do is try to get people to understand the data base in various areas well enough so they will know how they bear on exploration problems. You have to use exploration teams; you can't expect people to be expert in all fields, but people have to know enough about other fields to know how they will influence their own.

A problem is that people don't keep up with advances in other fields. For example, recent advances in the fields of stratigraphy and sedimentology are just outstanding in telling us how rocks are formed and how they change spatially . . . changes that are visible in the indirect measurements of rock properties that you're doing in geophysics. Every seismic section has a large number of anomalies, features that are different from the norm. But to identify and interpret the true anomalies, important to prospect generation, from false anomalies is vital to exploration success. Many people can't do this because there is, and always has been, fragmentation of material within companies. That shouldn't be. People are now accustomed to doing interactive work on a computer. Why shouldn't they think more about interacting with other sciences?

That sounds like a blanket endorsement for continuing education.

That's why you take CE, not only to learn about your own field but other fields too. I think that's the way the oil companies feel about it. They try to give young people a spectrum of courses — diverse geology, petroleum engineering, as well as geophysics.

The critical factor, though, is back at the office. When a young person comes back to the office after getting some advanced training, will the person that he or she is working for accept different approaches? In my opinion, there's a middle management problem in letting young explorationists do something in a new way. Of course, middle management is under time pressure to develop prospects and so most of the time some types of information we judge to be unimportant just get left out. This may be necessary but you have the feeling that the best scientific information in many areas is not being used in exploration.

I feel strongly that this is a major problem because new ideas are the fuel for exploration. I've seen drastic changes in exploration during my career and it follows that other changes, conceived by scientific teams, will lead to new prospects and new technology. For many years I taught a course called *History of Geologic Concepts*. In part, it was a study of the history of science because geology was important in the early development of scientific concepts. The history of science clearly records that at every point in time man has accepted as truth some incorrect concepts. Some of these were accepted to the point that they became unchallengeable. They were dogma. Some of the concepts in widespread use today are undoubtedly wrong; we just don't know which ones they are yet. The challenge to each new generation of scientists is to identify the incorrect models and change them. The challenge to each generation of management is to accept the change and not hinder the application of new scientific thought.

Successful exploration comes from people who have ideas and so I think it's obvious that to meet the challenge of the future people must welcome ideas, their own and those of others. The half-life of one's scientific knowledge has been described as eight

you can't establish a uniform standard and apply it everywhere. You have to look at the value of each element in each province.

I think that one way to reduce finding costs, which is the way you really get competitive, is to make the information available. That's a challenge to the entire exploration industry, but geophysics in particular has kept tight control over its information. How can we access that information? Is it better to have the information in a vault or have it readily accessible? If it were, would it lead to even more geophysical work? Or maybe to a breakthrough that would lower costs?

Back in the 1950s when I was on a seismic crew, we were working on Indian land and it was a requirement that the work had to be filed after project completion. So I went and looked to see if the tribal office had earlier work that would help our exploration effort. In that case I found data that were helpful. That may be a reason that information is so tight. Everybody looks at data differently so perhaps people are afraid that somebody will see something on their data that they didn't. Be that as it may, I think it's vital to get the vast amount of geophysical data more readily available, especially in the older producing regions that have been again abandoned by the majors.

Do you think there's still a split, as there has been historically, between geologists and geophysicists?

One of the big problems today is finding the proper scale and perspective to allow the integration of geologic models with geophysical data. I'm awfully impressed with the quality of stratigraphic information that's become available on geophysical years. Because the answers to problems keep changing, the education of the individual must continue — after university training — in formal or informal ways if that individual wants to maintain scientific competence. If people do not have the commitment to improve and expand their knowledge, they risk becoming scientifically and technically obsolete.

You can attempt to stay on top of the everchanging situation only by working at it — either by a do-it-yourself approach or by continuing education courses. I think that CE courses, if done right, make it a lot easier process.

This is an aspect of professionalism, a theme that you have stressed in the past because of your concern that it is vanishing. Do you still find that an appropriate description?

Yes, particularly because of the rapid changes in geology and geophysics. This is not a diplomatic thing to say for publication but I feel that we have had a lot of people come into exploration who lacked either the scientific background or the commitment to become truly professional. However, this isn't limited just to our fields. I see it as very widespread in society.

Jack Valenti, former president of the Motion Picture Association of America and special assistant to President Johnson, described very well what it means to be a professional. I'm not a particular fan of his but I think he was right on the money in this case. He said, "Let me sit in a discussion where decisions are made and I can quickly and accurately point out the professionals. They know the issues, they have untangled the crossing threads of logic and reaction, they understand the facts cold, and can, because they have done the necessary homework, come up with suggestions that may lack passionate intensity but usually make the most sense."

I see professionalism vanishing for four primary reasons. First, too many graduates in earth sciences have an inadequate education to enter a professional career. The second is the attitude of the individual. As I said before, I think many lack the commitment to do the hard work necessary to achieve and maintain competence in their field. One reason is that it's been easy for someone with some education to make a decent living without having to take the extra step needed to become a professional. Recent events may, however, reverse this trend. Third, there has been a change in values in many segments of society which, as continually illustrated by the news media, tends to measure success by material wealth and monetary gain alone. Recognition is diminishing for the person who does the job right for the sake of pride and accomplishment. Fourth, there is a trend to ignore or discredit the professional and his contributions to society. One form of this is the lack of recognition of extra performance and excellence by the organization for whom the professional works. Another is a condescending attitude toward the professional by legislative bodies. There are also special interest groups which constantly downgrade the achievements by professionals in scientific and related fields. These people are interested only in the outcome of a particular situation, no matter what the correct general solution might be.

As a result of this trend, I think it is vital that we promote professionalism in our field and others. This is perhaps more of a problem in geophysics than in some other sciences because people come into this field from so many different backgrounds. There is no obvious mechanism to assess what people are professional and whether or not they have broadened their perspective to keep pace with the tremendous explosion in earth science knowledge in recent years.

A lot of this is properly left to the incentive of the individual but management also plays a big role by encouraging people to become more and more professional and by rewarding them, in one way or another, when they do. **L𝙴**

Response to the Sidney Powers Medal Award

R. J. Weimer
Golden, Colorado

Much has been written about Sidney Powers and his unparalleled contributions to petroleum geology and to the AAPG. I encountered some of his genius during the course of work over the past several years in the Mid-Continent. The subject of his 1922 paper in *Economic Geology* was "Reflected Buried Hills and Their Importance in Petroleum Geology." After a lucid description of the buried hills, associated unconformities, and controls on reservoir distribution, Powers emphasized their importance by stating that 34% of total production in Kansas, Oklahoma, and north Texas has been in structures superimposed over buried hills. My current research uses different terms; for example, recurrent movement on basement faults, tectonic influence on deposition during sea level changes, and drape (force) folding over basement faults; but his paper contains the essence of these expanded concepts as we know them today, 62 years later.

I want to express a few random thoughts about the future. To maintain the strength of our nation, we are told that as much petroleum must be discovered in the next 30 years as has been found and used in the last 100 years. How will this goal be accomplished against the backdrop of the doomsday prophets of government and industry? The key is new ideas because they are the fuel for exploration. Drastic changes in exploration ideas have occurred during my 35-yr career, although some techniques and concepts remain applicable. So it follows that new ideas, conceived by scientific teams, will spawn new wildcats and new technology. For many years I taught a course called "History of Geologic Concepts." The course, in part, was a study of the history of science because, as a natural science, geology was important in the early development of scientific concepts. The history of science clearly records that at every point in time man has accepted as truth some incorrect concepts that became dogma. Some of the concepts in use today are wrong—we just do not know which ones. The challenge to each new generation of scientists is how to identify incorrect concepts and then change them. The challenge to each generation of management is to accept the change and not hinder the application of new scientific thought. These comments apply to theories in papers to be presented at this convention.

Exploration results from the ideas of people, aided by machines and, therefore, people must be prepared to meet the challenge of the future. The half-life of one's scientific knowledge has been described as 8 years. Because the answers to problems keep changing, the education of an individual, after university training, must continue in formal or informal ways to maintain scientific competence. Without a commitment to improve and expand their knowledge, people risk becoming scientifically and technically obsolete. When this happens, they lose the confidence of their supporters, investors, managers, or others, and one result is unemployment. Staying on top of the ever-changing answers, through an understanding of new ideas, requires work, whether you do-it-yourself or take courses through continuing education. However, continuing education courses, if done right, make it easier to update oneself. I have made a career commitment to continuing education programs, and I hope everyone else has as well, because better work will be needed if we are to fulfill the future goals of our society. Who among us can quantify the ultimate value of one new successful idea applicable in either old or new exploration areas.

More than new ideas and technology will be needed in the future; economic conditions must also be favorable. The economics of exploration will depend on the efficiency of scientists, their use of support technology, and the public's attitude. We must not make efforts so complicated and expensive that we price ourselves out of the market. And, as professionals we must more effectively convey our message of the importance of mineral resources to a technologically based, but scientifically ignorant, society. It is imperative that we fulfill our commitment to the public by better communications about what we do and why we do it. The public's erroneous image of the petroleum industry, as viewed through the eyes of the news media and the politician, must be corrected.

If ideas are the key to future exploration, then the university should be closer in activities to industry and government than in the past. After all, the role of the university is to generate ideas, test ideas, store ideas, and merchandise ideas in a free and open forum. The limited cooperative efforts of the past need to be expanded to assist in achieving the future projections.

Because Sidney Powers was one of the earliest true professionals in petroleum geology, I want to describe for you another concern of mine—vanishing professionalism. A profession pertains to a special occupation, often for monetary gain. A professional pursues as a business some vocation or occupation, but also develops an attitude that brings about a dedication of time and effort to acquire knowledge and to apply it for the benefit of mankind. Geologists, geophysicists, and geochemists practice in a profession, but too many are no longer dedicated to being professionals.

Jack Valenti, former president of the Motion Picture Association of America and a special assistant to President Lyndon Johnson, made this comment about professionals: "...let me sit in a discussion where decisions are to be made, and I can quickly and accurately point out the professionals. They know the issues, they have untangled the crossing threads of logic and reaction, they understand the facts cold, and can, because they have done the necessary homework, come up with suggestions that may lack passionate intensity, but usually make the most sense." To be the professional of this description should be the goal of all of us.

I see professionalism vanishing for four primary reasons. First, too many graduates in geology from universities or colleges have an inadequate education to enter a professional career. The second is the attitude of the individual. I believe many individuals today lack the commitment to do the hard work necessary to achieve and maintain competence in their field. For many years, it was too easy for an educated person in the country to make a living without that extra work to become a professional—so why do it? Events of the past two years may reverse the trend. Third, a change in values has occurred in many segments of society. Our society, as continually illustrated by the news media, tends to measure success by material wealth and by monetary gain alone. Recognition is diminishing for the person who does the job right, for the sake of pride and accomplishment, regardless of external considerations. Fourth, there is a trend to ignore or discredit the professional and his contribution to society. This may take one of two forms: the lack of recognition of extra performance and excellence by the organization for whom one works, or, a similar attitude toward the professional by legislative bodies who make the rules and regulations that guide society. Continued efforts are made by some special-interest groups in society to discredit knowledge, and with it, the achievements by professionals in scientific and related fields. These people follow the "win-the-case" syndrome regardless of what the correct solution may be to a problem. We must do more to promote professionalism in society, in our field and others.

Finally, I have experienced, as I am sure Sidney Powers did, one of the important factors in exploration—luck. Luck is a geologic factor that we don't understand. After being the recipient of luck, one should reinvest time and effort to determine why petroleum was found where it wasn't expected. If you know why you had luck, you have the best basis to look for more petroleum. An evaluation of the "luck factor" usually requires a more complete geologic model—which integrates all factors relating to source, migration, reservoir, and trap—than was used when the prospect was drilled. Data from a discovery well, and from subsequent wells, should be evaluated by a geologist, not a petroleum engineer. The model developed from this type work can then be used to better interpret geophysics and geochemistry and to improve the odds for success in exploration.

The name of the game is to make the exploration model more efficient to cut the finding costs per barrel of oil or mcf of gas. Efforts to develop more effective teams, and the use of their science in exploration, would then follow the example set by Sidney Powers more than 50 years ago.

I am most grateful to have derived the benefits of a professional society such as AAPG and to be the recipient of your highest award. Thank you!

Robert J. Weimer

Acknowledgement is given to the Indonesian Petroleum
Association for allowing the reprinting of this paper from
*Proceedings of the Fourth Annual Convention of the
Indonesian Petroleum Association*, 1975, v. 2, p. 261-268.

THE THREE R'S OF PETROLEUM EXPLORATION : RESOURCES, RISKS AND REWARDS

PROFESSOR MASON L. HILL *)

INTRODUCTION

Petroleum exploration comprises a many-faceted activity. It is a complex mix of science and technology, operating in a complex economic and political matrix. I will review some of the basic aspects of petroleum geology, economics and politics which influence this world-wide activity. I hope that such a review may be of some value to some of you, especially in viewing exploration in the context of changing world supplies and demands for petroleum.

Much petroleum has been generated and trapped in near-surface sedimentary rocks by unique geologic conditions, which have occured at only very localized places on the Earth. It has become exceedingly useful as an inexpensive and efficient source of energy for transportation, the generation of electricity, and a host of domestic and industrial purposes. However, since it is an exhaustible natural resource, it cannot last forever on this finite planet if per capita demand continues to increase. Of course, demand will decrease and use patterns will change as costs become exorbitant, and/or better sources of energy become available; therefore, some petroleum that could be produced will always remain in the ground. But until other economic energy sources are developed, it is desirable to maintain, or improve, the economic and political conditions which stimulate world-wide petroleum exploration, for the benefit of all mankind.

Obviously, manhours, materials and money (the 3 M's of exploration) are lost when exploration is unsuccessful. But what is the "risk" of loss and how great should the "reward" be for discovering petroleum "resources" (the 3 R's of petroleum exploration)? I maintain that economic and political factors are more important than science and technology in deter-

mining the amount of oil that will be discovered. I argue that exploratory risk and potential resources should not be expressed in mathematical terms. And I deplore some of the common misconceptions about petroleum exploration, shared by some economists, politicians, and even key individuals within the petroleum industry.

EXPLORATION PROCEDURES

It seems appropriate to begin this discussion with an outline of the customary sequence of exploratory activities which lead to test drilling in relatively unexplored areas. Those of you in the business will think of modifications and additions to the following six procedures:

(1)   Geologic research usually is the first step in exploration. This consists of a literature search for pertinent maps and reports and a review of the results of previous petroleum exploration, if any, in or near the prospective area. This preliminary survey ordinarily is relatively fast and cheap, but geologic judgment and intuition to choose the right regions, and make the right interpretations, relative to petroleum potential are extremely important. For example, favorable stratigraphy and structure for the accumulation of oil and gas may be indicated by studies of available geologic maps, or by ground and/or air reconnaissance. Before, or perhaps during, this study, economic and political analyses must be made to determine the feasibility of further exploration.

(2)   Reconnaissance geological and/or geophysical mapping comes next. At this stage some geological interpretations can be employed as an additional basis for appraising the petroleum potential of the area. And plans can be made for the kinds and sequences

*) U.C.L.A.

of detailed exploratory activities which will lead most directly to test drilling. For example, low-cost photogeology, flying magnetometer and/or a study of stratigraphic sections might be useful in preparation for later exploration.

(3) Contract negotiations or leasing is normally the next step, unless required earlier, in order to obtain rights and protect future expenditures. Economic and political considerations are paramount here, so that if petroleum exploration is successful a profit is possible, and rights are secure. This stage in the program can be quick and inexpensive in terms of manhours; or slow and expensive if long negotiations, including environmental issues, are involved (such as environmental impact reports, now required in the U.S.)

(4) Detailed geological and/or geophysical mapping is usually needed before good decisions about exploratory drilling can be made. Large expenditures in manhours, materials and money (the 3M's) are commonly made during this stage of exploration, if substantial seismic work and core-drilling are required. One of the keys to good exploration, particularly at this stage, is to do the *right* things in the *right* places, at the *right* times (e.g., if possible, do photogeology before seismic work — not the reverse as was once done in Egypt). Sound judgment (including intuition) is required to know when enough detailed mapping has been done to select an optimum location for test drilling.

(5) The location of a test well, the next and critical step in any exploration program, can be only as good as the interpretations of the stratigraphy and structure at depth. Uncertainties are involved, especially when translating geophysical data into geology (e.g., near misses in Cuyama and Alaska). Assuming drillable prospects are found, an early decision for test drilling is desirable, especially in a relatively unexplored area, to obtain information on the stratigraphic and structural habitat of the petroleum, if any. With this background, if favorable, better decisions can be made for further exploration. When deciding on a drilling site, consideration must be given to all abnormally high costs such as those resulting from difficult accessibility, adverse climatic conditions, deep water, etc. Obviously

some geologically successful projects can be financial failures if development and transportation costs exceed the value of the production. However, this outcome should be unlikely if good judgment has been employed in the previous stages of exploration. And geological considerations should usually outweigh logistic and mechanical ones in decisions on test well drilling sites.

(6) Drilling a test well is the only way to complete a petroleum exploration program. No shortcuts in drilling depth, logging or testing should be tolerated. No abandonment should be permitted until commercial production has been disproved. If a project is worth test drilling, it should be done right. This means that costs, including those due to mechanical problems, should not detract from the drilling objective, which is to appraise all the possibilities for commercial production. Unfortunately, the most expensive test wells are usually the failures. (e.g., in Alaska and North Sea.) Drilling costs are, for many reasons, continuing to increase, and therefore risk money for expensive wildcats *is* most likely to come from the larger oil companies. And high costs definitely reduce the number of test wells, which in turn restricts opportunities for finding new petroleum resources.

DISCOVERY AND DEVELOPMENT

If a test well makes a potentially economic discovery, confirmation wells are needed to delineate the size of the field. The purpose of this drilling is to determine, as soon as possible, approximate reserves and optimum producing rates. Then in the light of other economic and political conditions, the profitability of the project can be analyzed. If profitable production is unlikely, the operator may have to write off the loss and discontinue development work. Then, unless the loss of manhours, materials and money is too great, the company will pursue other exploratory ventures. Actually many oil companies do go out of business because of unsuccessful exploration (e.g., in California). Furthermore, even if successful exploration established substantial oil reserves to support development costs, financing may be difficult, even for large companies, if the oil cannot get to markets promptly (e.g., Alaska and North Sea).

RESOURCES, RISKS AND REWARDS

The total oil resources of the earth are unknown, although many estimates of the number of barrels yet to be found (potential reserves) have been published. (e.g., just this year, 45 billion barrels has been credited to Antarctica.) However, we do know the approximate number of barrels which have already been produced, plus the proved reserves which can be recovered, with current technology and economics. Actually, the potential oil reserves of the earth depend more on economic and political conditions than on the natural occurrences of oil in geologic strata and structures. That is to say, the degree of economic and political motivation directly controls the amount of exploration (number of wildcat wells), which largely determines the amount of new oil that is found. With sufficient motivation the conversion of potential resources into proved reserves is accomplished by the efficient (and perhaps fortunate) application of geology to petroleum exploration. Thus the continued discoveries of new oil has continued to disprove all past predictions of depletion. However, logic tells us that if we continue to find petroleum, and use it up at continually increasing rates, it cannot last forever. Therefore, it is not necessary to know "unknowable" potential resource figures to realize that we must conserve petroleum and look for other economically viable sources of energy.

Petroleum exploration is a high-risk activity in terms of the potential loss of man-hours, material and money; therefore, the rewards for success must be great. Certainly no one exploration entity (an individual, company or consortium) can continue to be unsuccessful in high-cost exploration and stay in business. Eventually the risk capital for exploration will shrink and additions to reserves will not meet future demands (already this has happened in the U.S.). To be successful in petroleum exploration, the financial rewards should be great enough to maintain, or expand, the rate of exploratory drilling. This means that good exploration companies will discover enough oil, enough times, to remain in business, although most of their wildcat wells result in total loss Many good exploration companies have gone out of business because of a long string of costly failures, and others, perhaps mainly by luck, have become prosperous. Only government oil companies, supported by taxes instead of profits, can continue to suffer expensive exploration failures.

In the U.S., at the present time, there is political pressure to retard exploration by increased taxation, by additional costs for environmental protection, and by not allowing exploration in the most attractive offshore areas. Elsewhere in the world the trend is toward higher taxes and costs, which weaken motivation for risky exploratory ventures. Noting that the equivalent of one 100 million barrel oil field only supplies the present world demand for two days, it seems obvious that exploration should be encouraged rather than discouraged.

EVALUATION OF RISK

Each exploration proposal must be analyzed in terms of potential loss (of the 3 M's) in case it should be unsuccessful. This is especially important, even for big companies, if the exploratory program, including test drilling, is relatively expensive. In most cases the decision involves choosing between several prospects; e.g., domestic or foreign, land or water, untested or producing area, small or large area, shallow or deep test, etc. The cost of obtaining drilling rights and kinds and security of production rights also are important factors in choosing a specific exploration project. And obviously, an evaluation of the risk of total loss compared to the potential for monetary gain (risk to profit ratio) must be made.

It is important to recognize that petroleum exploration, unlike many economic ventures, contains elements of risk that cannot be calculated in any conventional manner. For example, the risk to profit ratio for most manufacturing projects can be approximately determined before making large investments. Or a mining venture, such as for copper or oil shale, often can be appraised by measuring the volume and quality of the ore before making large investments. In contrast, nearly all petroleum prospects contain the unique element of risk that test drilling will not discover *any* of this valuable resource. This situation might be compared to the total loss of an agricultural crop except that a new crop can be planted on the same land next season with little risk of failure; whereas, the geologist must look for oil in some other place and risk another expensive failure. The principal reasons for this unique risk in petroleum exploration, beside underground concealment, is that times of accumulation, types of reservoir rocks and geometries of the traps are so variable that no two oil or gas fields can be alike.

It is true that enough similarities are found to exist between oil fields in maturely developed basins or trends to remove some of the exploration risk in new projects. Here, however, the remaining prospects tend to become poorer and smaller, which increases the economic risk. On the other hand, an untested sedimentary area must be evaluated in some different manner. In such situations only the potential for petroleum accumulations can be appraised on the basis of relatively unknown stratigraphic and structural conditions. In neither of the above cases can the possibilities of failure or success be quantified. And although most major oil companies crank in a "risk factor" with their computer programed economic analyses, and books have been written on mathematical solutions to exploration risk, I am sure that the practice is unsound in petroleum exploration. Therefore, instead of trying to quantify risk, I suggest that each prospect be qualitatively evaluated on the basis of geologic and geophysical data and interpretations. If such an evaluation indicates the potential for discovery of oil or gas, the project should be a candidate for test drilling. Thus, when a project is recommended by the explorationist (or exploration division) it should be assumed that a discovery can be made (instead of trying to calculate its percentage chance of failure).

Starting with the assumption of discovery success allows for a realistic comparison of available prospects on the basis of fairly quantifiable economic factors, such as development and transportation costs, and potential profits. Obviously, good geologic control on a large structure in an area of favorable stratigraphy should make an excellent target for test drilling, assuming the existence of reasonable economic and political conditions. This is because a big play is often about as inexpensive as a small one, and most of the world's oil is in giant fields (> 500 MM bbls). And this is why immaturely developed or untested offshore areas are now favored, especially in the U.S. where most of the land areas are already explored and developed. However, multiple discoveries in proven productive areas can result in as much profit as exploration in virgin areas. And a string of either profits or losses can make or break any exploration entity. Exploration failures, or losses due to other causes, often result in reduced exploration budgets, and the firing of explorationists. Although such actions may be of short-time advantage to managements, they can cause long-time harm to company stockholders.

POLITICS OF EXPLORATION

Occurrences of petroleum are unrelated in time and space to governments or to political boundaries. There have always been, and always will be, national inequities in the distribution of mineral resources, unless one body-politic takes, or is given, control of all the world's minerals. Currently, some countries are sellers and others are buyers (e.g., Saudi Arabia vs. Japan) but these situations can change with time (e.g., U.S. now buys and Britain will be selling). Governmental policies and legislation provide the climate to encourage or discourage exploration, and only by exploration can mineral resources be obtained.

Not long ago Australia became an oil-producing nation because it developed policy and legislation which encouraged exploration capital (the 3 M's) especially from the U.S. During this time (but not in the present political climate) the government even shared in some of the risk money in order to stimulate exploration activity. On the other hand, policies and legislation in the U.S. have managed to retard exploration by holding down prices for products (especially for gas), increasing taxes, increasing costs for environmental protection, and delaying offshore leasing.

Eight political factors influencing the places and paces of petroleum exploration are:

(1)   Nationalization by taking over interests held by foreign-based companies, or precluding foreign participation in exploration and development. Such actions tend toward worldwide

reduction in exploration activities. (2) Price increases, as those recently fixed by the Organization of Petroleum Exporting Countries, are conductive to increased petroleum exploration in other parts of the world, and to the development of other sources of energy. (3) Exploration contracts in which the risk of failure is taken by the operating company for a minor interest in, or a right to purchase some of, the discovered petroleum (if any). Such contracts tend to discourage exploration if the rewards for discovery are not sufficient to justify the risk of total loss. (4) A worldwide trend toward increasing taxes on, and oil payments by, the operating companies serves to increase the risk to reward ratio and thereby discourages exploration. (5) A worldwide trend to restrict areas available for exploration and to specify unreasonable environmental protective operations retards exploration (e.g., virtual shutdown of U.S. offshore exploration). (6) Bonus bidding, especially prevalent in the U.S., tends to eliminate all but the big exploration consortia and thus reduces competitive exploration. An alternative system of awarding leases on the basis of work obligations (the money going into exploration rather than into an acquisition bonus) would stimulate competitive exploration and could result in greater financiel benefit to governments from royalties and taxes. (7) An increase in the number of multinational oil companies, originating in a number of nations, could increase worldwide exploratory efforts, and also increase international understanding and prosperity. And (8) There is a trend, even in the U.S., toward complete nationalization of the oil industry. Such a trend, especially in risky mineral exploration, would surely lead to inefficiency, inactivity, increased prices, and lower living standards. (More on this subject later).

Knowing that petroleum is a depletable natural resource, it seems reasonable to conclude that governments should agree in principle to encourage: (a) exploration in order to keep supply approximately equal to demand, (b) conservation to extend the useful life of petroleum, and (c) development of other sources of energy (perhaps directly from the sun) to maintain, or perhaps improve, worldwide living standards. However, the economic and political facts of life, and the presently known distribution of petroleum on Earth, deter and defer any such logical international policy (even if possible in some individual countries).

SOME MISCONCEPTIONS

As expressed in my introduction, "Petroleum exploration is a complex mix of science and technology operating in a complex economic and political matrix". Therefore, geologists are challenged to explain the procedures and problems of exploration to others within the petroleum industry, to economists, politicians, resource and environmental agencies, and to the public, so that the best possible cooperative decisions can be made on exploration policies and programs. And, of course, economists and politicians need to explain their objectives and problems to petroleum geologists. As a partial summary of my previous remarks, I will now make recommendations for the removal of the following eight misconceptions about petroleum exploration:

Misconception (1): "Oil reserves of an intested area can be estimated without drilling." This is a common misconception. Here the pitfall is in not differentiating between "potential" and "proved" reserves. For example, the proved reserves in California (under present economics and technology) are approximately 3 billion barrels from knwon oil fields; whereas a figure of 14 billion barrels has been used for the oil reserves on the outer continental shelf off Southern California. The former figure is based on quantitative calculations of oil in place but the latter is only a subjective figure which can vary greatly, depending on whether the wishful thinking geologist (or some other less able person) is optimistic or pessimistic. Therefore, I believe potential reserves (possible resources) should not be expressed in mathematical terms because the figures only serve to confuse the public and cause politicians to make decisions based on the belief that this oil is available. Actually exploration and development, if economically producible oil is found, might take 10 or more years before a significant amount of oil could reach the markets. And much more time is required, depending on the intensity of exploration and development, before meaningful figures on the total reserves of any producing region can be determined.

Other examples of counting barrels before they are found are the estimates of 48 billion barrels of oil on the U.S. Atlantic offshore, and 33 billion barrels of oil on Naval Petroleum Reserve 4 (North Slope Alaska). In the former case, no test wells have been drilled and all that can be legitimately said about the potential oil resources is that some stratigraphic and structural interpretations favor the possibility of oil accumulations. However, until fields are found and the area is developed, any figure from zero to more than 100 billion barrels can be rationalized. The 48 billion barrel figure is continuing to generate policy arguments between the Federal Administration, Congress, State, local governments, and environmental groups (spurred on by the public's reaction to TV and press coverage). The U.S. Navy explored on the North Slope of Alaska in the 1940's without discovering economically producible reserves of oil or gas. However, since the Prudhoe Bay discovery, many miles to the east, the area has been reappraised by optimistic geologists and politicians. Here the 33 billion barrels of potential reserves now used for N.P.R. 4 is really less justified than the 48 billion barrels off the Atlantic Coast because rather extensive exploration has already been conducted. But these meaningless figures are being used as an argument for deferring exploration of California's outer continental shelf. Even if economic oil should be found on NPR 4, the lag time (10 – 15 years) would prevent that oil from easing any short-term supply crisis. Recently a figure of 45 billion barrels of oil in the Antarctic has been publicized with the comment that development costs might preclude its production, but it is not there until discovered and produced. (Does thunder exist if no one hears it?)

Misconception (2): "Exploration projects can be evaluated and compared on the basis of a mathematically expressed risk factor." This statement is wrong as exemplified by many surprised such as; (a) the "low-risk" Destin Anticline has so far been proved non-productive vs. the "high-risk" but productive Prudhoe Bay area (the former in the Gulf Coast oil province was acquired by high bonus bids, and the latter on the non-productive Alaska North Slope, was acquired for "peanuts" –

whereas the last one billion dollar sale there bought "peanuts"), or (b) the high-risk North Sea area, even if appraised after the Groenigen gas discovery, turns out to contain big reserves of both gas and oil, whereas the "low-risk" (as indicated by some bidders) Spanish Sahara, South Africa offshore, and many other promising areas have resulted in exploration failures. Furthermore, even in developed sedimentary basins, a comparative risk factor for discovery (or failure) cannot be mathematically determined because no two fields in a single basin are enough alike to determine the percentage chance for discovering another. Therefore, the chance of success or failure of a specific prospect should not be expressed in numerical terms.

Misconception (3): "Oil exploration and development will damage the environment." This is a widespread misconception, although it is based on many cases of poor operating practices and accidents which have caused some damage. Even the infamous Santa Barbara blowout is now known to have caused no permanent damage to the marine fauna and flora. (Hancock Foundation 1971). The oil companies cleaned up the mess, paid off property damages, and lost the good will of the public (especially in the city of Santa Barbara, where some people deplore the sight of an oil platform on esthetic grounds). Obviously, some damage has and can be done by oil operations (including tanker spills) but hardly any human activity is free from risk; whereas, the need for petroleum products during this interval of earth history justifies the continued efforts of oil companies to satisfy the public's requirements. Also recent improvements in operating practices and new legal regulations, further reduce possibilities for environmental damage.

Misconception (4): "Oil companies have more than enough money to do all the exploration that is needed." This misconception (probably not shared by bankers) has been evidenced recently by abandonments or delays of expensive industry projects, such as refinery construction or oil shale operations, because of the difficulty in obtaining capital. Even those companies with big reserves as collateral for loans cannot always afford high interest debts. Furthermore, exploration money comes mainly from net profits, so when profits are low, exploration budgets are cut. Of course, this is a poor long-term policy because an integrated oil company needs to replace its reserves by new oil to remain economically viable. And although stockholders expect quarterly dividends, sometimes that money might be better spent in exploration.

Misconception (5): "Governement can determine the value of petroleum resources before allowing exploration." This misconception is causing critical political problems in the U.S. Some politicians (perhaps due to public opinion — who leads whom?) believe the oil companies are getting too much profit at the expense of the public. Therefore, they think the government (in this case the U.S. Geological Survey) should undertake exploration to determine petroleum reserves so that only fair bonus bids and appropriate royalties will be accepted. The fallacy of this policy also stems from misunderstanding the difference between proved and potential reserves. In most situations, e.g., in bidding on offshore tracts, neither the oil companies nor government geologists can do more than guess whether any, some, or much oil will be found. Good evidence for this statement comes from the wide range of bids made for specific tracts, although many of the bidding oil companies have essentially the same geologic information.

In order for government geologists to appraise the potential for petroleum discoveries in offshore areas, they must make ocean bottom geologic maps and appropriate detailed geophysical surveys (this work can involve very expensive core drilling and reflection seismic work). However, the oil companies will probably do all of this preliminary work quicker and better than government, if given the opportunity. And; (a) in a competitive situation the oil companies will make their bonus bids, uninfluenced by any governmental estimates of value. Or (b) if competitive bidding on the basis of work obligations was allowed, many more exploration entities could be involved, thereby increasing competition and exploration, including the drilling of many more test wells. In either of these situations, much less governmental effort and money would be required. In the past, the role of the U.S. Geological Survey in mineral resource studies has been

to provide general geologic data and interpretations in order to stimulate exploration and development by private industry. However, in the present misinformed political climate, there is pressure on the survey to spend more money instead of less on offshore studies, even though their work is not needed to promote exploration.

Misconception (6): "Government should do some, or perhaps all, exploration and drilling in Federal waters." This is a misconception that could lead to government competing with private industry and gradually, with the use of nearly unlimited and unaccountable tax dollars, running the petroleum industry out of business. Many examples within the U.S. and elsewhere indicate that competitive private industry is, for many reasons, more efficient than government in the exploration and development of mineral resources. Furthermore, exploration of the same areas by several companies produces advances in exploratory technology and a wide range of interpretations on petroleum potential. On the other hand, exploration by government would generate no competition and would require taxation without measurable returns to the public.

Misconception (7): "A complete analysis of national (and world) reserves of petroleum should be made before a national energy policy is established." Some politicians have requested this even before the U.S. Department of Interior is allowed to issue any more offshore leases for exploration. And, at least by implication, it also means that the potential of all other energy sources to compete with, or replace, the uses of petroleum, should be determined before policy is set. Remembering my introductory statement about the complexity of petroleum exploration, and accepting my contention that it is impossible to determine the petroleum reserves of an area before test well drilling, field developments, and more test wells to disprove additional economic production, it is obvious that an energy policy cannot wait.

All a national energy policy really has to do is encourage (or perhaps discourage in some cases) petroleum exploration under reasonably controlled conditions. Of course this is easier said than done because the policy must be explicit and permanent, and yet flexible enough to meet changing local, national and international situations. Obviously, since petroleum is a non-renewable and mostly non-recyclible substance, government should encourage the development of other economic energy sources, including ways to use heat from the sun.

Misconception (8): "Multinational oil companies rape the oil producing countries and swindle their own governments." This statement, like the seven others described here, is not true, although there may be some cases to prove it. The reason there are multinational oil companies is that attractive geologic and economic prospects are available in many countries. A prime example now comprises the worldwide offshore areas where the chances for finding big reserves of petroleum are better than in the maturely developed land areas. U.S. based companies are especially interested in foreign offshore exploration because these areas are not widely available at home. Private industry wants, if possible, to explore in places where the potential for oil accumulations is judged to be the best, and petroleum occurrences are controlled by geology instead of political boundaries. This desire by industry is to maximize efficiency in competitive environment in order to keep stockholders from supporting the efforts of other companies, or other industries.

On the other hand, the investments made by multinational oil companies in foreign countries bring employment opportunities and monetary rewards to those countries. It can also improve educational opportunities, if needed, and raise living standards of some people. And perhaps the most important aspect of multinational companies, headquartered in any nation, is the long-term effect of increasing international people-to-people communication, and increasing the opportunities for cooperative endeavors, which are so important on this small "blue planet". The days of empire building are past, and the days for international agreements on long-term solutions to world resource and environmental problems are here.

CONCLUSIONS

Finally, I want to thank you for listening to these remarks about some of the geologic, economic and political aspects of petroleum exploration that seem important to a retired and provincial petroleum geologist from California. I have outlined the customary six steps involved in conducting an exploration program; I have emphasized that it takes *risk* to get *rewards* for finding petroleum *resources* (the 3 R's of exploration); I have argued contrary to custom, that the potential reserves of any area, or the exploration risk of a single project, should not be expressed by numbers; I have outlined eight political factors that influence petroleum exploration; and I have discussed eight misconceptions about the petroelum industry that should be banished.

I hope these remarks will help a little to provide better cooperation in petroleum exploration between people within industry, between industry and governments, and between governments, and thereby help in some way to solve present and future energy problems on our Earth.

Note: I wish to thank the Atlantic Richfield Company for giving me this opportunity of speaking to you. However, I want you to know that I am solely responsible for what has been said.

* * *

This paper, originally delivered at a convention of petroleum landmen, Vancouver, British Columbia, Canada, on September 11, 1986, is printed by permission of the author.

Winning

John A. Masters
Canadian Hunter Exploration Ltd.
Calgary, Alberta, Canada

Your speaker's committee recognized that Canadian Hunter wasn't reacting to the oil price depression in the same way as most companies. It appeared that we hadn't got the word. They asked if I would explain if we are just stupid, or do we know something the others don't.

I agreed to talk about "Winning." It is something we are determined to do, in any economic climate, under any political party.

My partner keeps a plaque on his desk that says: "Crises are Opportunities." Spare us from a thoroughly favorable red hot business climate. We discovered Elmworth in the pits of the middle 1970s, when you could farm-out 100,000 acres at a crack, when land prices were low, when there was room to move. We had tremendous exploration success this past winter, and we'll do it again next winter. I hate competition. I love it when everyone else is hunkered down.

To quote from an oil bulletin: "Talk to an oil man and he sees nothing for the future. There is no market, no possible expansion of the market, no chance for improvement, no visible ray of light." (*California Oil World*, September 14, 1914).

Most companies won't move unless everyone else is moving. Taxes have to be right, incentives right, royalties reduced, gas markets in place, oil price up. By the time all that is ready, you get run over by the stampede.

You win by being in early. And you miss if you're only good. You have to be first class. Excellent. It's as simple as that—and as difficult.

I don't go to the Petroleum Club anymore because all I hear there is overfed big shots complaining about all the things someone else should do to make their business profitable. You can work at this business from two different ends—the top or the bottom. Let me illustrate. And let me explain that I've been out of the country and did not hear about the Petroleum and Gas Revenue Tax (PGRT) until yesterday.

Start with $30 oil (Figure 1).

Take off PGRT, royalty, and operating, and you have $17 left. If you are an average company in the Canadian industry, according to the Canadian Energy Research Institute, your finding and development costs are $18. Presto, you're out of business. Even if PGRT is eliminated, considering the discount of future revenue, there is still no profit.

Most guys at the Petroleum Club moan that the trouble is at the top. Cut the taxes and the royalty, and I'll be alright, jack.

I just don't hear anyone talking about what he can do at the bottom end to make a profit.

Our friend Boone Pickens, who is here today, was one of the first people in the industry to realize this problem, call attention to it, and make a pile of money from it. He said, "Here, this is stupid. If you can't do any better than this, stop exploring with your profits from old oil and divide up the money." All the dinosaurs in the business said, "Boone you're a traitor, get off our backs and let us go on doing things the same old way, the government will bail us out, because we have to have oil supply security." They hurt Pickens feelings, and he is reported to have actually cried a little on the way to the bank.

Now, recently, the problem has become much more severe. Right now the price is about $20. And it looks like Figure 2.

With the atrocious finding and development costs of $18, the average company starts out by losing $7/barrel on the front end. Make that $4 now. So the moan rises to a high pitched scream for government help at the top. But still no one does very much about the bottom end of the problem.

Do you realize that the really good companies find and develop oil for about $3/barrel? Canadian Hunter's finding costs this winter were less than $1/barrel and development about $2. Coho Resources reports that they have found good production off the Heidelberg field in Mississippi for 50¢/barrel. That leaves plenty of room for a profit. Not as much as when the price was $30, but think of the other problems then: land prices were 2–3 times as high, drilling costs were higher, you couldn't get farm-outs, you didn't have a choice of prospects—you were lucky to get a piece of one.

Look what's happened to the price of oil in Canada

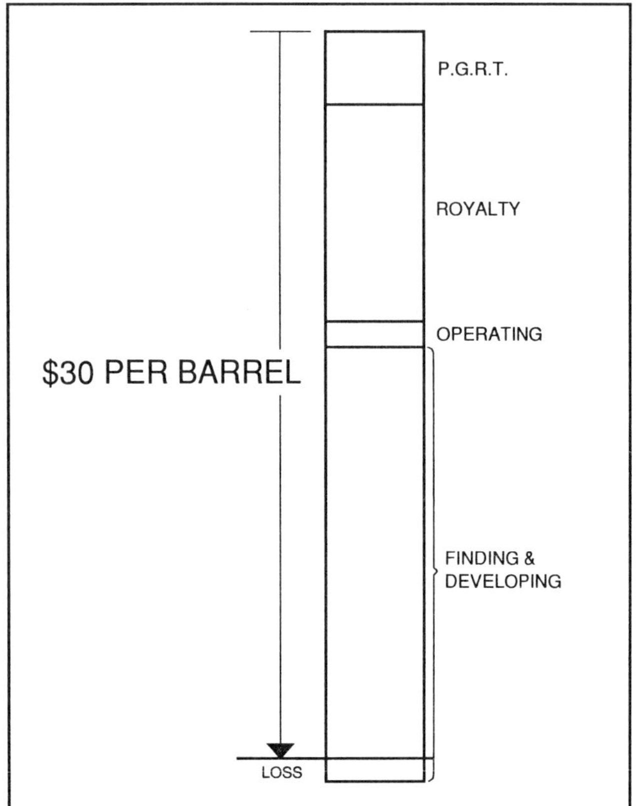

Figure 1.

Figure 2.

(Figure 3). For 20 years, it was flat at $2.50/barrel. You had to find oil cheap, and everyone did. In the 1970s, the economics of the industry changed overnight. For a period of 5 years, any idiot could make money in the oil business. A bunch of stockbrokers, and lawyers, and probably even some convicts and bankers, became oil men.

Everything was so easy and so exciting that staffs expanded (some of the majors had 1000 exploration people—terrible management, I call it), overhead exploded, the whole exploration process got clumsy and inefficient. The majors decided that the only place they could find oil with their bulldozer and sledge-hammer techniques was in the frontiers—so finding costs went out of sight.

Then 9 months ago, the price of oil, which was supposed to go up forever, went over the cliff.

Most of the industry is helpless in what they think is a new situation. They are like a bunch of fat men who suddenly find themselves entered in a 100-yard dash. Their reaction is epitomized by the bumper sticker I saw in Denver: "Please God let there be one more oil boom. I promise not to piss it all away this time."

Before you begin to think I am a twit talking about impossible remedies in an impossible situation, let me review a little history for you.

I was a kid in the "dirty thirties." Most of you weren't born yet. I remember tramps coming to the back door and my mother fixing them a sandwich. Look at some of the economic indicators (Figure 4). In America, there were 13,000,000 unemployed—25% of the whole work force—and no government assistance.

But look at the U. S. crude oil price index (Figure 5).

In 1980, the price was 50 times what the price was in 1930.

Things are not very good today in 1986, but they aren't the pits of the 1930s. We've only been in trouble for 9 months, not 10 years. We all believe it will end soon. We don't think our world is coming to an end as so many did in the 1930s. Now let me tell you something you don't know.

The 15 years from 1925 to 1940 saw the discovery of more oil reserves than any other period in the history of the U.S. (Figure 6). That isn't what you'd expect is it? The economists would have said it's related to the amount of drilling and availability of capital. But, that's not it. It's a function of available technology. Rotary rigs could drill to 5000 ft and the seismograph had just been developed and big structures were being found and everyone had to be in on it. Wallace Pratt, the best oil geologist who ever lived, built Humble into the biggest oil company in the U.S.—during

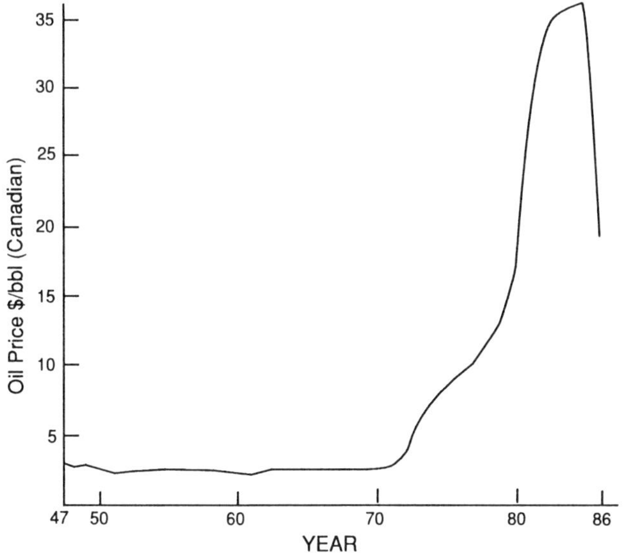

Figure 3.

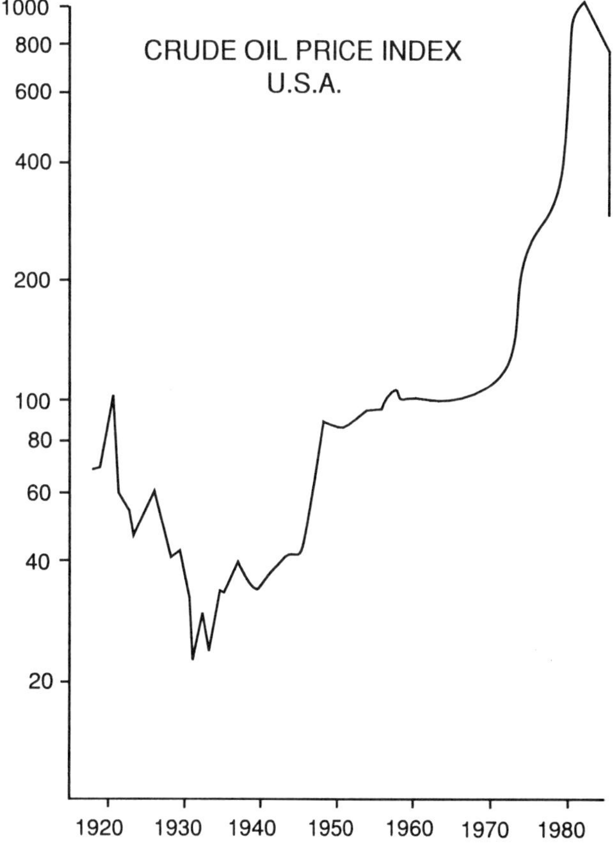

Figure 5.

Figure 4.

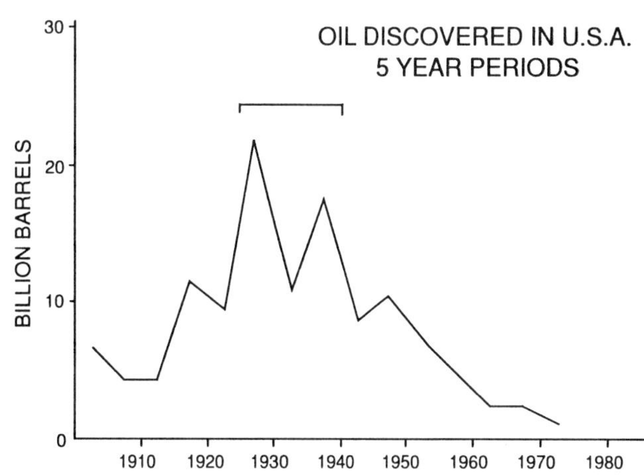

Figure 6.

the worst depression in history (Figure 7).

I want to quote from page 467 of Edgar Wesley Owen's *Trek of the Oil Finders*, the AAPG history of the oil business: "The producing business met its most excruciating test in the early 1930s, when the greatest excess productive capacity in the history of the domestic oil industry coincided with the most severe financial depression of the century." Sound familiar? "Perhaps the most momentous turn in Humble's fortunes was its decision during the period of financial depression and massive oversupply of oil in the early 1930s to conduct a large scale program of increasing its future reserves of crude oil." From 1930 to 1940—in the bottom of the depression—Humble tripled its reserves.

You don't find reserves when it's convenient. You find them when you can. One of the factors is competi-

Figure 7.

tion. When economics are favorable the big guys roll over you. If you're a small company *right now* may be the best time to explore. This may be your opportunity!

How do you do it?

Well, you have to have money, and you have to be a good oil finder.

First, money. The oil business is strapped, but the rest of the country is fat. You all say you can't get money, but how hard have you tried?

Have you covered eastern Canada, eastern U.S., Hong Kong, Europe? Have you talked to 20 companies, 40? If you want to be a winner, that's what you've got to do. I didn't say it would be easy. It's just simple. Canadian Hunter will use about $100 million of outside money in 1987 to fund our exploration programs.

To be a winner you have to have money—simple—and be a smart oil finder—not so simple.

Clearly, at today's oil price, you have to have low finding costs. That is the absolute key. How do you do that? Well, not with a staff of thousands. A few smart guys can move intelligently, faster, more imaginative-

ly than a large group. You can *find* oil with a few people. It takes a *lot* of people to produce it, but only a few to find it.

At Canadian Hunter, we have a staff of 12 exploration geologists and 3 geophysicists, but with a powerful support staff of log analysts, rock specialists, and engineers. This group drilled 42 exploratory wells last winter and made 30 discoveries. We think we found about 200 bcf of gas and 50 million barrels of oil. Our finding costs were less than $1/BOE.

This staff has been built *very carefully*. It is made up of young, eager, smart men and women. They are not geniuses and they do not use any secret technology. But every technical tool they use is applied with the most advanced skill available in the industry. These people just do the basics, but they are superb blockers and tacklers and passers and runners. Any one of them can break up a ball game! Every one of them, except the two youngest who are just out of school, have found oil and gas fields. They understand geology and economics, engineering and log analysis, and land and deal strategy. They are pros. They are highly paid and they get a net profit interest in their discoveries, so they'll be tough to hire away.

I will tell you a little more about our methods because I want to convince you that winning with lower finding costs is not a matter of luck, it's a carefully constructed program. First, you pick a niche in the oil business where you can compete. With a smaller size company it's not in the Frontiers, or in the Foothills, or in the Deep Devonian. It's not with regional seismic work. It has to be some geographic area, some part of the stratigraphic section, and has to be subsurface mapping of well control. So, you have to become the best at subsurface mapping.

The first thing you do is pick the best stratigraphic geologists you can find. But that doesn't put you very far ahead. A lot of companies have good geologists. However, most companies turn them loose to sink or swim. They are supposed to do their own log analysis, sample interpretation, core description, DST pressure interpretation, reservoir engineering, economics evaluation. You name it, the geologist is supposed to be able to do it. He's like the dog that Dr. Johnson saw walking on his hind legs. "He doesn't do it well, but the surprising thing is that he does it at all."

We got the bright idea that if we could provide the geologist with the most sophisticated, skillful, high tech measurements, he would be free to apply his skills to a maximum. So, we have the best log analysis experts in the industry equipped with computers, fancy equations to deal with clays, pore size, grain size, and hole diameter, and access to any consultant in the world for special problems. We have a team of the best specialists in the industry at sample and core description. They can take a tray of drill cuttings in sandstone, describe the grain size, clay content, cement, and mineralogy, estimate the porosity *and* the permeability, and then estimate within certain limits flow capacity of the reservoir after frac. No one else can do that. We have engineers who make DST pressure evaluations that tell us things about drill-stem tests which are nearly black magic. We have exploration *reservoir engineers* who work with every prospect to be sure we are taking full account of fluid data, pressure information, probable reservoir performance, etc. Our exploration geologists are young (with an average of only 8 years of experience), but they have enormous fire power.

This whole group works in close tandem with every other technical group in the company. In Canadian Hunter, *everyone* is in the exploration department. The drilling engineers design the wells with specific target formations in mind. The completion engineers work with the log analysts and the rock specialists to design the fracs.

In summary, we have teams—or, better, chains of specialists who develop our exploration prospects from conception to completion. Each link in the chain is vital, and no one man has ever been known to me who could do it all. There is nothing we do technically which isn't known to the major company research groups, but they have never been able to make the whole thing operational because they can't hold the teams together, and management isn't dedicated and committed to the concept. In our company,we almost don't have any turnover, so our teams have been working together, and learning together, for nearly 10 years. They are pretty hot shot groups. Several are run by geologists, one by an engineer, one by a landman. This winter, these teams had a 70% success rate in exploratory drilling. They personify what can be accomplished in achieving low finding costs. But, obviously, you can't put together such a group overnight.

However, the alternative is to go out of business. You simply cannot stay in with average finding costs.

When you go back to Calgary, what do you look for? You should encounter an exploration staff which spends about $2–4 million per year per geologist. If you're spending less than $1 million per geologist, you have too many. Which are the good ones? That's easy. Just like baseball players. How many home runs have they hit? The only reason to employ geologists is to make a profit from them.

Now, if you have high finding costs, a poor success ratio, and too large an overhead, whose fault is it?

Most companies have chosen to cut staff by 25–50%. What they ought to do is fire the V.P. Exploration. Get a good man there and let *him* cut out the ones who don't fit.

You hear a lot of people saying the companies are making a big mistake firing a lot of staff. In a few years, they'll be trying to hire them all back. Maybe. I don't think so. I believe the oil companies are finally getting the message that Mr. Pickens has been preaching for several years—that they can't go on running unprofitable businesses. Staffs of a thousand can't find oil effectively. They have to get smart, get small, and get efficient. Then they can win. The dinosaurs with small brains are eventually going to die.

It's part of the conventional wisdom that now is a good time to buy production. "You can buy it cheaper than you can find it." I believe that most people who have looked seriously for production to buy will tell you, wearily, that bargains are hard to find. It takes a lot of money and, of course, the sale is made to the buyer who will accept the lowest rate of return. Still, good deals are occasionally made and you should be alert for them. IPL says they made a wonderful buy of Home Oil at $4/barrel, and I can't argue with them.

A common idea going around is a floor price for oil, and some suggest a ceiling price sliding up with inflation. Well, that may save someone's tail now— but it's goodbye oil business. We're supposed to be the last of the free enterprisers. It makes me sick to see everyone running to the government for help. The next thing they'll want is a guaranteed income.

Wallace Pratt—the one and only Wallace Pratt— said "The very state of mind of the social order as well as of the individual is involved in successful oil finding." He would have grave doubts about the wisdom of making our industry into a farmer's co-op.

I think the most widespread and economically available opportunity is in exploration—if you know how to do it. If you can't, get into some partner deals with someone who can. And be willing to pay for the privilege. A lot of oilmen can only do the *arithmetic* on a deal and don't understand the values. They think a 50% deal is twice as good as a 25% deal. 50% of nothing is zero. You've got to pick winners.

Any good athlete will tell you winning is, finally, mental attitude. It is a dedication to excellence; and a grim, steely determination. Not easy—but kind of simple.

Jean Paul Getty said it all: "The meek *shall* inherit the earth—but *not* the mineral rights."

Exploration: A Game of Chance and Change

JAMES E. WILSON
President, The American Association of Petroleum Geologists
Shell Oil Company
Denver, Colorado

One of petroleum exploration's most illustrious pioneers was the late Everett L. DeGolyer. An aphorism of his, which I like to quote, goes like this:

> "It takes luck to discover oil. Prospecting is like gin rummy. Luck enough will win but not skill alone. Best of all are luck and skill in proper proportion, but don't ask what the proportion should be. In case of doubt," he said, "weight mine with luck."[1]

Nature created only a finite number of oil and gas accumulations, but we'll probably never know just how many do exist. What we do know is that with each one that is discovered, there is one less to be found—whether its discovery was by luck, skill, or a "proper proportion" of both.

In the 114 years since discovery of the Drake well, each decade has seen its advancement of the "state of the art" of exploration and its group of discoveries. The development of new exploration tools and the acquisition of new knowledge make an interesting and instructive study. I wouldn't minimize for a

---

[1] DeGolyer, "How Men Find Oil," Fortune 97 (Aug. 1948).

1

moment the part luck plays in the discovery game, and like DeGolyer, I'd like mine weighted with luck, for the uncertainties of geology do ineed make the search for petroleum a game of chance. Technological developments work to change the odds, and our exploration capability today is absolutely superb.

I'd like to review briefly our discovery experience and the history of some of the more important exploration tools.

The "Boot Days" of Exploration

Figure 1 shows the "giant fields" whose discovery was based on surface geology and other indications such as oil or gas seeps.[2] A "giant" field is by definition one whose ultimate production will exceed 100 million barrels of oil or one trillion cubic feet of gas. While all discoveries large or small all add up, I'd like to "key" on the "giants," for not only do they form the backbone of our reserves base, but they frequently gave a shot-in-the-arm to exploration.

In the seventy years following the Drake discovery in 1859, sixty-five "giant" fields were discovered in the United States, based on the most fundamental of all exploration methods. These were the "boot days" of exploration. The "tools of the trade" consisted primarily of a hammer, a Brunton compass or a plane table, and sturdy boots; and drilling equipment was relatively primitive. But so was there a lot of oil and gas in fairly obvious and shallow structures.

I have shown the names of only a few of the famous fields, and hope that your favorite hasn't been left off.

Aneth, in the Four Corners area of Utah, is the last recognized "giant" field credited to surface evidence. It might interest you to know that none of the large fields so far discovered in Alaska has been based on surface structure. Surface geology, however, has been very important in studying the regional stratigraphy and structural framework of the basins that have been successfully explored in Alaska.

As early as 1900, several oil companies had begun to make *systematic use of geology* in exploration, and toward the end of the twenties just about every surface anticline and fault struc-

---

[2] Halbouty, "Giant Oil and Gas Fields in United States," *Geology of Giant Petroleum Fields*, A.A.P.G. Memoir 14, Fig. 1, p. 93 (Nov. 1970).

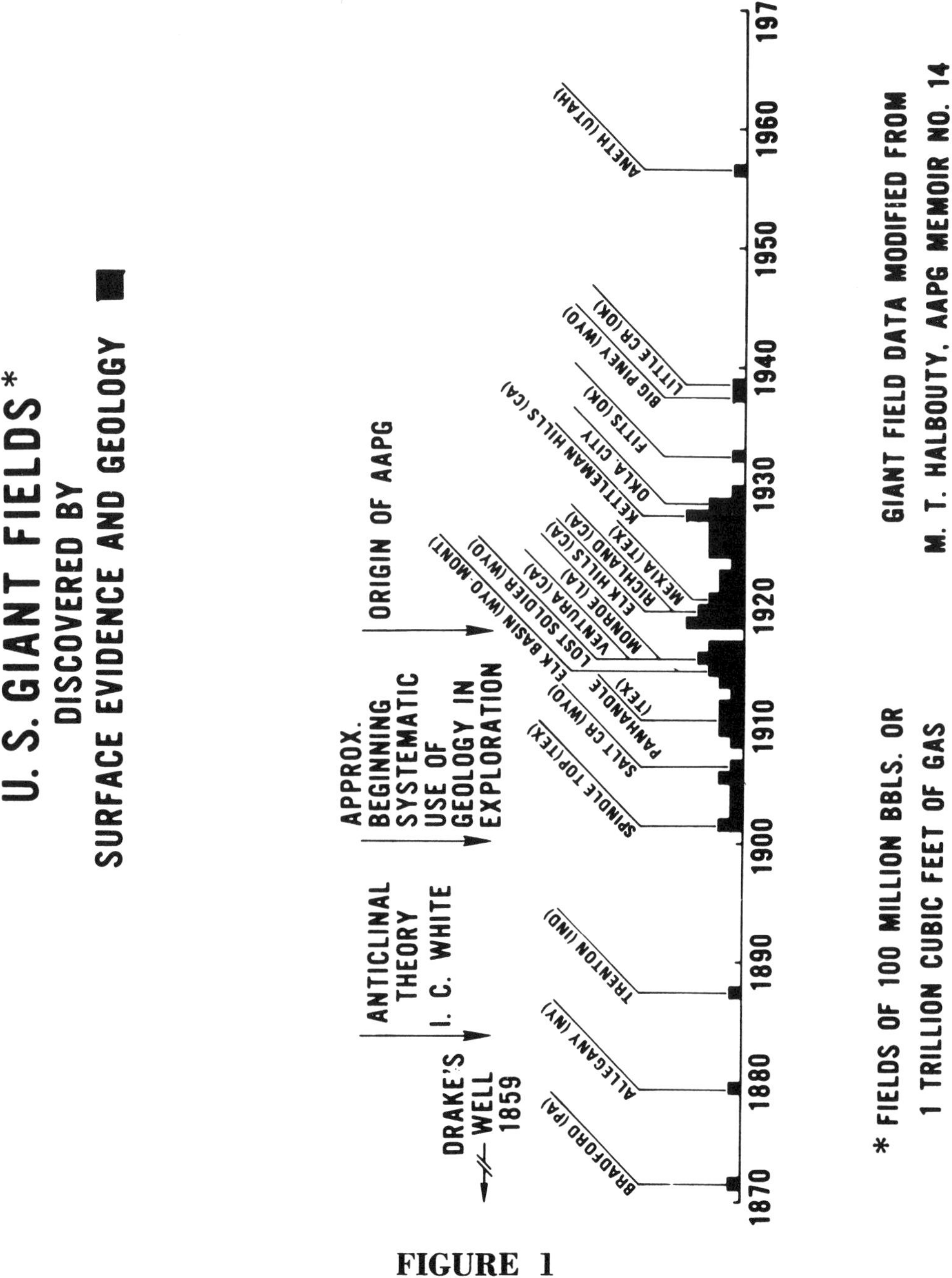

FIGURE 1

ture of any promise had been tested to the depth of the drilling capability for those days.

"Giants" by Random Choice

Figure 2 shows some new names and eighteen "giant" fields credited to what has been termed "random drilling." One of these "chance" discoveries was the super-"giant" East Texas Field

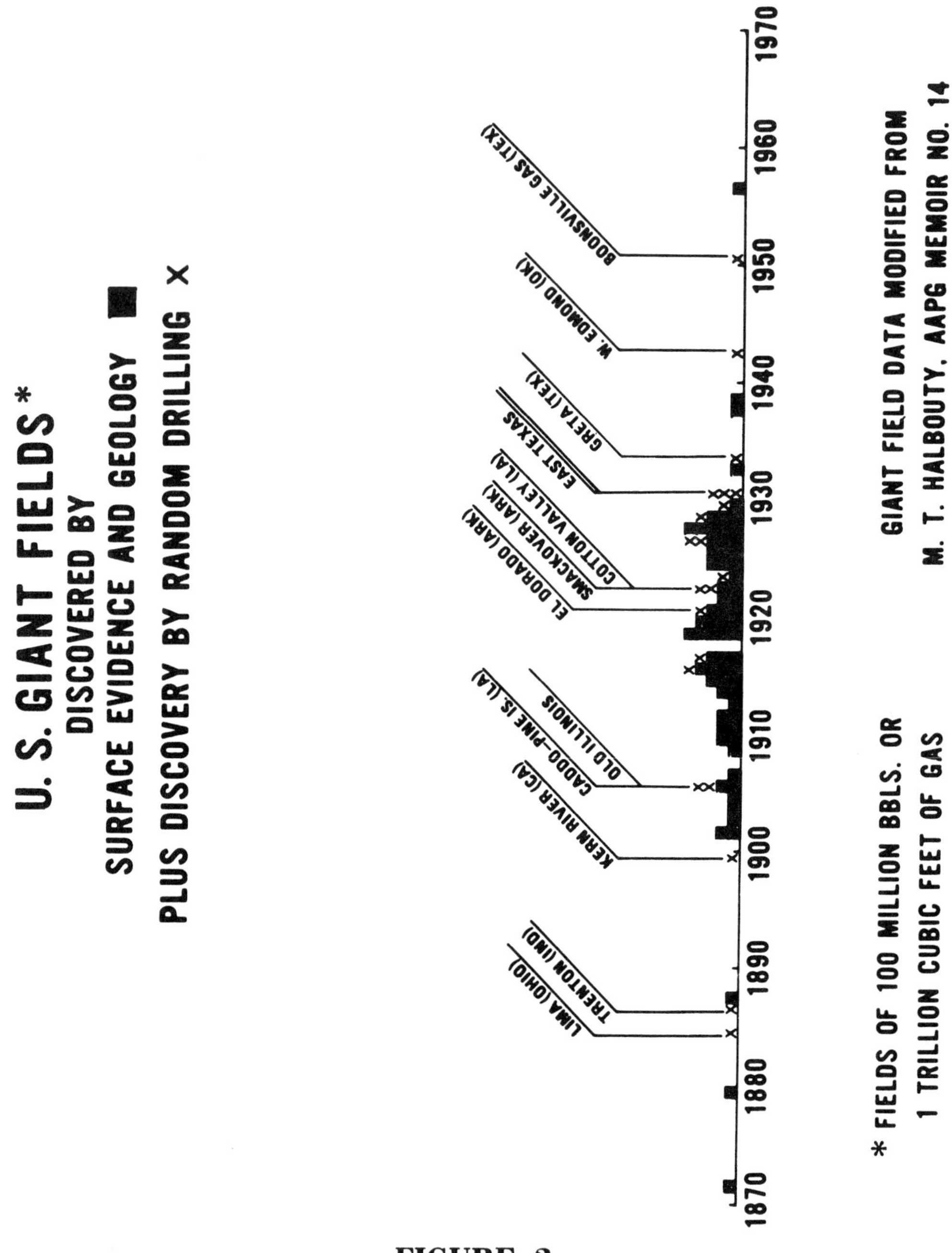

FIGURE 2

found in 1930 by C. D. Joiner. The story of "Dad" Joiner and the fabulous East Texas Field are, of course, legendary. What isn't often revealed is that Dad Joiner actually did have a geologist, Dr. A. D. Lloyd, who had recommended a location a couple miles northwest of the famous discovery.[3] As it later developed, Dr. Lloyd's recommended location would have resulted in dis-

---

[3] Rister, *Oil! Titan of the Southwest* 307 (Univ. of Okla. Press 1949).

covery of the field, but had Joiner moved his location one-quarter mile farther up dip he would have had a dry hole.

It is quite evident from the chart that 1930 was a pivotal year in petroleum exploration. The decade of the thirties saw the real beginning of instrument technology employed in exploration.

Technology Changes the Game

The reflection seismograph and electrical logging were introduced at about the same time in the very late twenties, and they began to change this game of chance quite significantly. We might first discuss the developments in electric logging. (Figure 3.)

Correlation by Curves

Electrical well-logging was conceived in France by Conrad Schlumberger and first tried out in this country in 1929.[4]

The *first electric log* was a simple single trace curve that recorded an "apparent resistivity" of the various strata to an impressed current produced by a device run in the borehole. Around 1934, a second curve known as the "spontaneous potential" was added to the logging service after it was noted that there was a natural potential difference between the fluids in the formations and the drilling mud in the uncased hold. The "SP" curve quickly came to be viewed as an indicator of porosity.

The Gulf Coast contains a thick sequence of sands and shales that all look very much alike in the drill cuttings. The new Schlumberger log showed that the various layers exhibited visible, although oftimes only slight, differences in electrical characteristics so that *"curve matching"* became a new, rapid, and acceptable *correlation tool.*

A significant thesis describing how identification of *formation fluids* could be made from electric logs was published in 1942 by G. E. Archie of Shell Oil Company.[5] Archie's concept was based on the fact that subsurface formations normally contain salt water and that salt water is conductive, whereas hydrocarbon

---

[4] API, *History of Petroleum Engineering* 522, 539 (1961).

[5] Archie, "The Electrical Resistivity Log as an Aid in Determining Some Reservoir Characteristics, 146 *Petroleum Development and Technology* 54-62, Transactions American Institute of Mining & Metallurgical Engineers (1942).

TECHNICAL DEVELOPMENTS
AND
PETROLEUM EXPLORATION

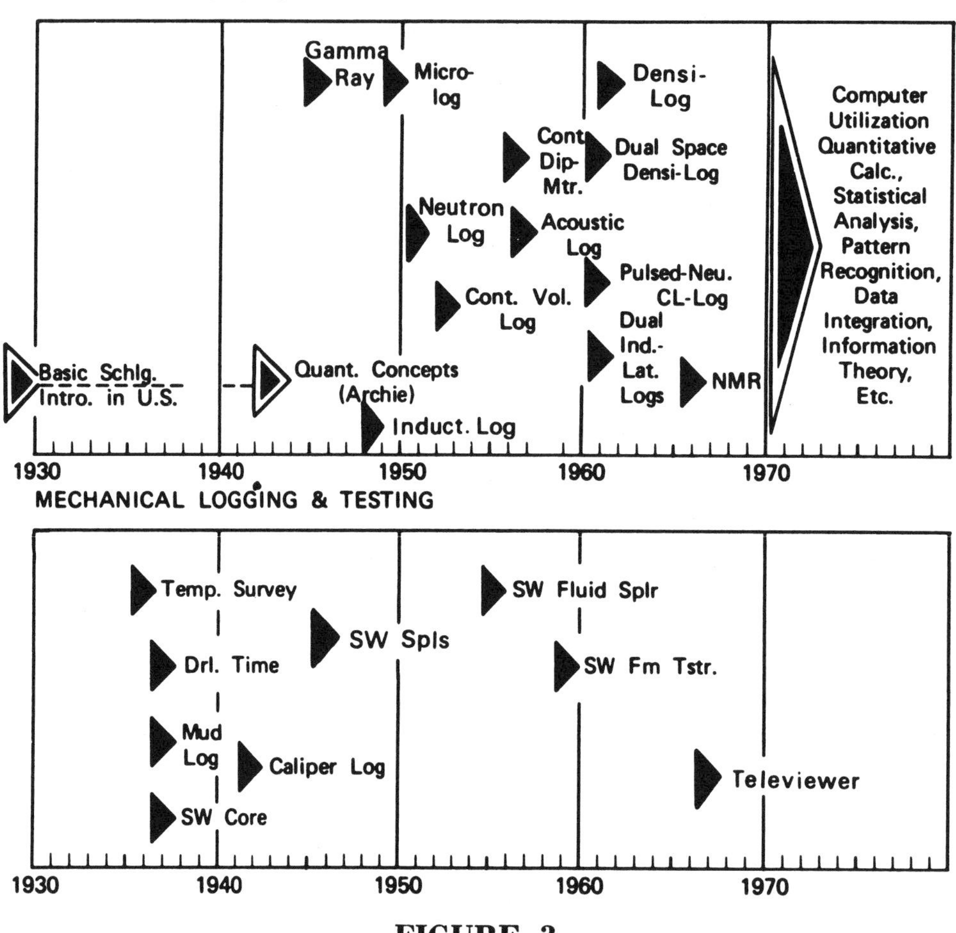

FIGURE 3

substances are less so. He pointed out that if one knew something of the salinities, and hence the conductivity of the fluids contained in the sediments, then it would be possible to calculate from the "apparent resistivity" the probable presence of oil or gas in particular strata. This "new dimension" greatly accelerated the use and demand for electric logging and opened the way for further *quantitative* use of the logs which had theretofore been used primarily for correlation work.

Chance Modified by Change—An Example

Shortly after World War II, Shell mapped with the seismograph a large anticline in the Anadarko Basin of western Oklahoma. Unfortunately, Continental Oil had already drilled a test on this structure, *but* their well had not reached the Springer sandstone, which was the first known objective and lay at a challenging depth for those days of around 13,000 feet. After assembling a lease block, Shell set out to drill a Springer test, west of the Continental old dry hole. Because of mechanical problems, this test also did not reach the Springer. No shows had been recorded during the drilling, and preparations were under way for abandonment. But because the well was deep and expensive, the electric log was sent to Gus Archie, Shell's log expert in Houston, just to "have a look," anyway. His analysis indicated some anomalous resistivities in certain zones of a "granite wash" section between 9,000—10,000 feet, and he recommended that tests be made of these formations. This meant additional expense, and the "granite wash" section had no productive history, but the decision was made to go ahead and do the testing anyway. The result was the discovery well of the Elk City Field—a "giant" gas-condensate field. As it turned out, the "granite wash" section around the Continental well was too "dirty" to have produced even if they had seen shows and attempted a completion. A footnote might be added that while Continental had surrendered its leases after their dry hole they *had* bought minerals on the structure and participated in the eventual development of the field.

Figure 4 shows that fifteen or more giant fields were added in the late thirties and early forties credited to *subsurface geology*. New or improved structural interpretations were made possible by electric log correlations as well as quantitative interpretations for the presence of hydrocarbons. So that discoveries by *subsurface geology* can largely be ascribed to the electric log.

The Borehole Now Gets the "Full Treatment"

Referring to Figure 3, we see that a panoply of new devices and improvements had been developed in both the fields of electrical and mechanical logging and testing.

The ability to evaluate productive potential has now reached a high degree of certainty in geologic provinces where pertinent subsurface data are available. An almost overwhelming array of physical and chemical parameters can be obtained with the logging and testing tools today.

Use of the computer permits analysis and integration that would not be possible otherwise. One of the most promising ap-

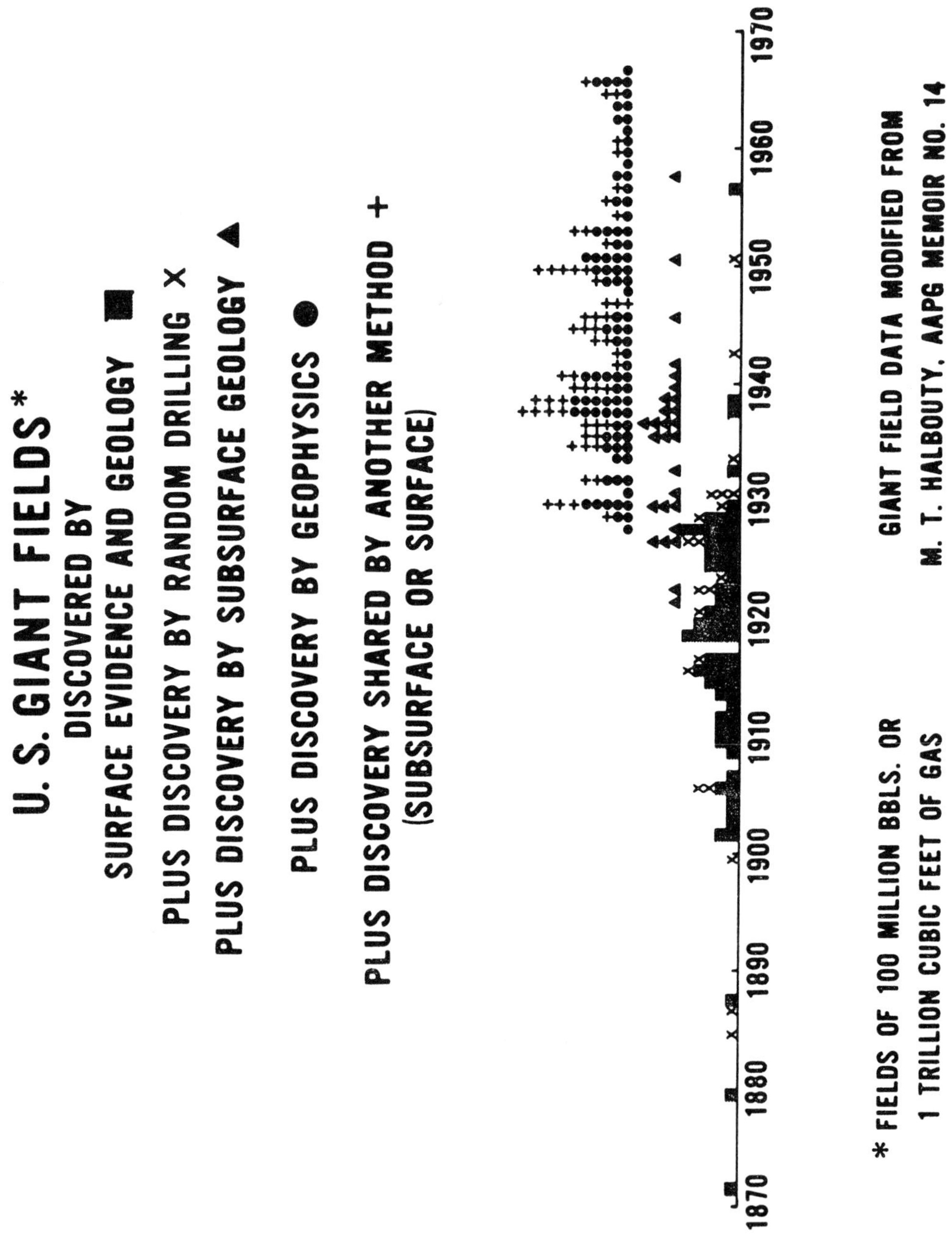

FIGURE 4

plications is the integration of borehole data with geophysical evidence; so it might be well now to go back and pick up the story on geophysics. (Figure 5.)

The "Doodlebug" Days

Although gravity and magnetic and refraction seismic work had been employed in petroleum exploration in increasing amounts

TECHNICAL DEVELOPMENTS
AND
PETROLEUM EXPLORATION

GEOPHYSICAL

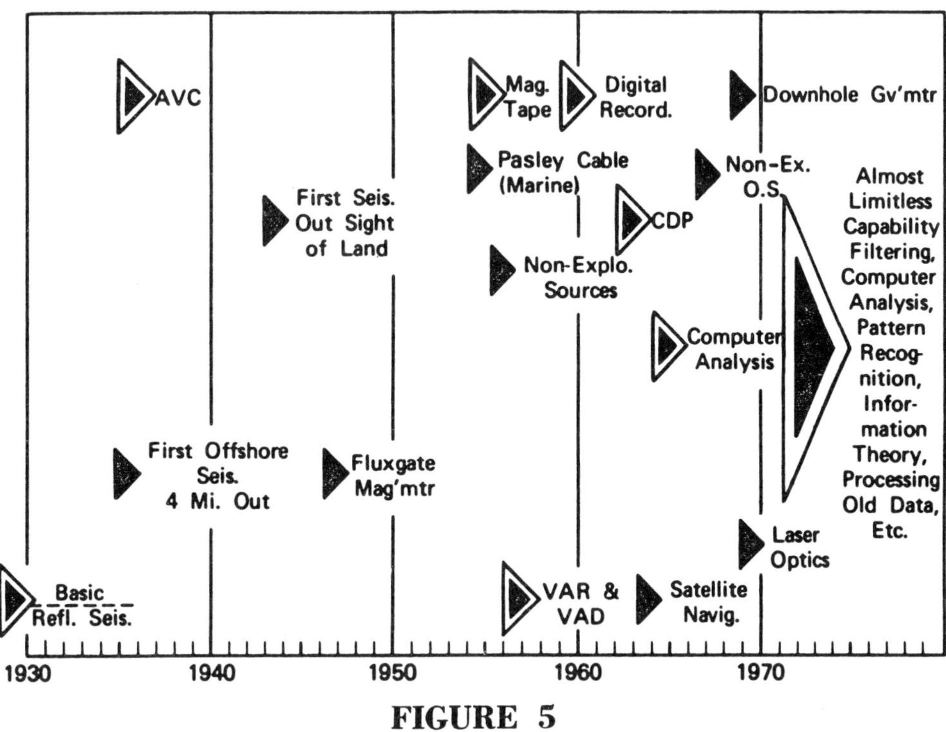

FIGURE 5

for almost a decade before 1930,[6] the *reflection seismograph* has been the principal geophysical tool for approximately forty years. The first reflection seismograph discovery was the Maud pool of Seminole County, Oklahoma, on work done in 1927.[7]

Good seismic reflecting horizons were found in Oklahoma, but in the lower Gulf Coast the records seemed a haphazard jumble of discontinuous reflections that defied interpretation. In 1929, the Geophysical Research Corporation developed a *dip method* of shooting that began to utilize the physics of sound waves in a more comprehensible fashion and greatly improved the "geologic sense" of the seismic data. The Geophysical Research Corporation,

---

[6] Historical Development of Exploration Geophysics, reprinted from 5 Geophysics No. 3, Pt. 1 (July 1940).

[7] Geophysical Society of Tulsa, Pamphlet *History of the Society of Exploration Geophysicists*, printed for the International Petroleum Exposition, Tulsa, Okla., May 15-22, 1948.

by the way, was organized by DeGolyer and others, and Eugene McDermott headed one of its first field crews.

Around 1935, *automatic volume controls* were developed to suppress any very strong seismic signals that tended to obscure or confuse succeeding reflections, and, in similar fashion, they amplified the weaker signals. For almost twenty-five years technological development in the seismic field was confined primarily to improvements in the recording apparatus and field equipment.

And, although there were no great new "breakthroughs," this new "tool," nevertheless, became exceedingly popular *and* successful.

The Rise and Fall of Seismic Effort

Figure 6[8] is a plot of the number of seismograph crews working per month in the United States since the early thirties.

Following World War II there was a pent-up need to explore. A proliferation of seismic crews resulted, and I don't suppose there was a basin in the lower 48 that escaped some sort of seismic invasion during this period. There was an almost exponential growth until it "peaked out" about 1952. Apparently, disappointment in discovery coupled with difficulties in solving shooting problems in a number of geologic provinces invoked rather abruptly the law of diminishing returns.

The steep down-turn was the compound effect of these two factors plus several others, including: a shift of exploration to foreign areas, a repression of economic incentives, and restrictive governmental policies.

In the late fifties and early sixties new technological developments gave encouragement that some of the difficult shooting problems might now be solved, and renewed attacks were made in numerous provinces. With this increasing technical sophistication came an arresting of this down-trend in seismic activity.

The Era of "Technical Sophistication"

Referring to Figure 5, we see that there were several important developments in the field of geophysics. Perhaps the most significant one was the recording of seismic signals on *magnetic tape*. This permitted later replay through selected electronic filters and

---

[8] Allen, "Geophysical Activity in 1969," 36 Geophysics No. 1 (Feb. 1971).

YEARS
1935
1940
1945
1950
1955
1960
1965
1970
SEISMIC ACTIVITY
CREWS PER MONTH PER YEAR
SOURCE: SOC. EXPLOR. GEOPH.
NAT. GAS ACT
W.W. II
FIRST OFFSHORE FIELD
KOREA
O.C.S. LAND ACT.
PHILLIPS DEC.
SUEZ CRISIS
IMPORT CONTROL
F.P.C. AREA RATE
DIGITAL RECORD
SUEZ CLOSED
TAX REFORM ACT
700
600
500
400
300
200
100
0
CREWS PER MONTH

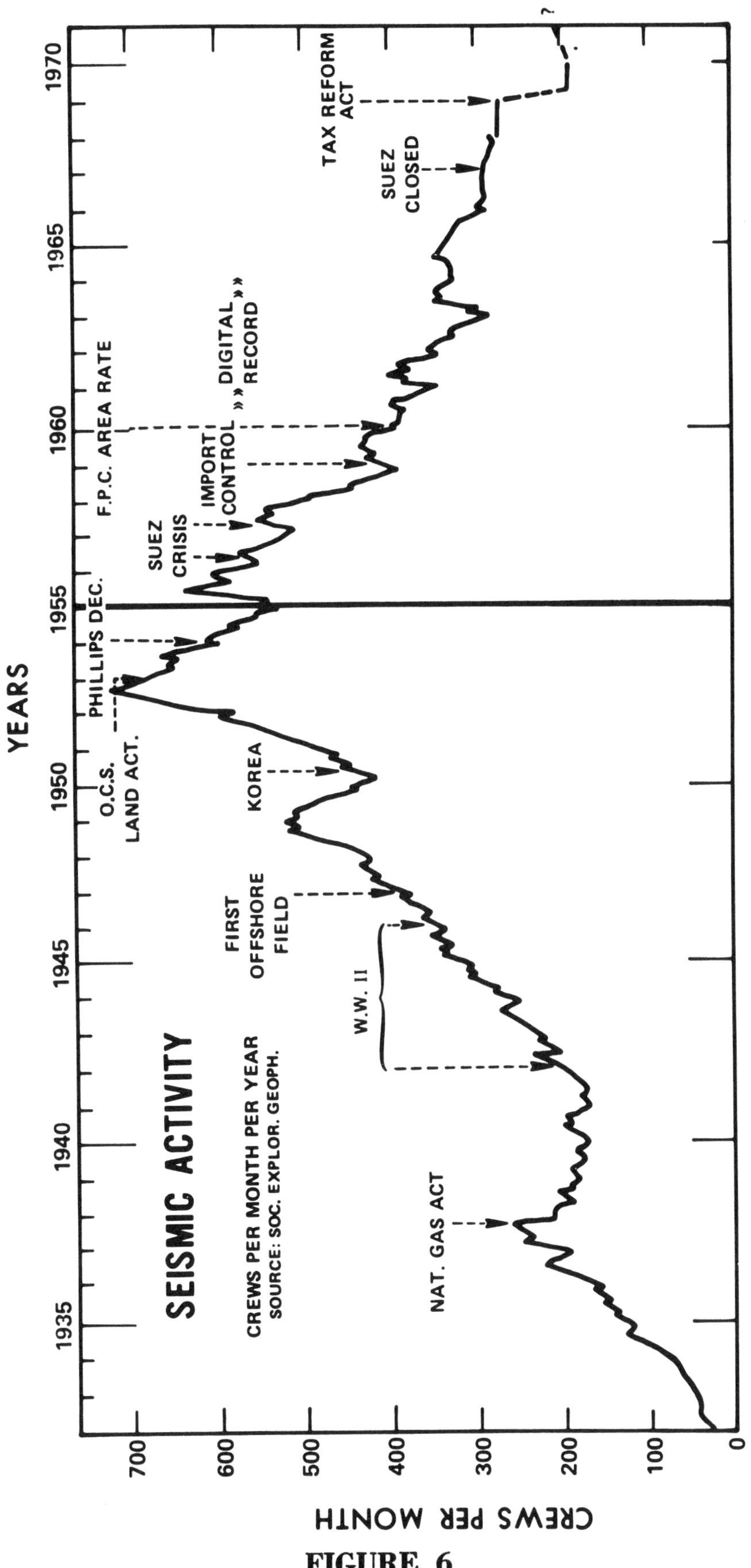

FIGURE 6

manipulation of visual display of the records. Closely following the use of tape came *digital recording* and today there is hardly a crew in the United States that does not use this technique. This form of seismic computation and "massaging of data" requires tremendous computer time, and while it is more costly, the flexibility and interpretative potential are likewise tremendous.

So, today, we have an almost limitless capability for computer analysis of seismic data. To make the fullest use of this seismic capability requires that physical parameters obtainable from borehole logging be combined into the most imaginative and accurate geologic "models." A new era of exploratory success lies within our technological capability, if economic and other political and administrative factors will operate more favorably.

In Figure 4, we see that some 80 "giants" were discovered by geophysics—primarily the reflection seismograph. It is somewhat unsettling that the Era of Increasing Sophistication beginning in the early sixties has not produced more major discoveries. One must keep in mind, however, that economic incentives and acreage availability have had a serious deterring effect.

In addition to the number of "unassisted" discoveries made by geophysics, it is also quite interesting to note the number of giants whose discovery was "shared" by more than one method.

The impact the combination of the seismograph and electric log had on exploration in the late thirties and early forties is very evident. Although Halbouty's plots stopped with 1966, there have been perhaps no more than five or six fields discovered that have been recognized as "giants" since that time.

While geophysics and electric logging did indeed play a leading role in technical exploration, there were important advancements on other fronts in both "software" and "hardware." Foremost among these was the rapid advancement of computer science, without which certainly seismic capability could not possibly have reached its present stage of sophistication. The digital computer has made possible the advancement of many technologies related to petroleum exploration.

Advancement on Other Fronts

A discussion of any of the topics shown in Figure 7 would alone make an interesting and lengthy article, but I shall give them only a "light brush" treatment.

TECHNICAL DEVELOPMENTS
AND
PETROLEUM EXPLORATION

DRILLING DEPTH RECORD

9,750' 12,780'	15,000' 16,650'	20,520' 22,570'	25,340'	25,600' 30,050'

GEOLOGIC CONCEPTS & MISC. DEVOL.

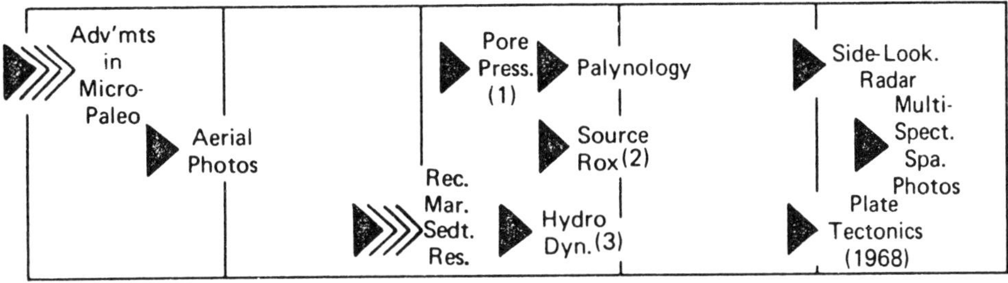

OFFSHORE CAPABILITY

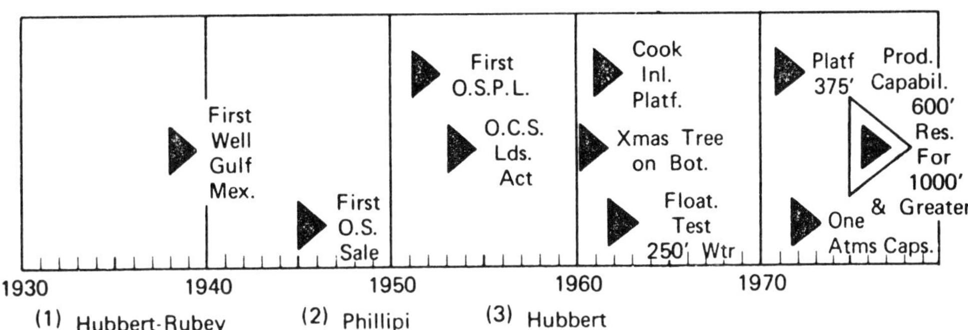

1930   1940   1950   1960   1970

(1) Hubbert-Rubey   (2) Phillipi   (3) Hubbert

FIGURE 7

Drilling Depth. The deepest well in 1930 was 9,750 feet; in 1972, the record was 30,050 feet. This is a depth record increase of about 5,000 feet per decade. Thirty thousand feet would penetrate the objective section of many basins, but what is needed is an ability to drill to such depths more routinely and cheaply.

Offshore capability has accelerated tremendously just in the past ten years. The first full-scale test drilled from a *floating drilling rig*, which could be controlled and completed as a producing well, was drilled in 1962 in 250 feet of water. This floating-drilling capability has now increased to the 1,000-foot water depth range. A bottom-supported *drilling and producing platform* has been installed in 375 feet of water in the Gulf of Mexico, and I understand that a prototype has been designed for 800 feet of water in the Santa Barbara Channel. Shell is currently testing a *one-*

atmosphere capsule system, where men can work under "shirt-sleeve" conditions in water depth of 1,500 feet and possibly much deeper.[9] The frontiers of water-depth operations are perhaps more a problem of economics than technology. That is, if the reward is sufficiently promising, the industry will solve the technical problems.

The *origin and migration* of oil has fascinated geologists from the earliest days. Everett L. DeGolyer once wrote that it wasn't as important to understand "origin" as "habit."[10] His analogy was that if one wanted to hunt quail it wasn't so critical to know how long it took them to hatch—or that they even hatched from eggs, but one had better know something about the *habits* of quail. Nevertheless, researchers have been working on *source rocks and the migration* of oil and gas for a long time in hope that there might be something in this alchemy that would aid in the discovery of significant accumulations. One of the earliest and now more accepted theories on source rocks was published by G. T. Philippi in 1957.[11] A better understanding of origin and migration is important to explain why one basin may be "richer" than another or why some areas are "gas prone" and others more "oil prone." It may also help explain the relation between oil and gas migration and the formation of a trap by geologic processes.

From the early days, geologists had looked *at* and described sedimentary features of recent origin such as deltas, offshore bars, river channels, etc., but in the late forties they began to study also the *anatomy* of these features. Logging devices for core holes and radiocarbon dating made it possible to obtain the third dimension as well as the time factor of typical deposits of recent geologic origin. A better knowledge of the structure of these recent sediments thus permitted a more accurate interpretation of the older geologic features of similar origin. This was the commencement of *modern stratigraphic geology* as applied to petroleum exploration. The importance of modern stratigraphic knowledge is that an accurate and imaginative "model" of geo-

---

[9] Collipp and Sellars, *The Role of One-Atmosphere Chambers in Subsea Operations,* Preprint OTC #1529, Offshore Technology Conference, May 1-3, 1972.

[10] See N.1 *supra* at 98.

[11] Philippi, "Identification of Source Beds by Chemical Means," Geologia del Petroleo 24-38 (Mexico D.F. 1957), 20th International Geological Congress Proceedings, Section III, Mexico City, 1956.

logic relationships must exist so that the seismic instruments can be "tuned" to bring out these relationships and thus define the "trap."

Several other *geological concepts* have been advanced that have been quite significant in exploration planning and evaluation.

Hubbert and Rubey[12] described in physical and mathematical terms the relationship which the *pressure of fluids* contained in the pores of rocks has on the strength of those rocks and, consequently, of their response to deforming earth stresses. Hubbert's comprehensive analysis of *hydrodynamics,* published in 1953, described the relationship of the movement of formation waters to the entrapment of oil and gas.[13]

The Future

It is ironic that at the time of our greatest need for domestic petroleum, our exploration is at the lowest level in some thirty years. The irony becomes almost a national disgrace when we recognize that our technological capability is absolutely superb, and what is sorely needed are economic incentives and an aggressive lease-sale policy by the Federal Government.

The uncertainties in geology will always be as the Creator made them. How the problems of geology are solved will depend on how well explorationists apply new ideas, new knowledge, and new technology. Petroleum exploration will always be an exciting and challenging game—a game of chance but also of change.

---

[12] Hubbert and Rubey, "Role of Fluid Pressure in Mechanics of Overthrust Faulting," 70 G.S.A. Bull. No. 2, pp. 115-205 (Feb. 1959).

[13] Hubbert, "Entrapment of Petroleum under Hydrodynamic Conditions," 37 A.A.P.G. Bull. No. 8, pp. 1954-2026 (Aug. 1953).

PETROLEUM EXPLORATION AND THE ROLE OF THE GEOLOGIST

Reprinted by permission of the Society of Exploration Geophysicists from *Geophysics: The Leading Edge of Exploration*, v. 7, no. 8 (August 1988), p. 26-28.

The changing role of the petroleum geologist

By JOHN F. BOOKOUT
President and C.E.O.
Shell Oil Company

At this assembly of the AAPG, the past, present and future roles of the geologist are represented. I am clearly part of the past, since my own career started almost 38 years ago. But today I am particularly interested in the future role of the petroleum geologist and I am optimistic about that future, despite the present state of the industry. It sometimes appears as though the glory days are behind us, since exploration activity peaked a few years ago. Yet I am reminded that for most of our industry's history, the future has been regarded as uncertain; and in those moments when the future was not regarded as uncertain, maybe it should have been.

I am optimistic about the future because I am mindful of the way technological advances have revolutionized this industry and this profession on many occasions. Geologists have seen their profession transformed into a hi-tech business. Along the way, the wonder of our science has always been that technology breakthroughs come as new thoughts and concepts that can be put to work immediately. We do not have to build huge plants or equipment. We just put our ideas directly into action. But, before discussing the future, let's review the history of what our science has accomplished to date.

We all know that the anticlinal theory made the science of geology vital to the oil industry. I. C. White's anticlinal theory was not only a scientific breakthrough that a lot of us should still use as a daily reminder, but he was able to put it right to work and to make money with it. Of course, it was an easy progression in our thinking from White's surface anticlines to anticlines and structural traps of all kinds hidden in the subsurface. I think back to the time when often the only tools

This article is an edited version of Bookout's remarks made to the 1988 Annual Meeting of the American Association of Petroleum Geologists in Houston on March 21.

available to geologists were the existing well logs.

For a long time, the geologist and geophysicist worked independently of each other. An important synergy developed between them as seismic technology progressed from point plot sections to record sections, to analog tape recording, to stacking with greater and greater multiplicity, to 3-D recording and imaging. The geologist and geophysicist were drawn closer and closer together.

We still have a long way to go in this essential field of trap mapping. For example, how soon will we be able to see salt/sediment interfaces clearly around salt domes, or get proper signals down through surface volcanics, glacial tills or shallow limestone beds? Our seismic imaging within complex structures is still very primitive. We often can only guess at the 3-D integrity of our fault traps. As we solve these problems, we will have more exploration opportunities.

Equally dramatic advances in the field of stratigraphy paralled this post-World War II explosion in descriptive structural geology. Stratigraphers had always done well vertically; now they started to think in terms of genetic stratigraphy. I believe that the breakthrough which expanded our thinking in environmental stratigraphy was the serious application of the fundamental principle, *the present is the key to the past.*

Until work in the recent, which really bloomed in the 50s, we simply contoured out reservoirs to zero in all directions from the well control, without any understanding of the genetic origin. Think what opportunities will emerge in this essential field of reservoir stratigraphy if we can solve the problems of postdepositional diagenesis. We have learned about the things that destroy our reservoirs — calcite cement, quartz overgrowths, vadose silts, salt plugging — but our ability to predict the distribution of these destructive barriers is still primitive.

I remember when, in my company, the

organic chemists and geologists met each other for the first time. Out of this marriage we learned to identify source rocks and to match common oils with their source rocks of origin. We now measure temperature histories of source rocks, allowing us to describe areas of immaturity, oil windows, gas windows and areas where temperatures have burned up all movable hydrocarbons.

Using these tools gives rise to a whole new set of questions that we didn't even know to ask before. Where did all the Woodford oil go in the Anadarko basin? When and where does the Monterey shale yield low gravity oils versus higher gravity oils? How did the Phosphoria oil get to the Minnelusa reservoirs in the eastern Powder River basin? Are we sure it is Phosphoria oil? Is there an oil window in the gas prone central Indus basin in Pakistan? We will find a lot more oil if we have the insight to figure out what the measurements are telling us.

How far will we be able to go in this essential field of oils and source rocks? Will we one day be able to measure with confidence all of the oil generated in a basin, trace its migration paths and capture all that is trapped? Will we learn to better understand migration and what surface seeps really mean? Actually, the whole field of surface hydrocarbon interpretation is still in its infancy.

Looking back, who would have guessed that our continental drift theories would bear such useful fruit? Plate tectonics, the interaction of lithospheric plates, genetically explains our basin origins. We now have a sensible, genetic classification system for the world's sedimentary basins. Even more important, the genetic origin of the basin is the vital key to the internal tectonic style of the basin. This is fundamental in understanding its traps and mapping them properly. I don't know how many years I thought the San Andreas was the only strike slip fault in the world and that Oklahoma was all vertical tectonics.

The late 60s brought us the seismic

velocity anomaly termed "bright spots." Since the bright spot was discovered, not invented, it came to different people at different times. Bright spots have helped the industry find a lot of oil and gas, but perhaps their more lasting contribution will be the spark they have set off in the whole field of stratigraphic geophysics. In technical sessions these days, it has become extremely difficult to distinguish the geologists from the geophysicists. The seismic record section is the common tool of both. Our answers are not complete until we can color our sections with the proper lithofacies and, hopefully, with a little porosity information. This is where seismic stratigraphy is making great contributions — and we haven't even scratched the surface.

What all this suggests to me is a promise for continuing progress in our science. With this in mind, let's consider two things: what the 1980s have taught us that our technology had not prepared us for, and the implications of this history on the future role of the geologist.

The 1979 world oil price escalation was followed by a burst of activity in the oil and gas industry; and, in fact, the US industry increased spending during 1979-82 by $85 billion in the lower 48 states. The $50 billion exploration portion of this translates into about 600 more seismic crews and 2000 more wildcats than the projected amount based on the industry's historical background. As a result of this extra effort, we added more oil and gas than our previous 10-year historical average; so for seven years, industry was able to halt the decline in oil and gas reserves and production that had been evident since 1970. Now, as we look over our shoulder at the financial performance of that activity, the accomplishment loses some of its luster. What made sense with oil prices at $33/bbl (and climbing) provides an unsatisfactory rate of return at today's price level.

What went wrong? We know that industry missed on its price premises. The dramatic oil price increases of 1979-80 raised long term price expectations to a level that hindsight shows to have been unrealistic. To a great extent, technical geologists should not be held responsible for the industry's price expectations. The other thing that went wrong was that the industry allowed its technical risk assessments to violate its economic risk capacity. We know there is no unique answer to a prospective oil field puzzle before it is drilled. But there is the issue of how, what and when one begins to drill. To illustrate, let me describe the well known three types of geologists that we tell our recruiters to look for.

• The Optimist: this fellow sees drill-sites everywhere. Predrilling, you cannot convince him that oil cannot be down there. Someone once estimated that every cubic mile of sedimentary rock, outside of Communist countries, had been recommended to me for lease purchase at one time or another.

• The Pessimist: he will not drill anywhere. The reasons are many: too small, too deep, not shot enough, land deal not quite good enough, etc. I remember after much argument over a drillsite with one such geologist, I asked him in exasperation, "Where would you drill?" His answer: "If we must drill, how about another infill well at Bay Marchand?"

• The Realist: you and me. The geologist who knows the perfect balance between economic risk capacity and the technical risk of the project. He is quick to act on new ideas and uses all the tools available but knows their limitations and never force fits them because they happen to be in vogue.

During the boom, I suspect we all became overly optimistic. To some extent, we forgot that even $100/bbl oil will not close

---

> *Taken together, these things should be reassuring to the geologist. There is plenty of oil to be found, and there will be a ready market for every barrel.*

---

a contour that is not there. Even $10/1000 gas may not take enough clay out of a sandstone's matrix to make it a reservoir.

However, I suspect the tide has gone too far in the other direction. Last year's average oil price of plus-or-minus $17 equates in real terms to the 1973 price level. That was a year when by our estimates industry earning power exceeded 5 percent real — perfectly satisfactory. Yet the current domestic industry activity level is well below that of 1973, when industry ran 50 more seismic crews and drilled 2000 more wildcats than in 1986. While I do feel that our current level of activity is too low, the boom experience has demonstrated convincingly that we cannot just "throw money at it."

There are two big unknowns in determining what this activity level should be: what is the future price of oil, and how much more of it can be found at a profit? If we could answer those questions accurately, good planning strategies and operating tactics would come easily. However, we all realize that we will never be able to answer those questions with perfect accuracy. Furthermore, the problem is even more complex because the answers are interrelated.

Let me address the last question first. (I think this might be the easier of the two.) Once a year my company asks its geologists and geophysicists to develop an industry volume forecast. We arm them with our most current price premises and ask for two volume forecasts: the next 20 years and the absolute ultimate.

The results are encouraging. With respect to volumes to be found in the free world, we estimate over 250 billion barrels of oil discoveries in the next 20 years. We conclude that there is a large enough resource base in the US to support discoveries in the next 20 years equal to the oil equivalent we found in the last two decades — and this will use up only half of what is yet to be found. This implies that US production, on the average, will slowly decline over the period. Taken together, these things should be reassuring to the geologist. There is plenty of oil to be found, and there will be a ready market for every barrel. There is no question about the exploration opportunities involved.

Now for the question of future oil prices. (Please do not lean forward in your seats in anticipation of my answer: I do not know what future oil prices are going to be.) Of course, we can calculate a rock bottom price with a little research into OPEC lifting and transportation costs. If we make a series of assumptions with regard to alternate fuel costs and timing, we can calculate a top price. The low price would make it difficult for the domestic industry to operate. But the experience of the last few years indicates that prices under $10/bbl appear unsustainably low. And we have learned, I believe, not to expect the high prices of the early 80s on a sustainable basis. Meanwhile, we have seen that even at today's $15-$20 oil, geologists can manage profitably.

What we can expect is that the price of oil will probably continue to be volatile, oscillating over the whole range I have just mentioned. For an exploration and development venture, we need to know the value of the reward out 10 years, which I use here as the half life of our average project. So what should we do?

One strategy for dealing with this outlook comes in two parts. The first is to work with more than one price outlook. With all the variables involved, it is not very difficult to make three price forecasts. More than three are possible but unmanageable. Less than three will not exercise enough options.

How will this be helpful? The US Department of Interior will not, of course, accept three checks with different amounts for the same tract in a sealed bid sale. I don't know any drilling contractor who will give me three different day rates depending on oil prices 10 years from spud date. But as one evaluates projects under different assumptions, a value order will emerge. Some will look better, some worse and some very sensitive to price, perhaps threatening a gambler's ruin.

The second part of the strategy is to have a well thought out long term investment plan, sensitivity-tested. That means resisting accelerated spending during price surges and having the conviction to avoid sharp contractions during downturns. We have seen plenty of both over the last 10 years.

I believe that future oil fields at our historical finding costs are likely to be profitable. With price volatility, it would help to find those new fields on a regular yearly basis. I think that is called "dollar averaging."

Let's turn now to the future role of the geologist. Throughout the history I have described, the geologist has remained the integrator and lead scientist in bringing forth exploration prospects and plays to management. But the requirements of this role have obviously changed over time. At the beginning of the period, the geologist was limited by the data he had available. As new tools developed, the availability of data grew exponentially. With this, the integration of the other sciences I have already mentioned has become an even greater imperative. As this trend continues, will the geologist be able to remain the ultimate integrator and playmaker? Will his role change?

Think for a moment about what we know: petroleum geologists will face an uncertain environment, just as we do now. But even in this environment, there are some givens. I have already indicated that there will be plenty of oil to be found. Geologists are not likely to be opportunity-limited.

Futher innovation is a given, if we do no more than answer some of the technical questions I have posed today. The industry is continuing to spend nearly half a billion dollars every year on E&P research and development, by our estimates. Perhaps the result of that effort will be a breakthrough that leads us in totally new directions. Or it may be more likely that the advances will push back the frontiers of the many disciplines now being used in petroleum exploration — including contributions from drilling, production and reservoir engineering. We do not yet know whether the advances will come from the large, integrated company research programs, from the smaller, more focused companies, or from the academic world. But while we may not know the nature of the change, we can count on advance, both inside and outside of our science.

We can also count on further integration of the sciences. Even now, expert systems are being built to deal with financial and risk assessments as well as technical data. It would seem to me that if the geologist of the future is to continue to play his leading role as the ultimate integrator, he will have to be comfortable with this new information technology. In fact, the total integration of all these systems may well be the next great breakthrough that again transforms the industry.

It seems to me the individual geologist will have two options. He can choose the necessary and important role of specialist, a contributor of knowledge to the decision process, or he can pursue his traditional role of facilitator and integrator, but to do this he will have to be able to handle an additional order of magnitude of information and know how to use it.

Whatever course he chooses, the geologist of the future will continue to have my profound respect. Indeed, I would tell the young man or woman contemplating a career in petroleum geology that I know of no work more satisfying; that I have always been glad I chose geology as my profession; that I would make the same choice today, without hesitation, if I had it to do again; and that I am proud to be associated with a discipline that has accomplished so much, not only for this industry but for all those whose needs we serve. **LE**

The American Association of Petroleum Geologists Bulletin
V. 70, No. 9 (September 1986), P. 1166-1168, 1 Fig.

Geologic Notes

The Future Petroleum Geologist[1]

ROBERT R. BERG[2]

INTRODUCTION

In July 1985, AAPG President William L. Fisher appointed a select committee to determine the capabilities that will be required of petroleum geologists in the future. His charge to the committee was based on the profound changes and uncertainties that were beginning to be felt in the industry and would surely affect the employment of geologists and their professional practice. These changes are well known: the supply of oil had exceeded demand, the price of oil was unstable, many companies were threatened by debt and buy-outs, and corporate restructuring was underway to meet changing economic conditions. All contributed to great uncertainty about the needs and requirements of geological employment and practice.

Specifically, President Fisher charged the committee (1) to distinguish those elements of recent times that are cyclic and those that are long-term in their effects; (2) to characterize the state of the industry for the next 25 years; (3) to predict the capabilities that the future petroleum geologist should possess to meet the challenges of the future; and most importantly, (4) to define the role of AAPG and its commitments to the membership under these changing conditions.

No one can foretell the future, but it is obvious that people with a variety of geological experience can sense the direction of change and the possible effects on our profession. It is in this spirit, one of opinion and judgment rather than fact and statistics, that 13 outstanding professionals accepted President Fisher's request to serve on this committee. The members are listed at the end of this report. I add my thanks for their hard work and contributions to this report, which is not my own but that of many others.

INDUSTRY TRENDS

The recent changes undergone by the oil and gas industry are profound, but one fact remains clear and unchanging: that oil and gas will continue to be essential for many years

to the growth and prosperity of the developed and developing countries. Given this premise, the short- and long-term changes can be predicted, and with this prediction can come a feeling for the effects on geological employment and practice.

The short-term changes will be in the price of oil and gas, and these changes will be cyclic in nature, which could be predicted even before the dramatic declines of early 1986. Fluctuations in price will cause fluctuations in the level of employment. Such has been the past and so will be the future.

The long-term changes will be felt primarily in the decline of U.S. oil production, and this decline will have an important effect on geologists. Exploration will continue, perhaps also with cyclic changes, and with decreasing size of discoveries, but exploitation of existing reserves and existing fields will be emphasized. The buzzword of a few years ago, Enhanced Oil Recovery, will be the byword of the future. The success of EOR will depend on imaginative engineering, and this is where geologists will contribute most. Their knowledge of the reservoir rocks and reservoir inhomogeneities will be essential to more efficient oil and gas production in the future.

EMPLOYMENT

The continued demand for oil and gas means continued employment, but as the price fluctuates, so will the level of employment. The numbers are difficult to predict, but a qualitative estimate can be made with some available data.

The AAPG membership represents at least a significant part of those geologists employed in the industry, and the history of membership can form a basis for prediction (Figure 1). In 1980, the total AAPG membership was about 25,000, a gain of 67% in a decade, but the distribution of members by age had not changed for years in its bimodal nature. For 1980, one mode had a mean age of 55 years while the other was 28 years; this distribution was the result of low employment levels in the 1960s.

By 1985, the AAPG membership had almost doubled to 44,000, principally by additions to the younger mode, including students, juniors, and active members, so that the younger mode had grown dominant and had a mean age of 29 years whereas the older mode was constant in number but had moved to 60 years.

[1]Report received, June 26, 1986. Report presented June 16, 1986, at the AAPG annual convention with SEPM, EMD, and DPA Divisions, Atlanta, Georgia, June 15-18, 1986.

[2]Texas A&M University, College Station, Texas 77843. Chairman, AAPG Select Committee on The Future Petroleum Geologist.

1166

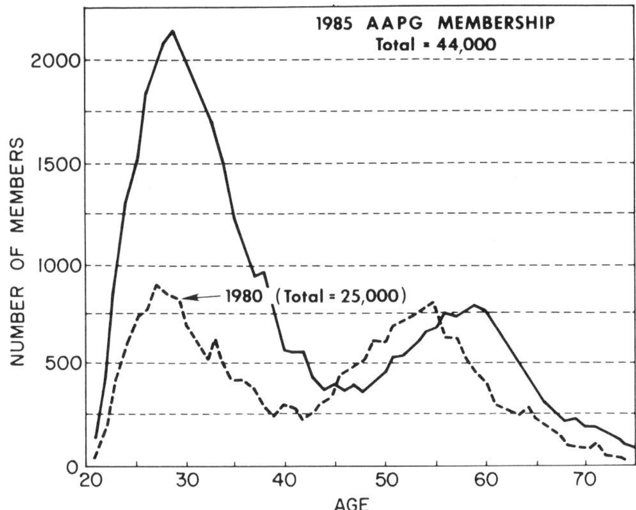

Figure 1—History of AAPG membership.

A variety of opinion exists about the meaning of this shift to a dominant younger geologist. To some, the ranks of employment had shifted to youth and imagination, which are qualities sorely needed in exploration; to others it meant inexperience and lack of training and judgment.

It is my opinion that a freshening of the industry was needed, and although it may be accompanied by some disadvantages, the result is an improvement in ideas and an increase in enthusiasm. As a consequence, exploration will benefit.

Then what about the future? The AAPG membership in 2000 will show another change, a smoothing of the distribution to a trimodal character. There will be a decrease in the present dominant mode by attrition, and certainly a normal attrition will affect the older mode. At the youngest end of the scale, however, the replacements will appear. A lag will be caused by present conditions, but eventually, the numbers will increase to near the total of 1980. As the oldest mode disappears, its place will be taken by the youngest mode of all. Beyond 2000 we expect that hiring will continue but also in a cyclic fashion.

The result of this conjecture is that hiring will rise, not to the levels of the early 1980s, but certainly to a level above that of today. There is a future for geological employment.

TYPE OF EMPLOYMENT

As oil is essential to economic well being, so will geologists be essential for the production of oil. Both short- and long-term changes in the industry will have similar effects on the types of geological employment. Short-term fluctuations in price will drive domestic producers to exploit known fields in order to maintain production without the expense of exploration. Long-term decline in U.S. production will further drive producers to more efficient methods of recovery by secondary techniques. Exploration will not disappear, of course, and will still demand the talents of the past, but the future emphasis will be on production.

In both efforts, the geologists' role is clear. They will be key professionals in predicting the size and shape of reser-

voirs for more efficient field development. They will be needed to predict the inhomogeneities within the reservoirs that are barriers to fluid flow. Many petroleum engineers today will admit that production problems are caused by the rocks and not the fluids. It is time also that geologists recognize the importance of fluids.

The imagination and foresight of the geological explorationist will have to focus on production. The geologist of the future will be forced to learn more about fluids just as the petroleum engineer has been forced to recognize the importance of rocks. Together, the geologist and engineer can find the answers to improved production from existing fields.

KNOWLEDGE REQUIREMENTS

Increased demand for production geology will require different skills for the future petroleum geologist. The problem is how these new skills will be obtained, whether by formal training, by experience, or both.

At most universities the geological curriculum is full. The essential requirements for academic training cannot be changed greatly. The future will require the same background as today: the fundamentals of geology, including field geology, as well as the physical sciences and mathematics will still be required. Even graduate studies, by definition a period of specialization, will not allow for much broadening into other subjects.

The additional fields of knowledge that will be required are easily classified by broad categories: geochemistry, geophysics, petroleum engineering, and economics. Obviously the educational requirements for geologists are already broad and must become even broader. There is no answer but continued learning beyond the years of formal education and during the years of practice. Self-study, short courses, experience, and to a lesser extent, company training will have to build on the formal training necessary for both exploration and production geology.

Professional development will become less an industry problem and more an individual responsibility. No exploration geologist today can be without a knowledge of geophysics and the chemistry of source rocks; in the same way, no geologist of the future can be without a knowledge of two- and three-phase fluid flow and transient-pressure analysis. These may be strange terms for some geologists but necessary to the understanding of production.

THE ROLE OF AAPG

The individual geologist's responsibility for learning can be aided in many ways by the AAPG. The structure and mechanisms have been established by past members and directors, and these can be maintained and expanded in the future. The mechanisms can be broadly grouped as (1) education, (2) professional development, and (3) professional activities.

Education

The AAPG educational activities have been successful for many years in providing members as well as nonmembers the opportunity to learn and relearn the principles of modern practice in petroleum geology. Schools, short courses, and field seminars have provided wide opportunity for learning at reasonable cost. Publications have also presented the latest developments in petroleum science as well as background for exploration. All of these activities can continue to serve the needs of the present members as well as those of the future. The Committee on Development Geology should play an important role in future continuing education programs.

Professional Development

Closely related to education is the need for professional contact, with other professionals as well as with the public. Through the Division of Professional Affairs, these essential activities can be expanded and directed to students and young professionals who must learn to deal with the issues of ethics, standards, and the responsibilities of practice.

Professional Activities

A myriad of other activities within the AAPG structure prepares geologists for interaction with the public and with future petroleum geologists. The Visiting Petroleum Geologist program can remain an essential tie between the knowledgeable professional of today and the professional of the future. Especially important in times of adversity is the spirit of professionalism generated within the Student Chapters. These student members are the professionals of the immediate future, and they need continued support and encouragement at this time as well as in the longer term future. Other youth activities are of critical importance to the continued health of the profession such as the Boy Scouts committees that are active in many areas.

CONCLUSIONS

Oil and gas will continue to be the economic foundation of the modern world for many years, and geologists will continue to be employed in exploration and production.

Employment opportunities will be cyclic, but they will be there.

Petroleum geologists must be prepared with broader knowledge, for there will be two roles to play: the explorationist and the production geologist. Both roles will require special skills, the ability to synthesize diverse concepts and data, and above all, imagination and initiative.

AAPG can play an important role in education beyond the years of formal academic training, but it will be the responsibility of the individual petroleum geologist to prepare himself for the future.

AAPG Select Committee on The Future Petroleum Geologist

Robert R. Berg, Chairman
Department of Geology
Texas A&M University
College Station, Texas 77843

Robert E. Boyer
College of Natural Sciences
Office of the Dean
W. C. Hogg Building, Room 108
University of Texas at Austin
Austin, Texas 78712-1199

Robert D. Cowdery
Petroleum Inc.
800 RH Garvey Building
Wichita, Kansas 67202

Gerald M. Friedman
Rensselaer Center of
　Applied Geology
P.O. Box 746
Troy, New York 12180

John D. Haun
Barlow and Haun Inc.
1238 County Road 23
Evergreen, Colorado 80439

Susan M. Landon
Amoco Production Company
P.O. Box 800
Denver, Colorado 80201

Rufus J. Le Blanc
Shell Oil Company
P.O. Box 481
Houston, Texas 77001

James O. Lewis, Jr.
10919 Wickwild
Houston, Texas 77024

John P. Lockridge
1444 Wazee Street
Suite 218
Denver, Colorado 80202

Marcus E. Milling
ARCO Oil and Gas Company
Geological Research
P.O. Box 2819
Dallas, Texas 75221

Max G. Pitcher
Conoco Inc.
Room 2906
P.O. Box 2197
Houston, Texas 77004

Peter R. Rose
Telegraph Exploration
P.O. Box 2
Telegraph, Texas 76883

John C. Shea
615 Patchester
Houston, Texas 77079

Donald L. Zieglar
Chevron Corporation
P.O. Box 7137
San Francisco, California
　94120-7137

Reprinted by permission of Exxon Company, U.S.A. (the successor company to Carter Oil Company) and by *Geotimes* from *Geotimes*, v. 3, no. 4 (1958), p. 6-7, 28.

THE OIL FINDER

by MERRILL W. HAAS [1]

Sometimes a geologist becomes convinced that oil cannot occur in an area and finally says he will drink all the oil found there.

When this happens, he should ask for a transfer. He has been in the area too long. Negative thinking never found a barrel of oil.

A good geologist should critically examine and challenge the statements and theories of others in his own mind. He should follow up with positive actions based on his ideas and sell them to management. By so doing, he will be a creator of fires instead of a fire fighter. It is the creator who has the most to gain and obtains the bulk of the oil reserves.

In the early 1930's pessimists estimated U. S. crude oil reserves at a 15 to 20-year supply. Their periodic warnings have continued during the last 25 years. Eventually they will be correct, but there is strong evidence demand will be met for many years to come.

What has been accomplished during the last 25 years? The industry has provided all needs, including the terrific demands of a World War. Accumulated production has amounted to more than three times the reserves of 25 years ago and remaining proved crude oil reserves have increased more than tenfold.

Many billions of barrels of oil are yet to be found in old densely drilled geologic basins and in the partially drilled basins. In some of the latter, well density averages about one well for each 250 square miles of basin area.

KNOWN BASINS WILL YIELD MORE

In these regions the subsurface control is so poor that the geologic maps are mainly conjecture. When sufficient information becomes available to better delineate the prospective areas for accumulation, these basins will add substantial volumes of oil to the reserve picture. The Williston Basin of North Dakota-Montana and the Paradox Basin in Utah and Colorado are recent examples which are now developing significant reserves.

Many basins throughout the world are in the reconnaissance stage of exploration now. Although some of these basins contain oil seepages, their development will be slow and far into the future. They are cited to stimulate imagination and to support the belief that a 1,000 billion barrel

target for free world oil reserves can be achieved by 1975. Present free world reserves are about 295 billion barrels.

Oil is hard to find and it has always been so. It is also expensive to find and the risks are great.

In the new and sparsely drilled basins' photogeology, surface geology and the various seismic tools will still continue to be the most successful exploration methods. In addition, the new mapping techniques, such as ratio and facies maps, are excellent means of delineating prospective areas from regional control. These will provide the guides for the use of more expensive exploration methods such as the core drill and seismograph.

In the densely drilled areas, such as eastern Oklahoma, the Illinois and Michigan basins, detailed subsurface studies will continue to be most effective. Although subsurface studies are inaugurated early in the development of a basin, this powerful exploration tool does not attain its full usefulness until thousands of wells are available to supply the necessary subsurface data.

NEW CONCEPTS AHEAD

New concepts about oil occurrence are certain to develop as more fundamental knowledge is obtained about source beds.

Since the anticlinal theory, many ideas about reservoiring oil have come forth, such as fault traps, salt domes, and stratigraphic traps. Even after the recognition of stratigraphic accumulations, who would be so bold as to state that a particular oil field was a stratigraphic occurrence and that the anticlinal structure was coincident with the stratigraphic trap? There must be accumulations today that are being explained by conventional ideas. When the true conditions are recognized, an entirely

---

[1] Merrill W. Haas is Vice President of Carter Oil Company. This article, which is reprinted from *The Link*, May-June 1958, was based on a talk which Mr. Haas presented to geologists attending a Carter training session.

6

GEOTIMES

new wave of exploration will sweep over the basin areas.

Adhering to the anticlinal theory, what geologist would recommend drilling for oil in a syncline? Yet large accumulations occur in them. Strictly following present ideas about the orgin of oil and source beds, who would drill several hundred feet into basement rock to search for petroleum? It has been done and a sizeable reserve is reservoired there.

Geologists are still "gun shy" at recommending drilling for hydrocarbons on a regional down-dip pinchout where the reservoir and source beds interfinger. The huge gas reserve in the San Juan Basin is an outstanding example of accumulations under these conditions. During the past several years in the United States major additions to reserves have been found by drilling around the edges of old oil fields. This will also happen in other regions. New concepts about oil occurrence will develop through research, imagination and sheer determination to test a theory where others have feared to tread.

Because of progress being made in well log interpretations geologists should examine periodically logs of unsuccessful wildcat wells and apply new interpretational techniques. Results will be astounding. Likewise examine logs in old producing fields. The number of shallower productive reservoirs that were missed during routine drilling operations will be surprising.

"Stratigraphic Era" Only Beginning

In many areas the stratigraphic era of exploration is only beginning. Because stratigraphic oil is hard to find, it provides the best hunting ground with respect to size and number of fields to be found.

Unfortunately, present exploration methods do not contribute substantially toward finding these accumulations. Deposition of sediments takes place in ascending order, yet oil men drill from the surface down. Few cross-sections exhibit a complete sedimentary section. They generally terminate with the bottom of well control. This has stifled geologic curiosity. It creates shallow thinkers. When imagination is applied to the undrilled section with the same vigor as to the shallower beds, large stratigraphic accumulations will be discovered.

The study of hydrodynamics or fluid pressures is a new approach in petroleum exploration. It is an aid in explaining the reservoiring of oil on the flanks of structure, which is a common occurrence in

(Continued on page 28)

THE OIL FINDER—*Continued from page 7*

the Big Horn Basin of Wyoming. It is also helpful from a regional viewpoint in determining areas where the hydrodynamic gradient is favorable or unfavorable for petroleum accumulations. The study of hydrodynamics is in its infancy, but is already contributing to the exploration effort. The integrated and team effort of the geologist and geophysicist is the best method yet devised in the field of exploration. These two great oil finding arts no longer can be separated. Blending of the two enhances the odds for success.

Fundamentally, both geologist and geophysicist must possess the same qualifications—enthusiasm, imagination, curiosity, daring and a basic knowledge of oil occurrence. The term "Explorationist" has been applied to individuals comprising these two groups. It is most appropriate and deserves common usage.

There is no substitute for hard work. A wildcat well demands the diligent watchfulness of an infant, and one does not raise children on a 40-hour week. Imagination evolved from sound geologic facts and tempered with practical judgment is also necessary. Competitors' data is equal to yours. The manner in which creative imagination is applied to interpret the data will determine the success of a venture.

Failures must not dampen enthusiasm. The odds are against explorationists from the beginning. When a venture is inaugurated the chance for a successful commercial completion is one in 44. Each defeat requires thorough analysis because it may be a link toward a future major success. Knowing that success may follow failures, this spurs the explorationist on to greater efforts.

Optimism must not go unbounded, because optimism cannot cure a dead horse.

Faith in Self Is Important

Intellectual honesty is a must. The risks are great and large sums of money are expended on the geologist's recommendations. But the geologist must never lose faith in his own capacity to locate oil. Above all he should adhere to the ethics of his profession and be an honorable citizen of his community.

Even though many in the profession are mature, experienced, and recognized as geologists, all geologists are not "oil finders." I estimate that the ratio of oil finders to geologists is about one in twenty. Fortunately the industry needs both types, but it is the "oil finder" who keeps his company alive. His ability is unique in that within limits he pinpoints areas of accumulation. He is regarded as having the magic touch. To become an "oil finder" one must develop the ability to apply his experience in all new situations. The non "oil-finders" seldom find the forest because of the trees.

BULLETIN OF THE AMERICAN ASSOCIATION OF PETROLEUM GEOLOGISTS
VOL. 49, NO. 10 (OCTOBER, 1965), PP. 1597-1600

MAXIMUM BRAIN POWER: NEW EXPLORATION BREAKTHROUGH[1]

MICHEL T. HALBOUTY[2]
Houston, Texas

All predictions of the nation's future growth indicate an increasing need for new petroleum reserves. In order for us in the United States to maintain an adequate reserve position for only the next 10 years, it is essential that we find a minimum of 45 billion barrels of oil and 275 trillion cubic feet of gas. These figures represent the equivalent of this nation's crude oil production for the past 19 years, and natural gas production for 30 years.

With these future requirements staring us in the face, it is ironical to note that in 1962 and 1963, successively, we found less crude oil reserves than we produced, and the statisticians tell us that in 1964 we barely discovered more reserves than we consumed.

In 1964 we drilled 9,250 exploratory wells and 36,200 development wells; in 1963, the figures were 8,607 and 35,471, respectively, and in 1962, there were 9,003 exploratory wells and 37,419 development wells. From these figures one fact emerges: it will be impossible for this industry to meet the reserve requirements of the future by drilling such a small number of exploratory wells.

In order to meet the demand for just the next 10 years we should be drilling approximately 25,000 exploratory wells and 35,000 development wells annually. This means a total of 60,000 wells a year beginning immediately.

It is obvious that 25,000 exploratory wells per year must require 25,000 drillable prospects per year, and this means a three-fold increase in the industry's exploration program.

Following the Spindletop discovery and into the early 30s, prior to the advent of the seismograph and the electric log, management relied primarily on the conceptual skills of its geologists to find drillable prospects. Imagination and vision were the prime guides and these were stimulated by use of at least four of the five senses. Thus, this era might be called the see, touch, taste, and smell period. The basic exploration technique was surface geology in which geologists

came into direct contact with the rock outcrop. Drillers, using these same sense skills developed by much exposure and experience, provided the log of the well; and, geologists, studying and analyzing the samples at the well site, filled in the detail of the driller's log.

With the development of the electric log and the seismograph, management came into possession of new instruments which, at first, seemed destined to replace geological brain power. Throughout history, man has always dreamed of a Utopian situation or has sought a panacea for all his problems—and exploration managers are no exception. Following a series of spectacular successes many management groups believed that these new geophysical exploration instruments had, at last, solved the problem of finding new reserves. During the 30s and 40s, any anomaly conceived through geological thinking was submitted to the seismograph and was classified as drillable only after the "black box" confirmed the geologist's concept.

This period could be termed the era during which the "black box" replaced the "brain." During this time, geophysicists, who were primarily mathematicians, physicists, or in some instances electrical engineers, performed as technicians, transcribing the data provided by the seismogram into a form which lent itself to examination and comprehension. When more and more seismic "structures" were found to be dry, management became disenchanted with the new "cure-all" exploration tool and many geophysicists found themselves presenting only those data which they could positively defend before and after the area was drilled. Naturally, as the original instrument began to fail more and more frequently, new innovations and improvements were developed. Finally, in the early 1950s, management realized that geophysical data are no better than the creative skill of the explorers who use them. This same thing can be said of structural, stratigraphic, and other subsurface geological data. With this realization, management began a concerted effort to encourage strong mutual cooperation and coordination between

---

[1] Manuscript received, June 10, 1965.

[2] Consulting geologist and petroleum engineer.

the geologists and geophysicists. It was only natural to assume that once the geologist and the geophysicist were able to exchange ideas and work together, the exploratory effort would be upgraded. During the last 15 years, many important discoveries have been made as the result of the joint work done by men in these two exploration activities, although there is still the need for a great deal of coordination to close the gap which yet exists.

At this point, however, the explorer must examine his own actions. He can not and must not stay in a "rut." Geologists and geophysicists are prone to "play it safe" by accepting precedent too readily and many use precedent as a means of dodging progressive thinking. It is too easy to be content, sit back, without "rocking the boat," and do what has already been done. With improvement in techniques, explorationists comforted themselves by anticipating that their new discoveries would be from conventional structures at depths which were previously unattainable or from virgin oil provinces at geographical localities which were previously inaccessible.

It is apparent from the industry's performance in recent years that the ability to drill to greater depths and to develop more sophisticated instrumentation will not assure the exploration success needed to meet future reserve demands. At least one warning flag has now been raised. Atwater and Miller (1965) have suggested that the reserves to be found at great depths may not prove to be as large or dependable as we now believe. Many additional data are needed before reaching a positive conclusion, but the available evidence indicates that, at least on the Gulf Coast, reservoir rocks lose favorable characteristics below 20,000 feet. If this initial observation proves to be true for provinces other than the Gulf Coast, then we must expect to find future new reserves in traps which we have missed or in traps different than those for which we have routinely explored in years past.

In order to find 25,000 drillable prospects per year in the years immediately ahead, the oil industry is going to have to employ maximum brain power. I would like to suggest several avenues by which this maximum exploration brain power can be achieved.

First, the management teams of exploration companies must create an atmosphere which generates maximum creative exploration thinking. Concurrently, management must re-appraise and take a long hard look at its own biased inclination to demand seismic confirmation of every sound geological anomaly presented by its oil finders. In addition, management must resist the temptation to play the role of Monday morning quarterback or an even more childish game, "pin the tail on the donkey." It is a basic law of life that creative thinking does not bloom readily in a stifling atmosphere clouded by second guesses and recriminations. Finally, management can assure an atmosphere conducive to creative work by amply rewarding those particular members of its exploration team who provide imagination and forward thinking. And in this connection, management also should appraise severely all exploration people early in their careers to weed out those who lack conceptual skills. Many sound geologists lack this streak of creativity but can still render their company good service in field development and data compilation activities.

Second, assuming a favorable atmosphere for creative work, the explorationists must strive to stimulate their own creative processes. We live in an age of specialization. Consider, for instance, that there was a time in the aeronautical field when one man built, flew, crashed, and repaired his own plane. Today it takes literally thousands of teams of specialists to put one man into orbital flight. Likewise, it is no longer reasonable to assume that one geologist can master all of the skills necessary to unravel and solve the mysteries of petroleum generation and accumulation. Teams of stratigraphers, log analysts, geochemists, geophysical interpreters, palynologists, paleontologists, structural subsurface experts, and others must pool their individual specialties to derive maximum benefit from all of the data available to them. With the advent of data-processing tools and the resulting ability to handle innumerable data rapidly, there is also room on the exploration team for mathematicians with strong geological backgrounds. Teams of experts such as these, working closely together, are bound to stimulate new thinking which would not be possible by any single group.

Explorationists also can stimulate their creative ability by a continuing program of self-education. This can be accomplished by several methods. It may sound naïve to suggest that a very simple method is to read papers published by other exploration geologists; however, much important new geological thinking with universal application is prepared and published but lost to the industry because the very people who should be reading and studying the data never do so. Another method of self-education can be effected through an organized continuous education program of short courses sponsored and approved by The American Association of Petroleum Geologists and its affiliated local societies or in conjunction with The Society of Exploration Geophysicists and the Society of Petroleum Engineers (AIME). The faculty for these courses could be drawn from both the academic and industrial fields, and the classes preferably should be held at selected regional universities or at seminars sponsored by local geological societies. In this manner those persons interested in furthering their education in petroleum exploration, and thereby becoming better acquainted with the ever-changing "state of the art," could avail themselves of the opportunity afforded by an approved curriculum of study, seminars, and discussions.

Third, in order to achieve maximum brain power, management should add a third member to its exploration team: the petroleum engineer. In his routine work in and around producing fields, the petroleum engineer develops many data which have unique exploration significance. Everyone who works in exploration will acknowledge the importance of changes in bottom-hole pressure, rates of decline, water contacts, *etc.*, as possible indicators of areas of new reserves. Nevertheless, in day-by-day practice the integration of significant engineering data seldom is employed by the geologists or geophysicists in their exploration thinking and work.

Geophysicists, geologists, and petroleum engineers have a great opportunity to find the much needed reserves of the future by working together in areas above, below, and around producing reservoirs. The exploration team must be "bold" —willing to take the risk in the realm of exploration or in the use of a new tool. Unorthodox thinking might well be the key to the discoveries of giant reserves. The earliest petroleum geologists looked for simple surface anticlines. Later, geologists and geophysicists combined their efforts to look for the less obvious subsurface anticlines. But the era of the perfect geophysical anticline is behind us (unless activity is conducted in a

virgin area) and the creative imagination of the exploration team—the maximum use of brain power—must be the new exploration "breakthrough tool."

The geologists, geophysicists, and petroleum engineers as an exploration team have a joint responsibility to develop new tools, new techniques, and innovations which will save time, money, and reserves. The petroleum industry is long overdue for some technical improvements in the areas of exploration, drilling, and recovery. As professional men, geologists, geophysicists, and petroleum engineers need to encourage each other to create these new tools and techniques for tomorrow. This can be accomplished only by a close-working, harmonious relationship. The responsibility for scientific invention and creative ideas must not be relegated solely to the research and development laboratories. In the immediate future, the industry will have to find new methods of improving discovery rates, new ways to reduce drilling costs, new approaches to stimulating flow from low porosity and permeability reservoirs, better methods of achieving more complete recovery, and better devices for accurate evaluation of potential reservoirs penetrated by the drill (Halbouty and Barber, 1964). No geologist, geophysicist, or petroleum engineer can do this alone, but all three working together as a unit—as a team—surely can do these things and more.

This maximum brain-power team must entertain new concepts concerning stratigraphic traps, shale oil development, thermal exploitation, deep-seated structure, and sedimentation. It must also generate more progressive thinking toward the unexplored basins such as those in Arizona, Nevada, and Idaho and in other states, and the still untested possibilities onshore and offshore along the entire eastern seaboard—from the Florida Keys to the tip of Maine.

Fourth, we must be bold in recommending the drilling of stratigraphic tests within small limits. The electric log is one of the most vital tools the petroleum geologist has ever been offered, but even today the log still is not fully used to achieve the maximum in exploration. For instance, we are reluctant to drill a well between two deep holes only a mile or so apart, especially if no structure or relief of any kind is apparent. But the updip pinch-out of a reasonably thick reservoir between the dry holes, ably revealed by the correlation of the electric logs of the two wells, could result in an elongate, highly prolific field. It is conceded that this kind of thinking is bold and of high risk, but it is this type of reasoning which will result in the big discoveries of the future.

Finally, a team of explorers using maximum brain power must also develop maximum resolution power. A dry hole should not stifle the courage to seek reasons to drill another hole on a prospect. Every geologist can name out of his own memory many fields in which one or more dry holes have been drilled before production was discovered. Consider High Island salt dome in Galveston County, Texas, which currently produces 10 million barrels per year. There have been approximately 100 dry holes drilled within or adjacent to the productive limits of the field. Admittedly, not all of these dry holes were drilled before the first production was obtained, but this theoretically would have been possible. Therefore, all who search for new petroleum reserves, from the geologist in the field to the chairman of the board, must develop a tenacity and bold vigor which are not weakened by one dry hole.

The oil industry will meet the reserve demands of the future. The most successful members of the oil industry in the future will be those who not only use wisely the scientific tools which are available but who also place maximum reliance on the conceptual skills of their own brain power.

It is now time for our industry to form this maximum brain-power exploration team. Our discovery rates will not improve unless we do— but will surely improve if we act positively with the knowledge that faith, determination, and optimism are the doors to scientific exploration success.

REFERENCES

API, Facts and figures, 1959–1963.

Atwater, G. I., and Miller, E. E., 1965, Deep Louisiana holes put a squeeze on payouts: Oil and Gas Jour., v. 63, no. 18, p. 71.

Halbouty, Michel T., and Barber, T. D., 1964, The responsibility of geologists and petroleum engineers in meeting exploration demands in the future: Jour. Petroleum Technology, v. 16, no. 3, p. 239–243.

U. S. Bureau of Mines, 1962–1963; and, API Annual Report, December 28, 1964.

Reprinted by permission of the South Texas Geological
Society from *South Texas Geological Society Newsletter*,
v. 6, no. 2 (1966), p. 4-11.

ESSENTIAL ABILITIES OF THE OIL-FINDING GEOLOGIST
Louie Sebring, Jr.

There are only two. The oil-finder must be able to locate and recognize a significant anomaly, and he must be able to get this anomaly tested. A significant anomaly is defined as a trapping anomaly, one that might trap hydrocarbons. So, the oil-finder has a dual job. He must be both a scientist and a salesman, a prospector and a promoter, a researcher and a hustler. He must locate the prospect, and, unless he is financially self-sufficient, he must persuade his management or his investors to test the prospect. One ability is not enough. To be a successful oil-finder he must have both abilities, although not necessarily in equal degree.

The truth is self-evident — all geologists are not equal in ability; either as trap locators or as prospect salesmen. Some have considerable ability to locate anomalies, but have great difficulty in persuading their management or investors to test their prospects. Others have difficulty in recognizing a prospect, but once one is located can easily persuade their management or clients to test it. And some, a truly blessed minority, can not only easily locate and recognize a significant anomaly but can, just as easily, persuade their management or investors to pursue the investigation of this anomaly to its logical conclusion.

In the past, those oil-finders who have excelled in selling prospects have been the most successful in their profession. This condition will undoubtedly continue. The super salesman may locate fewer prospects, although this is not necessarily so. He will certainly get a higher percentage of his prospects tested than will the geologist more delinquent in sales ability. Because of this, the salesman should find more oil. It behooves every petroleum prospector to develop his selling ability, his ability to write and speak persuasively, along with his exploratory techniques. A gifted anomaly-finder who is so incoherent, so unassertive, or so lacking in courage, that he cannot get an anomaly tested, regardless of its intrinsic merit; will never find oil.

Selling techniques will be doubly important to the geologist who has to sell his prospect to a manager or to an investor who is not technically qualified to judge the geologic merits of the prospect. The unqualified manager or investor will probably base his decisions on the eye appeal of maps and diagrams in the presentation, on the personal charm and persuasive ability of the prospect salesman, and on the apparent economics; rather than on the geologic worth of the prospect, which he is really not able to evaluate.

[1]Champlin Petroleum Company. The subject of this paper was suggested by W. B. Dansfiell. Mrs. Patty Fischer typed the manuscript. Mr. Sebring presented this paper before the South Texas Geological Society January 26, 1966.

4

There are only four ways by which the individual geologist or company can find more petroleum than their competition. These are: by doing better quality exploration work, by doing more exploration work, by being more aggressive than the competition, and by being luckier than the competition. It may take all four before a significant difference in results is noted. Of these four ways, the individual geologist can fully control only the first two. He can influence the third by his selling effort. The more prospects that he finds and sells the more exposure he will have to luck, and the luckier he is likely to be. Also, there appears to be considerable correlation between the number of lucky, pleasant surprises and the quality of the prospects and of the trend.

DESIRABLE TRAITS AND CHARACTERISTICS

The successful oil-finder will probably possess many of the personal characteristics, habits and traits which are described below. Some of these characteristics, which help him to find traps, will also help him to get them tested. Some will not.

It is essential that the oil-finder be optimistic. He must believe that there are commercial quantities of oil to be found and that he can find the prospects that might contain them. He must also believe that when he locates these prospects that he will be able to persuade his management or his investors to acquire an interest in these prospects. Thus a great deal of self-confidence is requried for both of his dual roles. This optimism must be buoyed by a sustained enthusiasm that will carry him through the many disappointments that blight the efforts of every prospector and every salesman.

The oil-finder will be a dedicated, effective worker who truly enjoys his exploration work. He had better like it more than just about anything he can do. If he doesn't, he can always find enough excuses not to do it. If he can't find enough excuses by himself; then his company, his community and his family will provide him with enough excuses in the form of routine paper work, meetings, training programs, visiting dignitaries, administrative and managerial problems, briefing sessions, civic good works, household tasks and yard work; so that he need not devote any time at all to the basic individual exploration effort that results in the discovery of petroleum. Unless he truly enjoys exploration work, he can quite easily rationalize himself into believing that he just doesn't have the time to do any of it at all.

To be effective, the oil-finder must spend his time at tasks most rewarding to his particular abilities. He should spend most of his time at exploratory jobs that only a person of his capabilities can do. If he is doing subsurface mapping, as are most of us in the United States, he should spend the bulk of his time correlating, examining and analyzing well logs, preparing maps, etc. The routine tasks of filing, spotting maps, drafting and plotting logs, which are just as essential, should be performed by persons not capable of performing the more demanding exploratory tasks.

5

A dedicated effort is required because oil and gas, at least in the United States, is becoming increasingly more difficult to find. This difficulty is caused by a shortage of obvious prospects at medium and shallow depths in proven petroliferous trends, brought about by the intense competition for reserves by the myriad companies and individuals who are searching for oil. Competition has resulted in a degree of development of our oil and gas resources unmatched anywhere else in the world. Each field found leaves one less to be found. Most of the more obvious prospects in the proven trends have been found. Most of the traps that remain are more complicated — deeper — less apparent — better hidden. This makes it necessary for the oil-finder to look at all of his data in a new, yet logical way. An all-out, dedicated search for these remaining traps is necessary. The day of the half-way exploration effort is over. If the company or individual searching for petroleum in the United States cannot dedicate themselves to their primary task, then they might as well give up right now. There are many intelligent people who are diligently searching for oil. The successful oil-finder must match or surpass their efforts.

One word of caution about this search for the more subtle, better hidden traps. The obvious traps should not be overlooked. An anticline will always constitute an excellent prospect, especially if it affects beds that have a productive history in the area. The geologist should not overlook such a trap just because his primary search is for a major oil field from a more appealing deep bed never before reached by the drill in the immediate area. Such a target has great charm simply because its potential is unknown and may be great. Original thinking, followed by aggressive action, will eventually pay off; if it is based on sound geological thinking. However, the odds are strongly against the success of any particular, single such venture.

The successful oil-finder will work just as hard at his selling job. How much of a selling effort is worthwhile? Whatever it takes to sell the prospect — and no more — is enough. Of course, in order to be sure that his prospect sells, the geologist will generally do more work in this connection than he really feels is necessary. Thus he will tend to overdo, to oversell his prospect. This is as it should be, since he soon learns that it is much more difficult to resell a prospect, regardless of its merit, once it has been turned down by the prospective buyer. His first pitch should be his best one — his high, hard one, so to speak.

It is patently impossible for any one person to be expert or even technically qualified in all of the exploration methods now being used. However, the successful oil-finder will be adept in at least one of the more reliable and useful exploration techniques. In the United States, he will probably be an expert subsurface geologist, seismologist or log analyst. Probably the first, since the data required by the subsurface geologist are more widely available than those required by other explorationists. The ready availability of well data, logs and

6

maps, his chief tools, allows the subsurface geologist to conduct an effective search almost anywhere in the United States. He can use almost any type of information to some purpose, i.e. old surface maps, core drill maps, seismic and gravity maps, etc., even if the information is fragmentary and of questionable reliability.

The successful oil-finder, whether he be primarily a subsurface geologist, a geophysicist or a log analyst, will have at least some practical knowledge of the techniques, advantages and limitations of the most important other methods used.

It is important for the prospect salesman to be technically qualified. Generally, although not always, he will be selling the prospect to a manager or client qualified to recognize its merits and its shortcomings. The salesman must know his product to sell to such a man. It is infinitely easier to sell a good product than it is to sell a poor one. Eventually, some of the prospects that he sells must produce. The name of the game is oil-finding. No matter what the individual's sales ability, failure after failure after failure of his prospects, unrelieved by any success, will eventually lose him the confidence of his company and his investors. Worse still, unless he is an absolute charlatan, he will lose his self-confidence, his enthusiasm and his optimism, all traits that he must have to conduct an effective search and sales effort. Successive failures will strain his courage, make him cautious when he should be bold, et cetera.

Because oil-finding is not yet an exact science, but rather more of an art or an educated gamble based on a knowledge of the odds and geologic rules, imagination is an essential trait. Levorsen, in his book "Geology of Petroleum" says that the most successful geologist is the one who can visualize the pool or locate the extension with the least advance information; i.e., he is the geologist who applies his imagination most effectively. Here the individual geologist's ability truly varies. One geologist may overlook several hundred feet of abnormal critical dip with no thought at all that here is primary evidence of a trap. Another may go into ecstasies of delight over the slightest change in the rate of dip or type of sedimentation. Another needs no excuse at all, but happily fills all of the blank spaces on his maps with untested anomalies whose size is determined by the amount of space available between his control points. Still another sizes and shapes his prospects to fit the available acreage.

The geologist who habitually overlooks the obvious anomalies may be just stupid or so lacking in courage that he dare not face his management and try to sell them a prospect. No matter, he is worthless as an exploration geologist.

The space-filler apparently assumes that all others who have prospected the area are geological morons who drill only in structurally low or stratigraphically normal areas, thus leaving only the fields already luckily found, or yet to be found, by some matchless exploration genius such as he. This man will eventually find oil, but it is quite possible that he could expend a major company's exploration budget before his efforts are crowned by success.

7

Every geologist must consider the possibility that each undrilled area that he inspects may contain an accumulation of oil or gas. Any exploratory program for prospects at commonly reached depths in the older producing districts, must be based on the possibility that the preceding programs have overlooked commercial oil and gas accumulations.

Wallace Pratt said, "Where oil is first found, in the final analysis, is in the minds of men." Wishful thinking, however, is not enough. Most undrilled areas don't contain a major oil field. Most exploratory wells are drilled by competent people looking for oil at locations that someone believes are prospective. When these wells are plugged most of them are not capable of producing commercial amounts of oil or gas. There are exceptions of course, and no exploration geologist should overlook the possibility that a key well was improperly tested, or that new production techniques might result in a commercial completion from zones where such a result was impossible only a short time ago.

There must be some indication of the presence of an anomaly — a show — a break in the dip, either a structural flattening or a structural steepening, an unusually high control point or unusually low point — a stratigraphic change; in other words, some variation that points to the presence of an anomaly. There should not only be the space for an anomaly, but the reasonable possibility, based on regional geological considerations, that such an anomaly can actually be present. The explorationist should be sharp enough, intelligent enough, to recognize the indications of an anomaly, and decide from the available evidence whether its presence is proven, probable, possible, improbable or impossible. If the anomaly's presence is only possible, then a more refined exploration tool, such as the reflection seismic method may be employed to determine its presence.

The oil-finder's imagination should be controlled by a good deal of applied judgment that involves a comparison and discrimination of evidence. He must have great powers of assimilation of quantities of data. Much of this data will be extraneous, of no economic use, and some may be unreliable. Some of his data will give conflicting indications. He must reject the unreliable and the extraneous and be able to weigh the relative importance of conflicting data in the light of his own experience. If the conflicting data are of equal reliability, he should accept the more optimistic data. He knows that the rewards for the discovery of a commercial accumulation of oil or gas are much greater than the penalties for a dry hole. Integrity of evaluation of the evidence is essential. He must give the most weight to the most reliable, the most pertinent and the most likely.

Imagination and good judgment are just as important to the salesman as they are to the oil-finder. An imaginative sales approach may succeed where a more prosaic approach might fail. The salesman must exercise a great deal of judgment, however. It is possible that he might alienate a competent evaluator if he gives the impression that he is trying to camouflage the prospect's weaknesses by too much windowdressing.

8

Integrity is essential to the salesman, too. If he disregards or alters his data, eventually he will be found out. Once an investor or a manager loses confidence in the honesty of the salesman or geologist, his usefulness is ended and the relationship should be terminated.

The search for oil and gas calls for a great deal of persistence both in exploration and in sales effort. The geologist will persist in his search for oil or gas in country that he believes is petroliferous until he locates evidence of a trap. He will then initiate a persistent selling effort in order to get the trap tested. If he is successful in getting a test well drilled and if it is dry, he will re-evaluate his data in the light of this new evidence. If he still believes that a trap exists, he will again renew his efforts to get the prospect tested. It is his duty to continue this until his efforts are crowned by success; until he is himself convinced that a trap capable of containing a commercial accumulation does not exist; or until he has exhausted all legitimate means of persuasion at his disposal.

One of the most difficult selling problems faced by the exploration geologist is encountered when he attempts to sell a prospect in an area where his supervisor or investor has detailed knowledge and a pre-conceived conception of the geology. Unless the salesman's prospect fits the evaluator's picture of the geology, he faces a monumental task. He must convince the supervisor or investor that his more recent interpretation is more reliable than the one now held by the evaluator. Both parties realize that essentially the geology doesn't change in our time — only our conception of the geology changes. The geologist must be quite resourceful here. If he gets more data, if he obtains additional interpretations that tend to verify his own interpretation, if he does more detailed work, maps more horizons, if he employs a new sales approach, etc.; it may be possible to change the evaluator's mind, but it's never easy. Generally, it is much easier for the independent explorationist to give up his sales effort to this particular client and submit the prospect to someone else. This, the honest company geologist cannot do. Persistence holds his only hope for success in such a venture.

Initiative is an essential quality of the oil-finding geologist. After he visualizes the prospect, it becomes his duty to develop the data on the prospect and decide whether he has enough data for a recommendation and a sales effort. If, in his judgment, the prospect needs more control or data, it is his job to take the necessary steps to see that the control is gathered. He cannot and should not wait for someone to tell him what to do.

The oil-finding geologist must have courage, intestinal fortitude, guts — if you will. After he has gathered and weighed all of the pertinent available evidence and convinced himself that the possibilities of the presence of an economic accumulation are worth the risk, he must commit himself. He must take a stand, express an opinion, and recommend an action. He will employ every legitimate and honorable resource at his command in this effort to persuade his management or investors to follow the course of action he recommends. He should not conceal the weaknesses that every prospect has, but

9

should endeavor to place the weaknesses and strengths of each prospect in their proper perspective. Ideally, the effect that success or failure of the recommended project might have on his own career should not be a consideration. Undue concern over this matter will affect his judgment of the merit of the prospect.

The oil-finding geologist will be responsible. He will be willing to accept the consequences of his own recommendations — be they good or bad. He will expect praise and rewards for a successful exploration effort, but he must be willing to be held accountable for the inevitable failures as well.

The oil-finder will have a good sense of economic judgment. He realizes that it is not enough for a prospect to produce oil or gas. It must produce enough oil or gas to make a profit. A successful exploration program must yield a reasonable profit above all the costs of finding, producing and transporting the oil and gas to the market. The oil-finder will constantly attempt to evaluate the economics along with the effectiveness of the exploration techniques that he uses. He will discard the ineffective ones, cut costs where he can and employ exploration innovations only after he is convinced of their economic merit. He will recommend the techniques that he believes will economically best reveal the essential elements of the trap.

The geologist who has the abilities, traits and characteristics that were described above, will develop, with up-to-date experience in a specified area, an almost intuitive feeling about the merits of individual prospects in that area. His discovery average will improve. He'll drill some dry holes, of course, but few foolish wells. His prospects will all be possibles or probables based on sound geological reasoning. The amount of exploratory risk that he is willing to assume will be keyed to the possible benefits to be received for a successful effort. Some of his discoveries will be uneconomic, but since he will stay in trends that he believes have good potential, some should be extremely successful both geologically and economically. He will on rare occasions be surprised by the discovery of a new type of trap or of a new trend in his area. However, when the occasion does arise, he will be alert enough, openminded enough, and nimble enough in his thinking, to launch an early search in this new trend or for this new type trap.

MANAGEMENT AND INVESTOR HELP

What can management or an investor do to help the oil-finder? Mainly, they can simplify the selling effort. They should require a minimum of extraneous data. Since even the best training in the latest exploratory techniques is useless if the geologist has no time to apply these techniques; after training, the geologist must be given time to search for prospects. Management should provide him with the tools he needs — logs, clerical and drafting help, good base maps and especially reliable well data. They can set realistic exploration goals and keep the geologist informed of these goals. Management can hope that their geologists will be more successful than their competition, but can expect their geologists only to match the best efforts and results of their competition. Quality of indigenous petroleum deposits cannot be improved by threats or inspirational talks. A transfer

10

of personnel has never changed the porosity and permeability of a rock. Any competent management will reward a successful exploration effort in order to provide incentive for future efforts. Any sufficiently aggressive exploration program is plagued by more individual failures than successes. Therefore, the logical explorationist who is not rewarded for his successes, will soon realize that all that faces him are likely penalties for his failures. Inevitably, he will conclude that it will serve his own self-interest best not to take any exploratory chances at all. This conclusion, if it is reached by a majority of the exploration people, will bring any exploratory program to a halt.

Management should not require too high a success average. Such a requirement can result in rejection of productive prospects because of a lack of stringent control that may be impossible to obtain. The amount of control required should be inversely proportional to the possible economic value of the prospect.

CONCLUSION

Management cannot find the prospects, that's the job of the exploration geologist. He has to find them, and convince his management or investors of their merit.

The exploration geologist who has exceptional oil-finding ability is an extremely valuable asset. His value to an exploratory effort is surpassed only by two even rarer birds. The regional geologist who has the ability and the opportunity to put his company or investor into a new and prolific province or trend before the pack arrives, is more valuable. So is the person with exceptional ability to select the most talented oil-finders from a group.

Exploration problems tend to diminish as the quality of exploration work improves. Therefore, the selection, care and coddling — the motivation — of exceptional oil-finders should be the main concern of exploration management and of investors in exploration projects.

11

The American Association of Petroleum Geologists Bulletin

V. 57, No. 1 (January 1973), P. 207–209

A Suggestion for Exploration Management[1]

ROBERT G. BEHRMAN, JR.[2]

Houston, Texas 77005

Abstract   For the geologist or explorationist to perform the continually more difficult task of finding petroleum reserves, a change in management philosophy will be necessary. This new philosophy is really only a modern adaptation of the oldest and most successful one—an individual must be charged with the responsibility for finding the reserves, and management must give him the freedom to invent and experiment. Management must create the environmental atmosphere within which an individual may practice his specialty.

FOREWORD

"If Moses had been a committee, the Israelites would still be in Egypt" (Friendly Chat, 1965) or, "If Dad Joiner had been a committee, the East Texas Pool would still be undiscovered."

As Levorsen admonished his geologist colleagues to ask themselves the question, "Would I discover the East Texas pool the way I am now exploring?," so should management ask itself the same question. He pointed out that while "Dad Joiner's recipe for the discovery probably came from an almanac" in 1930, the chief ingredients were well-known in 1919. The following concepts and suggestions are directed primarily to problems of domestic exploration.

INTRODUCTION

Many of the finest, most experienced of the industry's leading explorationists (Levorsen, 1964; Halbouty, 1965; and Cram, 1966) have expressed some valid "soul-searching" and detailed criticisms of exploration methods and management policies. However, few, if any, have been able to come up with a successful key for reversing the trend of the last 3 decades which has been finding less and less oil at a cost of more and more. No one who has been involved intimately in the search for petroleum would deny that the solution of the problem of exploration at a profit is the basis for the success of our industry. A major change in the trend might be accomplished by a relatively simple change in emphasis or philosophy. Any such change which might contribute to the desired result must be based on a correct analysis of the cause. When wildcat success ratios are still 1 out of 10 or more after 30 years, it seems logical that any management changes which have been made are a result of incorrect analysis of the cause.

No compliment to those of us who have been involved during this period is the fact that these odds have not improved, especially in view of the increase in experienced technicians and the availability of more sophisticated hardware. Current management philosophy seems to be based on the premise that these wildcat odds are to be *accepted as inevitable rather than improved.*

CAUSE

Is not the most probable cause that we simply have failed to admit to ourselves the real nature of the beast? Each exploration project, regardless of the amount of expert technical know-how and interpretive ingenuity in its conception, is essentially an experiment. The conception or recognition of a prospect requires the same process as an invention. The experiment is actually a bout between man and nature and the drilling of the first well is an integral part. The outcome is always in doubt, and many times until a completion attempt is made, the answer is not forthcoming. Our modern sophistication probably has tended to lull us into the false illusion that we can treat exploration for petroleum as more advanced than the experimental stage.

Management in recent years has spent much time in developing economic formulas for prospect evaluation, which in the last analysis have planted only more trees in the already cluttered forest of the explorationist. Although the aim of these formulas is highly commendable, they have not contributed the desired result. Most significant discoveries of today, just as in Dad Joiner's day, would not pass the test.

On inquiry, one proponent of the formula method said that it had worked well in his organization, but confessed that "the real purpose was to derive a form of communication between financial experts and exploration experts." One would suppose it also produced an exceptional arithmetic mental agility in the exploraltion expert. Hopefully, communication between experts

---

[1] Manuscript received, March 16, 1972; accepted, May 3, 1972.

[2] Petroleum consultant.

could be performed adequately by use of the language of the realm.

In any field of endeavor what experiments, most of which are infinitely less complicated than petroleum exploration, can be evaluated in realistic economic terms before they actually are conducted?

In a symposium conducted by the Geophysical Society of Tulsa, bringing together a panel of four company executives to discuss "What Management Expects of Explorations," one of the panelists, an exploration manager, best defined explorationist as "oil finder, combining the technical knowledge of geology, geophysics, paleontology, stratigraphy, geochemistry, mineralogy, oceanography, *etc.*, with PROFIT. Note the emphasis is on the word "profit" and not on "oil finder."

Any inventor, worth his salary, is already aware that his bread and butter depends on his employer's profits. Continued pressure from management to justify his experiments in dollars before they are conducted, simply reduces him gradually to a state of fear and frustration and, with luck, transforms him into a homemade economist.

SUGGESTED CHANGE

Why not begin the change with this definition of explorationist: *"Oil Finder: inventive, imaginative interpreter and applier of the technical knowledge of geology, geophysics, paleontology, stratigraphy, geochemistry, mineralogy, oceanography, petroleum engineering, and drilling technology"?* Successful results will take care of *profit.*

This definition suggests the following method of implementation: *put the oil-finding responsibility on one individual explorationist.*

No oil company is so large and rich that it should put the production of its very life's blood into a position of less than first place.

The largest single enterprise in the U.S. is its own government. Most of us still believe that this complex enterprise requires a single head with a minimum of a 4-year term. Is "oil finding" any less complex or less important to a petroleum enterprise?

In the histories of most successful petroleum companies, you will find that the periods of greatest advance in oil finding were marked by the leadership of one rugged individualist who was highly dedicated to one purpose and unhampered by a multitude of completely extraneous considerations, having nothing to do with whether or not a given piece of real estate had petroleum beneath it.

Probably one of the worst blocks for the modern explorationist is the time lag built into most management processes. If he can not move expeditiously into his potential discovery when it is conceived, he will lose it to the one who can.

CREATE THE POSITION OF EXPLORATIONIST

This will require the introduction of a new title in company organizations. It should eliminate the title, "district exploration manager." It will require separation of the exploration units of the company from the line organization. Budgetary and accounting rules may have to be changed. Even tax regulations may have to be revised. These are all "man-made" and can be changed, although the natural laws cannot.

Initially the foremost problem probably will be personnel selection for the position of "explorationist." It is doubtful that anyone associated with a company hierarchy can be classed as an explorationist under the above-proposed definition, although there may be sufficient numbers of suppressed potential explorationists in most companies. Certainly there are many of them outside the company organizations who have become qualified by experience. In the past and present systems of exploration management many are forced to leave the company organization to gain any experience as explorationists.

The position of explorationists is envisioned as the filling of the "wasteland," pictured by Levorsen (1964) in the center of the Petroleum Trap Triangle. The position would require that first loyalty be only to the disciplines of the three corners of the triangle and their objective interpretations. If this were only 75 percent accomplished the results should be astounding, as current practice rarely results in more than 2 percent utilization and proper evaluation of these specialties.

GIVE EXPLORATIONISTS SAME FREEDOM TO EXPERIMENT THAT YOU WOULD GIVE INVENTOR

First, deliver him from the restriction of a committee, and second, from the economic formula for evaluation of a single experiment. Use all the committees you like for determining the amount of money with which you allow him to experiment in a predetermined interval of time. Devise all the formulas you may for evalualtion of multiples of experiments. He must have room for some failures in order to reach the successful one.

PAY HIM WELL AND DEMAND RESULTS

Part of the explorationist's compensation could be a royalty interest in the oil he is responsible for discovering, just as an inventor sometimes is compensated. Other methods, such as net-profit interests, cash bonus for each discovery, or as in some companies, stock options would be merited. The creation of a new-pool discovery is just as significant today as it was when man made his first discovery of oil. Therefore, the finder of this most important commodity deserves compensation of a special nature, above and beyond the ordinary employee.

GIVE HIM AT LEAST THREE YEARS TO PERFORM

The period of trial may vary according to the annual amounts of money allowed the explorationist. The recommended 3-year minimum is based on fund availability to cover six to eight exploratory projects per year, or a total of 18–24 for the 3-year period. Review and adjustments by both management and the explorationist could be made annually, according to the results of each six to eight experiments.

REPLACE HIM IF HIS RESULTS ARE NOT SATISFACTORY

This would not necessarily mean terminating the explorationist from a company payroll. The man who fails to produce in the explorationist role may be a fine technician or specialist in a particular field and could continue to be a productive contributor to the company total effort.

One of the prime collateral duties of the explorationist would be to train his junior personnel in method and technological application for possible succession to the position of explorationist.

SUMMARY

In essence, past and current exploration-management practices have resulted in reducing the exploration effort to finding only "seismic" and "electrical" oil and gas.

The concepts and suggestions presented should allow a far greater application of the cumulative scientific knowledge of the nature of petroleum origin, generation, and entrapment to the recognition and development of an exploration prospect and the site selection for drilling and testing.

To most companies a reorganization along these general lines would be classed as an experiment, but the current state of domestic reserves should be ample justification for such an experiment.

REFERENCES CITED

Cram, I. H., 1966, The oldest is the newest: Am. Assoc. Petroleum Geologists Bull., v. 50, no. 5, p. 826-829.

Friendly Chat, 1965: Oil Slicks, Baker Oil Tools, Inc., v. 33, no. 8.

Halbouty, M. T., 1965, Maximum brainpower: new exploration breakthrough: Am. Assoc. Petroleum Geologists Bull., v. 49, no. 10, p. 1597-1600.

Levorsen, A. I., 1964, Big geology for big needs: Am. Assoc. Petroleum Geologists Bull., v. 48, no. 2, p. 141-156.

THE SCIENTIFIC METHOD

13.2 *Explanations: Scientific and Unscientific*

In everyday life it is the unusual or startling for which we demand explanations. An office boy may arrive at work on time every morning and no curiosity will be aroused. But let him come an hour late one day, and his employer will demand an *explanation*. What is it that is wanted when an explanation for something is requested? An example will help to answer this question. The office boy might reply that he had taken the seven-thirty bus to work as usual, but the bus had been involved in a traffic accident which entailed considerable delay. In the absence of any other transportation, the boy had had to wait a full hour for the bus to be repaired. This account would probably be accepted as a satisfactory explanation. It can be so regarded because from the statements that constitute the explanation the fact to be explained follows logically and no longer appears puzzling. An explanation is a group of statements or a story from which the thing to be explained can logically be inferred and whose acceptance removes or diminishes its problematic or puzzling character. Of course the inference of the fact as conclusion from the explanation as premiss might have to be enthymematic, where the "understood" additional premisses may be generally accepted causal laws,[6] or the conclusion may follow with probability rather than deductively. It thus appears that explanation and inference are very closely related. They are, in fact, the same process regarded from opposite points of view. Given certain premisses, any conclusion that can logically be inferred from them can be regarded as being explained by them. And given a fact to be explained, we say that we have found an explanation for it when we have found a set of premisses from which it can logically be inferred. As was indicated in our first chapter,[7] "*Q because P*" can express either an argument or an explanation.

Of course some proposed explanations are better than others. The chief criterion for evaluating explanations is *relevance*. If the tardy office boy had offered as explanation for his late arrival the fact that there is a war in Afghanistan or a famine in India, that would have been a very poor explanation,

---

[6]This complication will be considered further in Section 13.6.
[7]See pages 29–30.

or "no explanation at all." Such a story would have had "nothing to do with the case"; it would have been *irrelevant*, because from it the fact to be explained *cannot* be inferred. The relevance of a proposed explanation, then, corresponds exactly to the cogency of the argument by which the fact to be explained is inferred from the proposed explanation. Any acceptable explanation must be relevant, but not all stories that are relevant in this sense are acceptable explanations. There are other criteria for deciding the worth or acceptability of proposed explanations.

The most obvious requirement to propose is that the explanation be *true*. In the example of the office boy's lateness, the crucial part of his explanation was a particular fact, the traffic accident, of which he claimed to be an eyewitness. But the explanations of science are for the most part *general* rather than particular. The keystone of Newtonian Mechanics is the Law of Universal Gravitation, whose statement is

> Every particle of matter in the universe attracts every other particle with a force which is directly proportional to the product of the masses of the particles and inversely proportional to the square of the distance between them.

Newton's law is not directly verifiable in the same way as a bus accident. There is simply no way in which we can inspect *all* particles of matter in the universe and observe that they do attract each other in precisely the way that Newton's Law asserts. Few propositions of science are *directly* verifiable as true. In fact, none of the important ones are. For the most part they concern *unobservable* entities, such as molecules and atoms, electrons and protons, chromosomes and genes. Hence the proposed requirement of truth is not *directly* applicable to most scientific explanations. Before considering more useful criteria for evaluating scientific theories, it will be helpful to compare scientific with unscientific explanations.

Science is supposed to be concerned with facts, and yet in its further reaches we find it apparently committed to highly speculative notions far removed from the possibility of direct experience. How then are scientific explanations to be distinguished from those that are frankly mythological or superstitious? An unscientific "explanation" of the regular motions of the planets was the doctrine that each heavenly body was the abode of an "Intelligence" or "Spirit" that controlled its movement. A certain humorous currency was achieved during World War II by the unscientific explanation of certain aircraft failures as being due to "gremlins," invisible but mischievous little men who played pranks on aviators. The point to note here is that from the point of view of observability and direct verifiability, there is no great difference between modern scientific theories and the unscientific doctrines of mythology or theology. One can no more see or touch a Newtonian "particle," an atom, or electron than an "intelligence" or a "gremlin." What, then, are the differences between scientific and unscientific explanations?

There are two important and closely related differences between the kind of explanation sought by science and the kind provided by superstitions of various sorts. The first significant difference lies in the attitudes taken toward the explanations in question. The typical attitude of one who really *accepts* an unscientific explanation is *dogmatic.* The unscientific explanation is regarded as being absolutely true and beyond all possibility of improvement or correction. During the Middle Ages and the early modern period, the word of Aristotle was the ultimate authority to which scholars appealed for deciding questions of fact. However empirically and open mindedly Aristotle himself may have arrived at his views, they were accepted by some schoolmen in a completely different and unscientific spirit. One of the schoolmen to whom Galileo offered his telescope to view the newly discovered moons of Jupiter declined to look, being convinced that none could possibly be seen because no mention of them could be found in Aristotle's treatise on astronomy! Because unscientific beliefs are absolute, ultimate, and final, within the framework of any such doctrine or dogma there can be no rational method of ever considering the question of its truth. The scientist's attitude toward his explanations is altogether different. Every explanation in science is put forward tentatively and provisionally. Any proposed explanation is regarded as a mere hypothesis, more or less probable on the basis of the available facts or relevant evidence. It must be admitted that the scientist's vocabulary is a little misleading on this point. When what was first suggested as a "hypothesis" becomes well confirmed, it is frequently elevated to the position of a "theory." And when, on the basis of a great mass of evidence, it achieves well-nigh universal acceptance, it is promoted to the lofty status of a "law." This terminology is not always strictly adhered to: Newton's discovery is still called the "Law of Gravitation," whereas Einstein's contribution, which supersedes or at least improves on Newton's, is referred to as the "Theory of Relativity." The vocabulary of "hypothesis," "theory," and "law" is unfortunate, since it obscures the important fact that *all* of the general propositions of science are regarded as hypotheses, never as dogmas.

Closely allied with the difference in the way they are regarded is the second and more fundamental difference between scientific and unscientific explanations or theories. This second difference lies in the basis for accepting or rejecting the view in question. Many unscientific views are mere prejudices that their adherents could scarcely give any reason for holding. Since they are regarded as "certain," however, any challenge or question is likely to be regarded as an affront and met with abuse. If those who accept an unscientific explanation *can* be persuaded to discuss the basis for its acceptance, there are only a few grounds on which they will attempt to "defend" it. It is true because "we've always believed it" or because "everyone knows it." These all too familiar phrases express appeals to tradition or popularity rather than evidence. Or a questioned dogma may be defended on the grounds of revelation or authority. The absolute truth of their religious creeds and the absolute falsehood of all others have been revealed from on high, at various times, to Moses, to Paul, to Mohammed, to Joseph Smith, and to many others. That there are rival traditions, conflicting authorities, and revelations that contradict one another does not seem disturbing to those who have embraced an absolute creed. In general, unscientific beliefs are held independently of anything we should regard as *evidence* in their favor. Because they are *absolute,* questions of evidence for them are regarded as having little or no importance.

The case is quite different in the realm of science. Since every scientific explanation is regarded as a hypothesis, it is regarded as worthy of acceptance only to the extent that there is evidence for it. As a hypothesis, the question of its truth or falsehood is open, and there is continual search for more and more evidence to decide that question. The term "evidence" as used here refers ultimately to experience; *sensible* evidence is the ultimate court of appeal in verifying scientific propositions. Science is *empirical* in holding that sense experience is the *test of truth* for all its pronouncements. Consequently, it is of the essence of a scientific proposition that it be capable of being tested by observation.

Some propositions can be tested directly. To decide the truth or falsehood of the proposition that it is now raining outside, we need only glance out the window. To tell whether a traffic light shows green or red, all we have to do is to look at it. But the propositions offered by scientists as explanatory hypotheses are not of this type. Such general propositions as Newton's Laws or Einstein's Theory are not directly testable in this fashion. They can, however, be tested indirectly. The indirect method of testing the truth of a proposition is familiar to all of us, though we may not be familiar with this name for it. For example, if his employer had been suspicious of the office boy's explanation of his tardiness, she might have checked up on it by telephoning the bus company to find out whether an accident had really happened to the seven-thirty bus. If the bus company's report checked with the boy's story, this would serve to dispel the employer's suspicions; whereas if the bus company denied that an accident had occurred, it would probably convince the employer that her office boy's story was false. This inquiry would constitute an indirect test of the office boy's explanation.

The pattern of indirect testing or indirect verification consists of two parts. First, one deduces from the proposition to be tested one or more other propositions capable of being tested directly. Then, these conclusions are tested and are found to be either true or false. If the conclusions are false, any proposition that implies them must be false also. On the other hand, if the conclusions are true, that provides evidence for the truth of the proposition being tested, which is thus confirmed indirectly.

It should be noted that indirect testing is never demonstrative or certain. To deduce directly testable conclusions from a proposition usually requires additional premises. The conclusion that the bus company will confirm that its seven-thirty bus had an accident does not follow validly from the proposition that the seven-thirty bus did have an accident. Additional premises are needed, for example, that all accidents get reported to the company's office, that the reports are not mislaid or forgotten, and that the company does not make a policy of denying (or "covering up") its accidents. So the bus company's denying that an accident occurred would not prove the office boy's story to be false, for the discrepancy might be due to the falsehood of one of the other premisses mentioned. Those others, however, ordinarily have such a high degree of probability that a negative reply on the part of the bus company would render the office boy's story very doubtful indeed.

Similarly, establishing the truth of a conclusion does not demonstrate the truth of the premisses from which it was deduced. We know very well that a valid argument may have a true conclusion even though its premisses are not all true. In the present example, the bus company might confirm that an accident happened to the seven-thirty bus because of some mistake in their records, even though no accident had occurred. So the inferred conclusion might be true even though the premisses from which it was deduced were not. In the usual case, though, that is highly unlikely; so a successful or affirmative testing of a conclusion serves to corroborate the premisses from which it was deduced.

It must be admitted that every proposition, scientific or unscientific, that is a relevant explanation for any observable fact has *some* evidence in its favor, namely, the fact to which it is relevant. Thus the regular motions of the planets must be conceded to constitute evidence for the (unscientific) theory that the planets are inhabited by "intelligences" that cause them to move in just the orbits that are observed. The motions themselves are as much evidence for that myth as they are for Newton's or Einstein's theories. The difference lies in the fact that that is the *only* evidence for the unscientific hypothesis. Absolutely no other directly testable propositions can be deduced from the myth. On the other hand, a very large number of directly testable propositions can be deduced from the scientific explanations mentioned. Here, then, is *the* difference between scientific and unscientific explanations. A scientific explanation for a given fact will have directly testable propositions deducible from it, other than the one stating the fact to be explained. But an unscientific explanation will have no other directly testable propositions deducible from it. It is of the essence of a scientific proposition to be empirically verifiable.

It is clear that we have been using the term "scientific explanation" in a quite general sense. As here defined, an exploration may be scientific even though it is not a part of one of the various special sciences like physics or psychology. Thus the office boy's explanation of his tardiness would be classified as a scientific one, for it is testable, even if only indirectly. But had he offered as explanation the proposition that "God willed him to be late that morning, and God is omnipotent," the explanation would have been unscientific. For although his being late that morning is deducible from the proffered explanation, no other directly testable proposition is, and so the explanation is not even indirectly testable and hence is unscientific.

13.3 *Evaluating Scientific Explanations*

The question naturally arises as to how scientific explanations are to be evaluated, that is, judged as good or bad, or at least as better or worse. This question is especially important because there is usually more than a single scientific explanation for one and the same fact. A person's abrupt behavior may be explained either by the hypothesis that the person is shy or by the hypothesis that the person is unfriendly. In a criminal investigation, two different and incompatible hypotheses about the identity of the criminal may equally well account for the known facts. In the realm of science proper, that an object expands when heated is explained by both the caloric theory of heat and the kinetic theory. The caloric theory regarded heat as an invisible weightless fluid, called "caloric," with the power of penetrating, expanding, and dissolving bodies, or dissipating them in vapor. The kinetic theory, on the other hand, regards the heat of a body as consisting of random motions of the molecules of which the body is composed. These are *alternative* scientific explanations that serve equally well to explain some of the phenomena of thermal expansion. They cannot both be true, however, and the problem is to evaluate or choose between them.

What is wanted here is a list of conditions that a good hypothesis can be expected to fulfill. It must not be thought that such a list of conditions will provide a *recipe* by whose means anyone at all can construct good hypotheses. No one has ever pretended to lay down a set of rules for the invention or discovery of hypotheses. It is likely that none could ever be laid down, for that is the *creative* side of the scientific enterprise. Ability to create is a function of imagination and talent and cannot be reduced to a mechanical process. A great scientific hypothesis, with wide explanatory powers like those of Newton's or Einstein's, is as much the product of genius as a great work of art. There is no formula for discovering new hypotheses, but there are certain rules to which acceptable hypotheses can be expected to conform. These can be regarded as the criteria for evaluating hypotheses.

There are five criteria commonly used in judging the worth or acceptability of hypotheses. They may be listed as (1) relevance, (2) testability, (3) compatibility with previously well-established hypotheses, (4) predictive or explanatory power, and (5) simplicity. The first two have already been discussed, but we shall review them briefly here.

1. Relevance

No hypothesis is ever proposed for its own sake but is always intended as an explanation of some fact or other. Therefore it must be *relevant* to the fact it is intended to explain; that is, the fact in question must be *deducible* from the proposed hypothesis—either from the hypothesis alone or from it together with certain causal laws that may be presumed to have already been established as highly probable, or from these together with certain assumptions about particular initial conditions. A hypothesis that is not relevant to the fact it is intended to explain simply fails to explain it and can only be regarded as having failed to fulfill its intended function. A good hypothesis must be *relevant*.

The chief distinguishing characteristic of scientific hypotheses (as contrasted with unscientific ones) is that they are testable. That is, there must be the possibility of making observations that tend to confirm or disprove any scientific hypothesis. It need not be directly testable, of course. As has already been observed, most of the really important scientific hypotheses are formulated in terms of such unobservable entities as electrons or electromagnetic waves. As one research scientist has written,

> A physicist of this century, interested in the basic structure of matter, deals with radiation he cannot see, forces he cannot feel, particles he cannot touch.[8]

But there must be some way of getting from statements about such unobservables to statements about directly observable entities such as tables and chairs, or pointer readings, or lines on a photographic plate. In other words, there must be some connection between any scientific hypothesis and empirical data or facts of experience.

3. Compatibility with Previously Well-Established Hypotheses

The requirement that an acceptable hypothesis must be compatible or consistent with other hypotheses that have already been well confirmed is eminently reasonable. Science, in seeking to encompass more and more facts, aims at achieving a system of explanatory hypotheses. Of course, such a system must be self-consistent, for no self-contradictory set of propositions could possibly be true—or even intelligible. Ideally, the way in which scientists hope to make progress is by gradually expanding their hypotheses to comprehend more and more facts. For such progress to be made, each new hypothesis must be consistent with those already confirmed. Thus Leverrier's hypothesis that there was an additional but not yet charted planet beyond the orbit of Uranus was perfectly consistent with the main body of accepted astronomical theory. A new theory must fit with older theories if there is to be orderly progress in scientific inquiry.

It is possible, of course, to overestimate the importance of the third criterion. Although the ideal of science may be the gradual growth of theoretical knowledge by the addition of one new hypothesis after another, the actual history of scientific progress has not always followed that pattern. Many of the most important new hypotheses have been inconsistent with older theories and have in fact replaced them rather than fitted in with them. Einstein's Relativity Theory was of that sort, shattering many of the preconceptions of the older Newtonian theory. The phenomenon of radioactivity, first observed during the last decade of the nineteenth century, led to the overthrow—or at least the modification—of many cherished theories that had almost

---

[8]Lloyd Smith, "The Bevatron," *Scientific American*, Vol. 184, No. 2, February 1951.

achieved the status of absolutes. One of these was the Principle of the Conservation of Matter, which asserted that matter could be neither created nor destroyed. The hypothesis that radium atoms undergo spontaneous disintegration was inconsistent with that old, established principle—but it was the principle that was relinquished in favor of the newer hypothesis.

The foregoing is not intended to give the impression that scientific progress is a helter-skelter process in which theories are abandoned right and left in favor of newer and shinier ones. Older theories are not so much abandoned as corrected. Einstein himself always insisted that his own work was a modification rather than a rejection of Newton's. The Principle of the Conservation of Matter was modified by being absorbed into the more comprehensive Principle of the Conservation of Mass-Energy. Every established theory has been established through having proved adequate to explain a considerable mass of data, of observed facts. And it cannot be dethroned or discredited by any new hypothesis unless that new hypothesis can account for the same facts as well or even better. There is nothing capricious about the development of science. Every change represents an improvement, a more comprehensive and thus more adequate explanation of the way in which the world manifests itself in our experience. Where inconsistencies occur between hypotheses, the greater age of one does not automatically prove it to be correct and the newer one wrong. The *presumption* is in favor of the older one if it has already been extensively confirmed. But if the new one in conflict with it *also* receives extensive confirmation, considerations of age or priority are definitely irrelevant. Where there is a conflict between two hypotheses, we must turn to the observable facts to decide between them. Ultimately our last court of appeal in deciding between rival hypotheses is experience. What our third criterion, compatibility with previous well-established hypotheses, comes to is this: the totality of hypotheses accepted at any time should be consistent with each other,[9] and—other things being equal—of two new hypotheses, the one that fits in better with the accepted body of scientific theory is to be preferred. The question of what is involved in "other things being equal" takes us directly to our fourth criterion.

4. Predictive or Explanatory Power

By the predictive or explanatory power of a hypothesis is meant the range of observable facts that can be deduced from it. This criterion is related to, but different from, that of testability. A hypothesis is testable if *some* observable fact is deducible from it. If one of two testable hypotheses has a greater number of observable facts deducible from it than from the other, then it is said to have greater predictive or explanatory power. Thus, Newton's hypothesis of universal gravitation joined together with his three laws of motion

---

[9]Scientists may, however, consider and even use inconsistent hypotheses for years while awaiting the resolution of that inconsistency. This was for many years the situation with respect to the wave and the corpuscular theories of light.

had greater predictive power than did either Kepler's or Galileo's hypotheses, because all observable consequences of the last two were also consequences of the former, and the former had many more besides. An observable fact that can be deduced from a given hypothesis is said to be explained by it and also can be said to be *predicted* by it. The greater the predictive power of a hypothesis, the more it explains, and the better it contributes to our understanding of the phenomena with which it is concerned.

Our fourth criterion has a negative side that is of crucial importance. If a hypothesis is inconsistent with any well-attested fact of observation, the hypothesis is false and must be rejected. Where two different hypotheses are both relevant to explaining some set of facts and both are testable, and both are compatible with the whole body of already established scientific theory, it may be possible to choose between them by deducing from them incompatible propositions that are directly testable. If H_1 and H_2, two different hypotheses, entail incompatible consequences, it may be possible to set up a *crucial experiment* to decide between them. Thus if H_1 entails that under circumstance C phenomenon P will occur, while H_2 entails that under circumstance C phenomenon P will *not* occur, then all we need do to decide between H_1 and H_2 is to produce circumstance C and observe the presence or absence of phenomenon P. If P occurs, this is evidence *for H_1* and *against H_2*, while if P does not occur, this is evidence *against H_1* and *for H_2*.

This kind of crucial experiment to decide between rival hypotheses may not always be easy to carry out, for the required circumstance C may be difficult or impossible to produce. Thus the decision between Newtonian Theory and Einstein's General Theory of Relativity had to await a total eclipse of the sun—a situation or circumstance clearly beyond our power to produce. In other cases the crucial experiment may have to await the development of new instruments, either for the production of the required *circumstances,* or for the observation or measurement of the predicted phenomenon. Thus proponents of rival astronomical hypotheses must often bide their time while they await the construction of new and more powerful telescopes. The topic of crucial experiments will be discussed further in Section 13.6.

5. Simplicity

It sometimes happens that two rival hypotheses satisfy the first four criteria equally well. Historically, the most important pair of such hypotheses were those of Ptolemy (fl. 127–151) and Copernicus (1473–1543). Both were intended to explain all of the then known data of astronomy. According to the Ptolemaic theory, the earth is the center of the universe, and the heavenly bodies move about it in orbits that require a very complicated geometry of epicycles to describe. Ptolemy's theory was relevant, testable, and compatible with previously well-established hypotheses, satisfying the first three criteria perfectly. According to the Copernican theory, the sun rather than the earth is at the center, and the earth itself moves around the sun along with the other planets. Copernicus's theory too satisfied the first three criteria perfectly. With respect to the fourth criterion, that of predictive power, there was not a great deal of difference between the two theories. But with respect to the fifth criterion there was a very significant difference between the two rival hypotheses. Although both required the clumsy method of epicycles to account for the observed positions of the various heavenly bodies, *fewer* such epicycles were required within the Copernican theory. The Copernican system was therefore simpler, and this contributed greatly to its acceptance by all later astronomers.

The criterion of simplicity is a perfectly natural one to invoke. In ordinary life as well as in science, the simplest theory that fits all the available facts is the one we tend to accept. In court trials of criminal cases, the prosecution attempts to develop a hypothesis that includes the guilt of the accused and fits in with all the available evidence. Opposing the prosecuting attorney, the defense attorney seeks to set up a hypothesis that includes the innocence of the accused and also fits all the available evidence. Often both sides succeed, and then the case is usually decided—or *ought* to be decided—in favor of that hypothesis that is simpler or more "natural." Simplicity, however, is a very difficult term to define. Not all controversies are as straightforward as the Ptolemaic–Copernican one, in which the latter's greater simplicity consisted merely in requiring a smaller number of epicycles. And, of course, "naturalness" is an almost hopelessly deceptive term—for it seems much more "natural" to believe that the earth is still while the apparently moving sun really does move. The fifth and last criterion, simplicity, is an important and frequently decisive one, but it is difficult to formulate and not always easy to apply.

13.4 *The Detective as Scientist*

Now that we have stated and explained the criteria by which hypotheses are evaluated, we are in a position to describe the general pattern of scientific research. It will be helpful to begin by examining an illustration of that method. A perennial favorite in this connection is the detective, whose problem is not quite the same as that of the pure scientist, but whose approach and technique illustrate the method of science very clearly. The classical example of the astute detective who can solve even the most baffling mystery is A. Conan Doyle's immortal creation, Sherlock Holmes. Holmes, his stature undiminished by the passage of time, will be our hero in the following account.

1. The Problem

Some of our most vivid pictures of Holmes are those in which he is busy with magnifying glass and tape measure, searching out and finding essential clues that had escaped the attention of those stupid bunglers, the "experts" of Scotland Yard. Or those of us who are by temperament less vigorous may think back more fondly on Holmes the thinker,

. . . who, when he had an unsolved problem upon his mind, would go for days, and even for a week, without rest, turning it over, rearranging his facts, looking at it from every point of view until he had either fathomed it or convinced himself that his data were insufficient."[10]

At one such time, according to Dr. Watson,

He took off his coat and waistcoat, put on a large blue dressing-gown, and then wandered about the room collecting pillows from his bed and cushions from the sofa and armchairs. With these he constructed a sort of Eastern divan, upon which he perched himself cross-legged, with an ounce of shag tobacco and a box of matches laid out in front of him. In the dim light of the lamp I saw him sitting there, an old briar pipe between his lips, his eyes fixed vacantly upon the corner of the ceiling, the blue smoke curling up from him, silent, motionless, with the light shining upon his strong-set aquiline features. So he sat as I dropped off to sleep, and so he sat when a sudden ejaculation caused me to wake up, and I found the summer sun shining into the apartment. The pipe was still between his lips, the smoke still curled upward, and the room was full of a dense tobacco haze, but nothing remained of the heap of shag which I had seen upon the previous night.[11]

But such memories are incomplete. Holmes was not always searching for clues or pondering over solutions. We all remember those dark periods—especially in the earlier stories—when, much to the good Watson's annoyance, Holmes would drug himself with morphine or cocaine. That would happen, of course, between cases. For when there is no mystery to be unraveled, nobody in his right mind would go out to look for clues. Clues, after all, must be clues *for* something. Nor could Holmes, or anyone else, for that matter, engage in profound thought unless he had something to think about. Sherlock Holmes was a genius at solving problems, but even a genius must have a problem before he can solve it. All reflective thinking, and this term includes criminal investigation as well as scientific research, is a problem-solving activity, as John Dewey and other pragmatists have rightly insisted. There must be a problem felt before either the detective or the scientist can go to work.

Of course, the active mind sees problems where the dullard sees only familiar objects. One Christmas season Dr. Watson visited Holmes to find that the latter had been using a lens and forceps to examine "a very seedy and disreputable hard-felt hat, much the worse for wear, and cracked in several places."[12] After they had greeted each other, Holmes said of it to Watson, "I beg that you will look upon it not as a battered billycock but as an intellectual problem."[13] It so happened that the hat led them into one of their most interesting adventures, but it could not have done so had Holmes not seen a problem in it from the start. A problem may be characterized as a fact or group of facts for which we have no acceptable explanation, that

[10]A. Conan Doyle, *The Man with the Twisted Lip.*
[11]Ibid.
[12]A. Conan Doyle, *The Adventure of the Blue Carbuncle.*
[13]Ibid.

seem unusual, or that fail to fit in with our expectations or preconceptions. It should be obvious that *some* prior beliefs are required if anything is to appear problematic. If there are no expectations, there can be no surprises.

Sometimes, of course, problems came to Holmes already labeled. The very first adventure recounted by Dr. Watson began with the following message from Gregson of Scotland Yard:

> Mr. Dear Mr. Sherlock Holmes:
> There has been a bad business during the night at 3, Lauriston Gardens, off the Brixton Road. Our man on the beat saw a light there about two in the morning, and as the house was an empty one, suspected that something was amiss. He found the door open, and in the front room, which is bare of furniture, discovered the body of a gentleman, well dressed, and having cards in his pocket bearing the name of "Enoch J. Drebber, Cleveland, Ohio, USA." There had been no robbery, nor is there any evidence as to how the man met his death. There are marks of blood in the room, but there is no wound upon his person. We are at a loss as to how he came into the empty house; indeed, the whole affair is a puzzler. If you can come round to the house any time before twelve, you will find me there. I have left everything in statu quo until I hear from you. If you are unable to come, I shall give you fuller details, and would esteem it a great kindness if you would favour me with your opinion.
>
> Yours faithfully,
Tobias Gregson[14]

Here was a problem indeed. A few minutes after receiving the message, Sherlock Holmes and Dr. Watson "were both in a hansom, driving furiously for the Brixton Road."

2. Preliminary Hypotheses

On their ride out Brixton way, Holmes "prattled away about Cremona fiddles and the difference between a Stradivarius and an Amati." Dr. Watson chided Holmes for not giving much thought to the matter at hand, and Holmes replied: "No data yet. . . . It is a capital mistake to theorize before you have all the evidence. It biases the judgment."[15] This point of view was expressed by Holmes again and again. On one occasion he admonished a younger detective that "The temptation to form premature theories upon insufficient data is the bane of our profession."[16] Yet for all of his confidence about the matter, on this one issue Holmes was completely mistaken. Of course one should not reach a *final judgment* until a great deal of evidence has been considered, but this procedure is quite different from *not theorizing*. As a matter of fact, it is strictly impossible to make any serious attempt to collect evidence unless one *has* theorized beforehand. As Charles Darwin, the great biologist and author of the modern theory of evolution, observed,

> . . . all observation must be for or against some view, if it is to be of any service.

---

[14]A. Conan Doyle, *A Study in Scarlet.*
[15]Ibid.
[16]A. Conan Doyle, *The Valley of Fear.*

The point is that there are too many particular facts, too many data in the world, for anyone to try to become acquainted with them all. Everyone, even the most patient and thorough investigator, must pick and choose, deciding which facts to study and which to pass over. One must have some working hypothesis for or against which to collect relevant data. It need not be a *complete* theory, but at least the rough outline must be there. Otherwise how could one decide what facts to select for consideration out of the totality of all facts, which is too vast even to begin to sift?

Holmes's actions were wiser than his words in this connection. After all, the words were spoken in a hansom speeding toward the scene of the crime. If Holmes really had no theory about the matter, why go to Brixton Road? If facts and data were all that he wanted, any old facts and any old data, with no hypotheses to guide him in their selection, why should he have left Baker Street at all? There were plenty of facts in the rooms at 221-B Baker Street. Holmes might just as well have spent his time counting all the words on all the pages of all the books there, or perhaps making very accurate measurements of the distances between each separate pair of articles of furniture in the house. He could have gathered data to his heart's content and saved himself cab fare into the bargain!

It may be objected that the facts to be gathered at Baker Street have nothing to do with the case, whereas those awaiting Holmes at the scene of the crime were valuable clues for solving the problem. It was, of course, just this consideration that led Holmes to ignore the "data" at Baker Street and hurry away to collect those off Brixton Road. It must be insisted, however, that the greater relevance of the latter could not be *known* beforehand but only conjectured on the basis of previous experience with crimes and clues. In fact, a *hypothesis* led Holmes to look in one place rather than another for his facts, the hypothesis that there was a murder, that the crime was committed at the place where the body was found, and that the perpetrator had left some trace or clue. Some such hypothesis is always needed to guide an investigator in the search for relevant data, for in the absence of any preliminary hypothesis, there are simply too many facts in this world to examine. The preliminary hypothesis ought to be highly tentative, and it must be based on previous knowledge. But a preliminary hypothesis is as necessary as the existence of a problem for any serious inquiry to begin.

It must be emphasized that a preliminary hypothesis, as here conceived, need not be a complete solution to the problem. The hypothesis that the man was murdered by someone who had left some clues to his identity on or near the body of the victim was what led Holmes to Brixton Road. This hypothesis is clearly incomplete: it does not say who committed the crime, or how it was done, or why. Such a preliminary hypothesis may be very different from the final solution to the problem. It will never be complete: it may be a tentative explanation of only part of the problem. But however partial and however tentative, a preliminary hypothesis is required for any investigation to proceed.

3. Collecting Additional Facts

Every serious investigation begins with some fact or group of facts that strike the investigator as problematic and thus initiate the whole process of inquiry. The initial facts that constitute the problem are usually too meager to suggest a wholly satisfactory explanation for themselves, but they will suggest—to the competent investigator—some preliminary hypotheses that lead to the search for additional facts. These additional facts, it is hoped, will serve as clues to the final solution. The inexperienced or bungling investigator will overlook or ignore all but the most obvious of them; but the careful worker will aim at completeness in the examination of those additional facts to which the preliminary hypotheses had led. Holmes, of course, was the most careful and painstaking of investigators.

Holmes insisted on dismounting from the hansom a hundred yards or so from their destination and approached the house on foot, looking carefully at its surroundings and especially at the pathway leading up to it. When Holmes and Watson entered the house, they were shown the body by the two Scotland Yard operatives, Gregson and Lestrade. ("There is no clue," said Gregson. "None at all," chimed in Lestrade.) But Holmes had already started his own search for additional facts, looking first at the body:

> . . . his nimble fingers were flying here, there, and everywhere, feeling, pressing, unbuttoning, examining. . . . So swiftly was the examination made, that one would hardly have guessed the minuteness with which it was conducted. Finally, he sniffed the dead man's lips, and then glanced at the soles of his patent leather boots.[17]

Then, turning his attention to the room itself,

> . . . he whipped a tape measure and a large round magnifying glass from his pocket. With these two implements he trotted noiselessly about the room, sometimes stopping, occasionally kneeling, and once lying flat upon his face. So engrossed was he with his occupation that he appeared to have forgotten our presence, for he chattered away to himself under his breath the whole time, keeping up a running fire of exclamations, groans, whistles and little cries suggestive of encouragement and of hope. As I watched him I was irresistibly reminded of a pure-blooded, well-trained foxhound as it dashes backward and forward through the covert, whining in its eagerness, until it comes across the lost scent. For twenty minutes or more he continued his researches, measuring with the most exact care the distance between marks which were entirely invisible to me, and occasionally applying his tape to the walls in an equally incomprehensible manner. In one place he gathered up very carefully a little pile of gray dust from the floor and packed it away in an envelope. Finally he examined with his glass the word upon the wall, going over every letter of it, with the most minute exactness. This done, he appeared to be satisfied, for he replaced his tape and his glass in his pocket.

---

[17]A. Conan Doyle, *A Study in Scarlet.*

"They say that genius is an infinite capacity for taking pains," he remarked with a smile. "It's a very bad definition, but it does apply to detective work."[18]

One matter deserves to be emphasized very strongly. Steps 2 and 3 are not completely separable but are usually very intimately connected and interdependent. True enough, we require a preliminary hypothesis to begin any intelligent examination of facts, but the additional facts may themselves suggest new hypotheses, which may lead to new facts, which suggest still other hypotheses, which lead to still other additional facts, and so on. Thus having made his careful examination of the facts available in the house off Brixton Road, Holmes was led to formulate a further hypothesis that required the taking of testimony from the constable who found the body. The man was off duty at the moment and Lestrade gave Holmes the constable's name and address.

> Holmes took a note of the address.
> "Come along, Doctor," he said: "we shall go and look him up. I'll tell you one thing which may help you in the case," he continued, turning to the two detectives. "There has been murder done, and the murderer was a man. He was more than six feet high, was in the prime of his life, had small feet for his height, wore coarse, square-toed boots and smoked a Trichinopoly cigar. He came here with his victim in a four-wheeled cab, which was drawn by a horse with three old shoes and one new one on his off fore-leg. In all probability the murderer had a florid face, and the fingernails of his right hand were remarkably long. These are only a few indications, but they may assist you."
> Lestrade and Gregson glanced at each other with an incredulous smile.
> "If this man was murdered, how was it done?" asked the former.
> "Poison," said Sherlock Holmes curtly, and strode off.[19]

4. Formulating the Hypothesis

In any investigation the stage will be reached, sooner or later, at which the investigator—whether detective, scientist, or ordinary mortal—will begin to feel that all the facts needed for solving the problem are at hand. The investigator has the "2 and 2," so to speak, but the task still remains of "putting them together." At such a time Sherlock Holmes might sit up all night, consuming pipe after pipe of tobacco, trying to think things through. The result or end product of such thinking, if it is successful, is a hypothesis that accounts for all the data, both the original set of facts constituting the problem and the additional facts to which the preliminary hypotheses pointed. The actual discovery of such an explanatory hypothesis is a process of creation, in which imagination as well as knowledge is involved. Holmes, who was a genius at inventing hypotheses, described the process as reasoning "backward." As he put it,

[18]Ibid.
[19]Ibid.

Most people if you describe a train of events to them, will tell you what the result would be. They can put those events together in their minds, and argue from them that something will come to pass. There are few people, however, who, if you told them a result, would be able to evolve from their own inner consciousness what the steps were which led up to that result.[20]

Here is Holmes's description of the process of formulating an explanatory hypothesis. Whether his account is right or wrong, when a hypothesis has been proposed, its evaluation must be along the lines that were sketched in Section 13.3. Granted its relevance and testability, and its compatibility with other well-attested beliefs, the ultimate criterion for evaluating a hypothesis is its predictive power. As a more recent writer has put it,

The formation of hypotheses is the most mysterious of all the categories of scientific method. Where they come from, no one knows. A person is sitting somewhere, minding his own business, and suddenly—flash!—he understands something he didn't understand before. Until it's tested the hypothesis isn't truth. For the tests aren't its source. Its source is somewhere else.[21]

5. Deducing Further Consequences

A really fruitful hypothesis will explain not only the facts that originally inspired it, but will explain many others in addition. A good hypothesis will point beyond the initial facts in the direction of new ones whose existence might otherwise not have been suspected. And, of course, the verification of those further consequences will tend to confirm the hypothesis that led to them. Holmes's hypothesis that the murdered man had been poisoned was soon put to such a test. A few days later the murdered man's secretary and traveling companion was also found murdered. Holmes asked Lestrade, who had discovered the second body, whether he had found anything in the room that could furnish a clue to the murderer. Lestrade answered, "Nothing," and went on to mention a few quite ordinary effects. Holmes was not satisfied and pressed him, asking, "And was there nothing else?" Lestrade answered, "Nothing of any importance," and named a few more details, the last of which was "a small chip ointment box containing a couple of pills." At this information,

Sherlock Holmes sprang from his chair with an exclamation of delight.
"The last link," he cried, exultantly. "My case is complete."
The two detectives stared at him in amazement.
"I have now in my hands," my companion said, confidently, "all the threads which have formed such a tangle. . . . I will give you a proof of my knowledge. Could you lay your hands upon those pills?"
"I have them," said Lestrade, producing a small white box.[22]

[20]Ibid.
[21]Pirsig, *Zen and the Art of Motorcycle Maintenance.*
[22]A. Conan Doyle, *A Study in Scarlet.*

On the basis of his hypothesis about the original crime, Holmes was able to predict that the pills found at the scene of the second crime must contain poison. Here deduction has an essential role in the process of any scientific or inductive inquiry. The ultimate value of any hypothesis lies in its predictive or explanatory power, which means that additional facts must be deducible from an adequate hypothesis. From his theory that the first man was poisoned and that the second victim met his death at the hands of the same murderer, Holmes inferred that the pills found by Lestrade must be poison. His theory, however sure he may have felt about it, was only a theory and needed further confirmation. He obtained that confirmation by testing the consequences deduced from the hypothesis and finding them to be true. Having used deduction to make a prediction, his next step was to test it.

6. Testing the Consequences

The consequences of a hypothesis, that is, the predictions made on the basis of that hypothesis, may require various means for their testing. Some require only observation. In some cases, Holmes needed only to watch and wait—for the bank robbers to break into the vault, in the *Adventure of the Red-Headed League*, or for Dr. Roylott to slip a venomous snake through a dummy ventilator, in the *Adventure of the Speckled Band*. In the present case, however, an experiment had to be performed.

Holmes asked Dr. Watson to fetch the landlady's old and ailing terrier, which she had asked to have put out of its misery the day before. Holmes then cut one of the pills in two, dissolved it in a wineglass of water, added some milk, and

> . . . turned the contents of the wineglass into a saucer and placed it in front of the terrier, who speedily licked it dry. Sherlock Holmes's earnest demeanor had so far convinced us that we all sat in silence, watching the animal intently, and expecting some startling effect. None such appeared, however. The dog continued to lie stretched upon the cushion, breathing in a laboured way, but apparently neither the better nor the worse for its draught.
>
> Holmes had taken out his watch, and as minute followed minute without result, an expression of the utmost chagrin and disappointment appeared upon his features. He gnawed his lip, drummed his fingers upon the table, and showed every other symptom of acute impatience. So great was his emotion that I felt sincerely sorry for him, while the two detectives smiled derisively, by no means displeased at this check which he had met.
>
> "It can't be a coincidence," he cried, at last springing from his chair and pacing wildly up and down the room: "it is impossible that it should be a mere coincidence. The very pills which I suspected in the case of Drebber are actually found after the death of Stangerson. And yet they are inert. What can it mean? Surely my whole chain of reasoning cannot have been false. It is impossible! And yet this wretched dog is none the worse. Ah, I have it! I have it!" With a perfect shriek of delight he rushed to the box, cut the other pill in two, dissolved it, added milk, and presented it to the terrier. The unfortunate creature's tongue seemed hardly to have been

moistened in it before it gave a convulsive shiver in every limb, and lay as rigid and lifeless as if it had been struck by lightning.

Sherlock Holmes drew a long breath, and wiped the perspiration from his forehead.[23]

By the favorable outcome of his experiment, Holmes's hypothesis had received dramatic and convincing confirmation.

7. Application

The detective's concern, after all, is a practical one. Given a crime to solve, he has not merely to explain the facts but to apprehend and arrest the criminal. The latter involves making application of his theory, using it to predict where the criminal can be found and how he may be caught. He must deduce still further consequences from the hypothesis, not for the sake of additional confirmation but for practical use. From his general hypothesis Holmes was able to infer that the murderer was acting the role of a cabman. We have already seen that Holmes had formed a pretty clear description of the man's appearance. He sent out his army of "Baker Street Irregulars," street urchins of the neighborhood, to search out and summon the cab driven by just that man. The successful "application" of this hypothesis can be described again in Dr. Watson's words. A few minutes after the terrier's death,

> . . . there was a tap at the door, and the spokesman of the street Arabs, young Wiggins, introduced his insignificant and unsavoury person.
>
> "Please, sir," he said touching his forelock, "I have the cab downstairs."
>
> "Good boy," said Holmes, blandly. "Why don't you introduce this pattern at Scotland Yard?" he continued, taking a pair of steel handcuffs from a drawer. "See how beautifully the spring works. They fasten in an instant."
>
> "The old pattern is good enough," remarked Lestrade, "if we can only find the man to put them on."
>
> "Very good, very good," said Holmes, smiling. "The cabman may as well help me with my boxes. Just ask him to step in, Wiggins."
>
> I was surprised to find my companion speaking as though he were about to set out on a journey, since he had not said anything to me about it. There was a small portmanteau in the room, and this he pulled out and began to strap. He was busily engaged at it when the cabman entered the room.
>
> "Just give me a help with this buckle, cabman," he said, kneeling over his task, and never turning his head.
>
> The fellow came forward with a somewhat sullen, defiant air, and put down his hands to assist. At that instant there was a sharp click, the jangling of metal, and Sherlock Holmes sprang to his feet again.
>
> "Gentlemen," he cried, with flashing eyes, "let me introduce you to Mr. Jefferson Hope, the murderer of Enoch Drebber and of Joseph Stangerson."[24]

[23]Ibid.
[24]Ibid.

Here we have a picture of the detective as scientist, reasoning from observed facts to a testable hypothesis that not only explains the facts but also permits a practical application.

ART. XXVII.—*The Inculcation of Scientific Method by Example, with an illustration drawn from the Quaternary Geology of Utah;* by G. K. GILBERT.   (With a map, Plate VIII.)

Presidential Address read before the American Society of Naturalists at Boston, December 27, 1885.

Mr. President and Gentlemen:—This is an association of teachers of science and investigators. Those of us who are primarily engaged in investigation have come here more especially as educators. It is our function to discuss, not our results, nor the subject matter of the several sciences with which we are concerned, but our methods of investigation, our methods of publication or promulgation, our methods of teaching.

It is fitting that this, one of the first formal addresses before the Society, should deal with some of the most general considerations affecting methods. In the statement of these considerations it is impossible to avoid that which is familiar, and even much that is trite. Indeed all expectation of entertaining or edifying you with the original or the new may as well be disclaimed at the outset. I shall merely attempt to outline certain familiar principles, the common property of scientific men, with such accentuations of light and shade as belong to my individual point of view.

The teacher's work is susceptible of a logical division into two parts. He stores minds, and he trains them. The modern educator believes the second function to be the higher, because the trained mind can store itself. Nevertheless the two go hand in hand and are in great part inseparable. The effort of the intelligent teacher is to employ such methods in storing the minds of his pupils with knowledge that they shall acquire at the same time the best training.

In that particular department of teaching which is called scientific, there is the same logical duality, and to a great extent there is a practical unity; but in this case there is a pre-determined classification of those who fall under the teacher's instruction, which has the effect of practically dividing his methods. A portion of his pupils are preparing to engage in the work of research, and look to a scientific career. Another portion are to be occupied with business or in other pursuits not implying research, at least in the ordinary sense, and desire to obtain, as a part of a liberal education, an acquaintance with the materials and results of science. The first demand a training in methods, the second consciously ask only for a store of knowledge. Nevertheless, the general student can best accomplish his purpose with the aid of a certain amount of training in method, while to him who proposes a career of investigation, there is an equal necessity for a large amount of positive knowledge.

Before proceeding to amplify these propositions it seems best to give consideration to the essential nature of scientific research—to restate, for the sake of a common understanding, the process by which science advances.

Scientific research consists of the observation of phenomena and the discovery of their relations. Scientific observation is not sharply distinguished from other observation. It may even be doubted whether there is such a thing as unscientific observation. If there is a valid distinction, it probably rests on the two following characters. Scientific observation, or the observation of the investigator, endeavors to discriminate the phenomena observed from the observer's inference in regard to them, and to record the phenomena pure and simple. I say "endeavors," for in my judgment he does not ordinarily succeed. His failure is primarily due to subjective conditions; perception and inference are so intimately associated that a body of inferences has become incorporated in the constitution of the mind. And the record of an untainted fact is obstructed not only directly by the constitution of the mind, but indirectly through the constitution of language, the creature and imitator of the mind. But while the investigator does not succeed in his effort to obtain pure facts, his effort creates a tendency, and that tendency gives scientific observation and its record a distinctive character.

Scientific observation is moreover selective and concentrated. It does not gather facts indiscriminately, but, recognizing their classification, it seeks new facts that will augment established groups. The investigator, by restricting his observation to a limited number of groups of phenomena, is enabled to concentrate his attention, and thus sharpens his vision for the detection of matters that are unnoticed by the ordinary observer.

The superficial relations of phenomena are discovered by induction—by the grouping of facts in accordance with their conspicuous common characters—or, in other words, by empiric classification. Such empiric classification is a preliminary work in all sciences. It is a convenient and temporary sorting of our knowledge, and with the increase of knowledge it is perpetually remodeled. But it is more than a mere convenience; it is a stepping-stone to a logical, or rational, or, more strictly, relational classification; for it leads to the understanding of those deeper relations which constitute the order of nature.

Phenomena are arranged in chains of necessary sequence. In such a chain each link is the necessary consequent of that which precedes, and the necessary antecedent of that which follows. The rising of the sun is consequent on the rotation of the earth. It is the logical antecedent of morning light. Morning light is in turn the consequent of sunrise and the antecedent of numerous other phenomena. If we examine any link of the chain, we find that it has more than one antecedent and more than one consequent. The rising of the sun depends on the position of the earth's axis as well as on its rotation, and it causes morning heat as well as morning light. Antecedent and consequent relations are therefore not merely linear, but constitute a plexus; and this plexus pervades nature.

Relational classification may be considered as of two sorts, first linear, and second, coördinate as determined by linear, that is to say, phenomena are linearly arranged in chains of sequence, and they are coördinately arranged in natural classes. A natural class is a group of coördinate facts having the same antecedents.

It is the province of research to discover the antecedents of phenomena. This is done by the aid of hypothesis. A phenomenon having been observed, or a group of phenomena having been established by empiric classification, the investigator invents an hypothesis in explanation. He then devises and applies a test of the validity of the hypothesis. If it does not stand the test he discards it and invents a new one. If it survives the test, he proceeds at once to devise a second test. And he thus continues until he finds an hypothesis that remains unscathed after all the tests his imagination can suggest.

This, however, is not his universal course, for he is not restricted to the employment of one hypothesis at a time. There is indeed an advantage in entertaining several at once, for then it is possible to discover their mutual antagonisms and inconsistencies, and to devise crucial tests,—tests which will necessarily debar some of the hypotheses from further consideration. The process of testing is then a process of elimination, at least until all but one of the hypotheses have been disproved.

In the testing of hypotheses lies the prime difference between the investigator and the theorist. The one seeks diligently for the facts which may overthrow his tentative theory, the other closes his eyes to these and searches only for those which will sustain it.

Evidently, if the investigator is to succeed in the discovery of veritable explanations of phenomena, he must be fertile in the invention of hypotheses and ingenious in the application of tests. The practical questions for the teacher are, whether it is possible by training to improve the guessing faculty, and if so, how it is to be done. To answer these, we must give attention to the nature of the scientific guess considered as a mental process. Like other mental processes, the framing of hypotheses is usually unconscious, but by attention it can be brought into consciousness and analyzed.

Given a phenomenon, A, whose antecedent we seek. First we ransack the memory for some different phenomenon, B, which has one or more features in common with A, and whose antecedent we know. Then we pass by analogy from the antecedent of ·B, to the hypothetical antecedent of A, solving the analogic proportion—as B is to A, so is the antecedent of B to the antecedent of A.

Having thus obtained an hypothesis, we proceed to test it. If the hypothetical antecedent is a familiar phenomenon, we compare its known or deduced consequents with A, and observe whether they agree or differ. If it is unfamiliar, we ascertain its consequents by experiment or some other form of observation; and in the selection of the particular experiments or observations to serve as tests, we are guided once more by analogy, inverting the previous formula.

The question, whether or not the function of the mind in devising hypotheses and the tests of them is creative, is foreign to the present purpose. It suffices that we recognize the process as analogic, requiring for its success a preliminary knowledge of numerous instances of consequential relations. The consequential relations of nature are infinite in variety, and he who is acquainted with the largest number has the broadest base for the analogic suggestion of hypotheses. It is true that a store of scientific knowledge cannot take the place of mental strength and training, i. e. of functional ability inherited and acquired, but it is nevertheless a pre-requisite of fertility in hypothesis.

The great investigator is primarily and preëminently the man who is rich in hypotheses. In the plenitude of his wealth he can spare the weaklings without regret; and having many from which to select, his mind maintains a judicial attitude. The man who can produce but one, cherishes and champions that one as his own, and is blind to its faults. With such men, the testing of alternative hypotheses is accomplished only through controversy. Crucial observations are warped by prejudice, and the triumph of the truth is delayed.

Returning now to the subject of education, take first the case in which the student is to become an investigator. He is to observe phenomena, he is to frame and test hypotheses. As a matter of course, in order to learn to do these things he must do them. Sooner or later he must be sent directly to nature, out of doors or in the laboratory, and must in her presence train his faculties by practice. But before he undertakes this, the teacher can aid him by imparting methods. It is probably not best to offer them in the abstract until he has become well acquainted with them in the concrete. Typical investigations should be described in detail, illustrating the varied phases of the method of hypothesis, and not omitting to show how its successes are achieved through series of failures. The history of at least one science should be developed, with the rise and fall of its successive theories. These educational factors are directed to the training of his mind, but his mind needs also to be stored with scientific knowledge, which shall serve as a foundation for analogies. If he would explain some feature of nature, he must depend on the explanations others have reached for other features; and he needs large resources of knowledge of the relations of phenomena.

The course of training for the apprentice of science should give him, in the study room and in the class room, a varied acquaintance with the laws of nature that have been discovered by research. It should not needlessly burden his memory with empiric classifications, for these belong to the humbler walks of science, and it is unwise to impress on the novice the high importance of that which the master regards as provisional. It should teach observation by actual practice,—practice rigorously restricted to selected groups of phenomena. It should illustrate with varied reiteration—by books, by lectures, by demonstrations in the laboratory—the method of discovery by the aid of hypotheses. It should assign him actual investigation and subject his methods to criticism.

Students whose projected careers are not scientific, but who are unwilling to ignore so important a subject, naturally wish to cover a wide field in a short time. Their teacher, imbued with the vastness of science, is tempted to give them a maximum number of facts, with such order and classification as best favor their rapid statement. If he yields to the temptation, there is reason to fear that a permanent misapprehension is established, and the essence of science is not communicated. In my judgment he will do better to contract the phenomenal, and enlarge the logical scope of his subject, so as to dwell on the philosophy of the science rather than its material. For such students laboratory work may or may not be expedient, but they can at least accompany the pupils who look forward to careers in research in some of the descriptive illustrations of methods of scientific achievement.

The investigator becomes an educator when in giving his work to the world he describes the route by which his end was reached. It is not denied that the publication of sound conclusions is in itself educational, but it is maintained that the publication of the concrete illustration of a good method is educational in a higher sense. It is not insisted that all skillful investigations should be published *in extenso;* it is only affirmed that the number of such publications is far too small. We need for educational purposes more narratives of good work in all departments of research. Let the discoverer of a new principle recite every hypothesis that occurred to him in the course of his search, telling, if he can, how it was suggested. Let him lay bare the considerations which rendered it plausible, the tests that were conceived, and those which were applied. Let him show in what way the failure of one hypothesis aided in the invention of another. Let him set forth not only the tests which verify his final hypothesis, but the considerations which leave a residuum of doubt as to its validity. And finally let him indicate, if he can, the line or lines of research that promise to throw more light.

By so doing he will accomplish many things. He will guard himself against an overestimate of the strength of his uneliminated hypothesis, and he will thus diminish his self-conceit. By conscious attention to his methods he will improve them. He will therefore educate himself.

He will inspire the young investigator by his example, and even his experienced compeer will take courage from the success that after many failures finally crowns his efforts. He will give to every investigator who reads his paper a lesson in method— a good lesson if his method is good, and not necessarily a bad one if his method is bad. He will therefore educate his fellow workers.

If his work admit of popular presentation, he may be a missionary as well as a teacher, for he may help to dissipate the wide-spread impression that there is something occult in the ways of science. He will at least aid in showing that, whenever a theory is created and tested, knowledge is the gainer, whether the theory itself stands or falls; and that the demolition of hypotheses, instead of testifying to the futility of research, is the method and condition of progress. His educational influence will thus extend to that great lay member, the general reader.

In making this plea for education by example, it would be unfair to ignore another point of view. Not all are willing to be educated, not all need be; the majority of those who examine an essay seek only to learn its conclusions and have time for nothing more. For their use there should be appended or prefaced a concise summary of results.

And, on the other hand, it should be observed that the service rendered to science by one who describes his course of investigation is not educational merely. Rejected hypotheses have a positive value in the domain of the subject to which they belong, and he who makes them public gives to his

Am. Jour. Sci.—Third Series, Vol. XXXI, No. 184.—April, 1886.
19

fellow-workers in the special field the fullest advantage of his material. Some steps of his progress, which did not prove suggestive to him, will find fertile ground in the mind of another and bear fruit. This consideration places the progress of knowledge before the glory of the individual, and is opposed by a natural egoism; but it is only the man of small calibre who has no ideas to spare, and secretiveness in matters of science is ordinarily a confession of weakness.

It was intimated a moment ago that precept unsupported by example could not be depended on to infuse method—and the dictum applies even to the burden of this discourse. I am persuaded that my meaning will be better apprehended if I supplement my disquisition by an outline of an investigation of my own. The seeming egotism must be condoned, for it is manifestly impossible for me to trace out the actual course of observation and reasoning in the case of another's work.

To guard against possible misapprehension it is necessary to emphasize the fact that the following discussion contains an outline merely of its subject. Its subject is a certain geologic uplift that has been observed in Utah. To render it intelligible to those who are unacquainted with the literature of the geology of Utah, it will be introduced by a short account of Lake Bonneville.

The basin of Great Salt Lake lies in a region of mountains; to picture its character to your mind, conceive a plain the surface of which is embossed by parallel ridges of moderate length, from fifteen to twenty-five miles apart, and from 2,000 to 6,000 or 7,000 feet high. Conceive further that portions of this plain are uplifted, together with their mountain ridges, so as to enclose a basin 150 miles in either dimension, and you have the general structure of the district in question. The debris washed down from the mountains has for ages accumulated in this depression, so that the central-lying mountain ridges are nearly buried; indeed there is reason to suspect that some of them are quite buried, a plain of fine silt being spread smoothly over them. Great Salt Lake itself lies on the east side of the basin; the western half, which is only a few feet higher, is a saline desert.

In the last geologic epoch—the Glacial Epoch—the lake expanded so as to fill the basin to overflowing. The water surface was then very much larger, and as its area included the basins of several lakes now independent, it has been given a separate name. Lake Bonneville was very irregular in form; the mountain ranges of the basin ran long peninsulas from its north and south shores, and projected from its surface in numerous islands. The Quaternary winds, playing on its surface, dashed waves against its shores, and the spits and beaches and cliffs wrought by these waves remain in a high state of preservation to testify to the position of the ancient water margin. Through the greater part of its extent this shore-line forms a conspicuous feature in the topography of the country, and is readily traceable. It has been actually traced out, surveyed, and mapped with much care, and our knowledge of the old lake is in many other respects definite and full.

Its width from a few miles east of Great Salt Lake to the west side of the Great Salt Lake desert was 125 miles. From the mountains on the north of the desert to those on the south its expanse was about the same, but it did not terminate with the southerly mountains. It extended through them in several straits and formed beyond a second and much smaller body of water. The main body was 1,000 feet deep, the minor body about 500 feet.

Manifestly, when this old shore-line was made, all parts of it lay in the same horizontal plane, with no other curvature than that which belongs to the figure of the earth ; that is, all parts of it were level. If it is not now level—if some parts are higher than others, it seems equally manifest that there have been local elevations or subsidences of the land. It is to such differences of level in the shore-line as it stands, and to their interpretation, that I desire to call your attention.

As far as the eye can judge, the shore is still level, and so long as no measurements were made its horizontality remained unquestioned. The two geologists who were probably the first to measure its height recorded their results in language implying no suspicion that more than one determination was necessary.

It happened that in the year of my first exploration of the Bonneville area I saw the shore-line not only in the Salt Lake basin but in the more southerly basin, and that I passed from one basin to the other by a route that did not reveal their connection. In doubt whether two lakes were under observation or only one, I sought to answer the question by determining the height of the shore-line in each, my instrument for the purpose being the barometer. The verdict of the barometer was that the southerly shore-line was somewhat higher than the northerly, but the computations necessary to deduce it were not made until the mutual continuity of the two shore-lines had been ascertained by direct observation. The barometric measurement was therefore superseded as an answer to the original question, but it answered another which had not been asked, for it indicated that the ancient shore at one point had come to stand higher than at another. The postulate of horizontality was thus overthrown.

An hypothesis immediately took its place. It is one of the

great inductions of geology that as the ages roll by the surface of the earth rises and falls in a way that may be called undulatory.   I do not now refer to the anticlinal and synclinal flexures of strata, so conspicuous in some mountainous regions,but to broader and far gentler flexures which are inconstant in position from period to period.   By such undulations the Tertiary lake basins of the Far West were not only formed but were remodeled and rearranged many times.   By such undulations the basin of Great Salt Lake was created.   As to their cause, geology is absolutely ignorant, and she is almost absolutely silent.   When it was ascertained that the Bonneville shore-line attwodistant points had not the same height, the first hypothesis to suggest itself merely referred the difference to this gentle undulatory movement of the crust.   As other hypotheses are to be mentioned, it will be convenient to christen this one the hypothesis of unexplained undulation.

A few years later the discovery was made that a fault had occurred along the western base of the Wasatch range of mountains since the Bonneville epoch.   This range lies just east of Great Salt Lake, and the Bonneville shore is traced across its western face.   The effect of the fault was to lift the mountain higher, with reference to the lake bottom, and to carry that part of the old shore-line upward.   The amount of the uplift varied in different parts of the fault from ten to fifty feet.

This discovery was something more than the finding of a post-Bonneville fault; it was the discovery also of a new method of recognizing faults—of a peculiar type of cliff produced by faulting, which, though by no means obscure, had previously been overlooked by geologists.   It gave rise to a new tentative explanaticn for the displacement of shore-line discovered by barometer, namely, that it arose by faulting; and it opened a new line of observation.

Two hypotheses were now under consideration, but they were not strictly alternative.   Perhaps it would be better to say that the origination of the second not only gave an alternative but also modified the first.   The Wasatch fault must affect the height of the shore-line, and wherever it crossed the shore-line that line must be discontinuous and exhihit two levels.   The admission of disturbance by faulting was therefore compulsory, but faulting might or might not be sufficient alone.   If it was not sufficient, then undulation might complement it.   The modified first hypothesis was, undulation and faulting combined, the second, faulting alone—both undulation and faulting being themselves unexplained.

It was not difficult to devise tests.   An instrumental level line might be carried along the old beach so as to ascertain whether it rose or fell in regions where no faults occur; or its height might be carefully measured at two points and the difference of altitude compared with the total throw of the intervening post-Bonneville faults. The second of these tests was applied, and with success. By means of the surveyor's level the height of the old shore above the water surface at the shore of Great Salt Lake was measured at two points twenty miles apart. One of these points is on the Wasatch range near Salt Lake City; the other is on the next range west, the Oquirrh. The only post-Bonneville fault between them is that at the base of the Wasatch, and its throw is there about fifty feet, the west side having gone down. If then faulting is alone responsible for shore-displacement, the beach on the Oquirrh range should be fifty feet lower than that on the Wasatch. The measurement however showed it to be twenty-eight feet higher, and thus demonstrated a difference of seventy-eight feet to be referred to undulation.

Thus a step was made in advance, but the resulting position was not final, for the inquiring mind could find no satisfaction in the knowledge that crust undulation and crust faulting were conjointly efficient, so long as both these remained without explanation. No new hypotheses were at once invented, but it was determined to continue observation until the solitary phenomena at command were expanded into a group, and to seek new light in the classification of this group. As the basin was traversed in the conduct of the general investigation of the old lake, a search was made for the records of recent faults and at every opportunity the height of the shore was accurately measured. Six such measurements were made in the immediate vicinity of Great Salt Lake, the lake affording a common datum plane. Ten others were made on the lines of railways, where the leveling data of the railway engineers could be utilized. At some points the height was found greater than on the Oquirrh, at others less than on the Wasatch, the range from highest to lowest being 168 feet.

Faults were discovered at the bases of numerous mountain ranges, but none of them are so great as that along the Wasatch, and nearly all are very small. None were found associated with the half-buried mountains of the center of the desert, and yet on these same mountains are the highest shore records to which measurement was carried.

In general it was found that the displacements recorded by the shores have been much larger than the displacements demonstrated by faults, so that faulting can be appealed to in explanation of shore displacement only to a small extent. It was found that the throw of the faults was in some cases opposed in direction to the total deformation on the shore-line and in other cases coincident. For these reasons faulting was provisionally regarded as a disturbing factor merely, and the deformation demonstrated by the measurements of shore-height was treated as simply flexural or undulatory.

To classify the shore-heights, as a basis for further hypothesis, they were platted on a map, and their grouping was compared with geographic features. It appeared that the highest measured points lay within the area of the main body of Lake Bonneville, that the lowest points lay at the extreme north and at the extreme south, that the eastern shore of Lake Bonneville in the vicinity of Great Salt Lake was intermediate in height, and that the single point determined on the western shore of the old lake agreed in height with the eastern shore. Unfortunately, the distribution of the measurements, which had been largely determined by the distribution of railroad lines, was not equable throughout the basin of the lake, and nearly the whole of its western shore was undetermined in altitude. When lines of equal altitude were drawn among the figures representing measurements, after the manner of the isobars on a Signal Service weather map, it was found that they were not fully controlled by the determined points; but when they had been given the most satisfactory adjustment, they contoured a figure of deformation which may be characterized as a low, broad dome, having its crest over the center of the main body of Lake Bonneville, and extending a subordinate member to the region of the southern body of the lake. One half of this figure was fairly inferred from the data of observation; the remaining half was imaginary and its drawing merely gave graphic expression to the hypothesis suggested by the incomplete contours—the hypothesis that the deformation stands in some necessary or causal relation to the lake and its disappearance.

Now it has been independently determined that the cause of the lake and the cause of its disappearance were climatic; it was not drained by the wearing down of its outlet, nor emptied by the unequal uplift of portions of its rim, but it was dissipated by evaporation. If then the disappearance of the lake and the deformation of the land are connected in a causal way the change in the lake was the cause, and the change in the land was the effect. How can we suppose the drying up of the water to have produced the up-arching of the plain on which the water lay?

In the attempt to answer this question three tentative explanations were suggested, and these will be stated in the order of their origination.

It is well known to geologists that in several instances a great formation thousands of feet in thickness consists wholly or chiefly of shore deposits. To account for them it is necessary to suppose that the sea floor locally sank down as rapidly as the sediments were added. Conversely there is reason to believe that the adjacent continent, which by erosion furnished the sediment, rose up as rapidly as its surface was degraded. It is a favorite theory—at least with that large division of geologists who consider the interior of the earth as mobile—that the sea-bottom sinks in such cases because of the load of sediment that is added and that the land is forced up hydrostatically because it is unloaded by erosion. A similar theory might explain the up-arching of the desiccated bed of Lake Bonneville, for the unloading of 1000 feet of water from an area more than one hundred miles across would give to the supposed liquid interior an irresistible uplifting force. This was the first explanation to suggest itself.

The second suggestion did not spring from any geological theory consciously retained in memory, but I have since suspected that the germ of the idea may have been caught from a passage in Croll's 'Climate and Time.' It is this: The geoid of which the ocean's surface is a visible portion is not an ellipsoid of revolution, but differs from that symmetric surface by undulations which depend on local inequalities in the density and in the superficial configuration of the earth. The water level is everywhere normal to the plumb-line, but the plumb-line, as geodesy has shown, is subject to local deflection. Now the ocean itself is one of the attracting factors, and if the ocean were to be removed, the geoid would thereby be modified. The surface of Lake Bonneville was part of a geoid at a higher plane than that of the ocean surface, and the removal of the water of the lake unquestionably modified the local form of the geoid. Only at first blush the cause seems too small for the effect observed.

The third suggestion relates to the distribution of temperatures beneath the surface of the earth. It is well established that the inner parts of the earth are extremely hot. The outer surface is relatively cool, and in the intermediate region there is a gradation of temperatures. The isogeotherms, or planes of equal temperature, are not even surfaces, but undulate in response to variations of conductivity and of superficial temperature. At the poles, where the external surface of the crust is exceptionally cold, the isogeotherms lie lower down than in warmer latitudes; and if a portion of the earth's surface undergoes a permanent change in temperature, the influence of this change is propagated slowly downward through the crust, and the isogeotherms are locally raised or lowered. Where they are raised, the crust is locally expanded, and its surface is uplifted; where they are depressed the surface of the crust subsides. If, therefore, it can be shown that the temperature at the bottom of Lake Bonneville was raised in connection with the dessication of the lake, we have a true cause of upward movement, and if we can show furthermore that the temperature of the surrounding region was not equally raised, we have at least a qualitative explanation of the differential uplift, the phenomenon to be accounted for.

It is now several years since these explanations were first suggested, and subsequent reflection has developed no others. While all of them appear perfectly rational, only a very slight inspection was necessary to raise a doubt as to the quantitative sufficiency of the second and third. It was therefore determined to ascertain as accurately as possible the maximum change which might be ascribed to each of the three suggested causes, and to compare it with the actual change. The actual change is susceptible of various statements. If we consider only the measurements on the margin of the main body of water, and in its center we find a difference of 100 feet; by including observations on outlying bays we get a maximum difference of 168 feet; and a study of the peripheral slopes of the uplift suggests that they extend somewhat beyond the boundaries of the lake. Crude extrapolation gives 200 feet as a maximum estimate of the height of the crustal dome.

Take first the hypothesis that the crust of the earth, floating on a molten nucleus, rose up in the region of the basin when its weight was locally diminished by the removal of the water of the lake. The weight of the load removed is measured by the depth of the water before evaporation, 1000 feet. The theory supposes that as the crust rose there flowed in beneath enough molten rock to replace the weight of the evaporated water. If the rock was very heavy, a layer of moderate depth was necessary; if it was less heavy, more was required; but in any event the thickness of the introduced layer must be equal to the amount of the superficial uplift. It is known that the density of the earth's material increases downward, for the mean density of the earth, expressed in terms of the density of water, is about 5·5, while that of the upper portion of the crust is about 2·7. Nothing is known however of the law under which the density increases, and nothing is known as to the depth of the zone at which matter is sufficiently mobile to be moved beneath the Bonneville basin. We may, however, indicate limits, and I think this is fairly done by assuming that the density of the introduced matter was not less than 3, nor more than 5·5. If it was 5·5, the uplift consequent on the evaporation of 1000 feet of water would be 182 feet. If it was 3, the uplift would be 333 feet. Now it has already been stated that the greatest value observation suggests for the amount of the uplift is 200 feet. The postulate is therefore abundantly competent in a quantitative way.

To evaluate the effect produced under the second hypothesis (the hypothesis, that is, that the geoid represented by the water surface of Lake Bonneville has been deformed by the withdrawal of the attraction exerted by the water itself) it is necessary to employ mathematical analysis of a high order. As my schooling in mathematics did not qualify me to undertake this, I submitted the problem to an eminently competent colleague, who has solved it rigorously and deduced for the deformation of the geoid within the area of Lake Bonneville a maximum amount of two feet. The second explanation is therefore eliminated from consideration, because quantitatively insufficient.

It remains to consider the rise of the isogeotherms, and to evaluate the resulting elevation of the basin. It has been established by numerous observations that in all lakes having a depth as great as 1000 feet, the temperature at the bottom is about 39° F. This depends upon the fact that water at that temperature is heavier than at any other, and having once reached the bottom of a deep lake, it is withdrawn from the circulation to which the upper layers are subject, and remains undisturbed. The meteorological records show that the mean annual temperature of the desiccated basin of Lake Bonneville at the present time is 52°. The change from a humid to an arid condition has therefore raised its temperature 13°. The temperature of the surrounding regions has at the same time undergone a change, of which we have no precise estimate. The epoch of Lake Bonneville was the Glacial Epoch, and the local climate was then in all probability cooler. If it was 13° cooler, the isogeotherms would be no more affected at the center of the basin than at its margins, and there would be no differential elevation. If it was cooler by less than 13° a differential uplift would occur. For the sake of giving this uplift a maximum value, we will assign a very small figure to the general change of temperature, namely 3°, and assume that the differential change with respect to the basin was 10°. A formula devised by Fourier enables us to estimate the rise of the isogeotherms, if only we know the conductivity of the material of the earth, and the time which has elapsed since the Bonneville shore line was carved. Then, if we know additionally the rate of expansion of rock for a degree of temperature, we are able to estimate the upheaval. Sir William Thompson has determined experimentally a coefficient of conductivity. The late Prof. Bartlett, of West Point, has determined the coefficient of expansion for several building stones, which may be assumed to represent the crust beneath the Bonneville basin. We do not know how these coefficients are affected by high temperatures and great pressures, such as exist deep in the crust, and an element of uncertainty attaches for that reason.

In order to obtain a maximum result despite this uncertainty, I have made an extreme assumption in regard to time. The shore line of Lake Bonneville is in a wonderfully perfect state of preservation. While one stands upon it, it is easy to believe that it is but a few centuries old, and the geologist, accustomed as he is to the contemplation of eons of time, hesitates to estimate its antiquity in greater units than thousands or at most tens of thousands of years. When therefore we postulate its antiquity at one hundred millions of years, we pass so far beyond the range of probability as to protect ourselves against a possible underestimate. With these data the computation has been made, and it has been ascertained that a maximum uplift of 36 feet* can thus be accounted for. Since observation shows an uplift of not less than 100 feet, the thermal explanation is shown to be entirely inadequate; and if we were able to substitute for our imperfect data the actual data, we should probably find the computed uplift too small to be taken into consideration.

If therefore we admit that the removal of the water of the lake was the cause of the upheaval of the lake-bottom, there seems no way to avoid the conclusion that the efficient *modus operandi* was an upbending of the solid crust of the earth, caused by hydrostatic pressure communicated through a mobile substratum. But we are far from being forced to that admission. The coincidence in locus of the uplifted dome and the Quaternary lake may have been fortuitous; or there may even have been no coincidence, for the contoured figure of deformation was in part supplied by the imagination; and in either of these cases we can fall back on the agnostic hypothesis of unexplained undulation. In the present state of observation and inference the hypothesis of the hydrostatic restoration of equilibrium by the underflow of heavy earth-matter is the only explanation which explains, and none of the observed facts antagonize it; but the alternative hypothesis is not barred out.

To reach a satisfactory conclusion more observation is necessary, and this discussion of the subject would be premature were it not that the necessary observation is very expensive, and there is no immediate prospect that it will be supplied. It is fitting, however, that the desirable lines of research be pointed out.

The undertaking that promises most is an exhaustive hypsometric survey of the Bonneville shore line, including all bays and islands. If this were executed, it would be possible to deduce

---

* In the original paper, as read, 12 feet instead of 36 were erroneously given. A friend has since pointed out that the estimate of 12 feet includes expansion in the vertical direction only, whereas the coincident horizontal expansion would almost necessarily be converted into uplift.

a much more satisfactory expression of the shape of the uplift, and to determine either that it is intimately related to the form of the body of water removed, or that it is not so related. If the relation were demonstrated, the observations might so far indicate its nature as to render possible an evaluation of the rigidity of the earth's crust.

Another profitable method of continuing the inquiry would be to make a similar investigation of the shore of another extinct lake, for example, the one to which Clarence King has given the name of Lahontan, and which ranks second to Lake Bonneville among the Quaternary lakes of the Great Basin. If in a second instance the center of the desiccated lake were found to be the locus of upheaval, the hydrostatic theory would be practically established.

It is hardly necessary for me to assure you that my personal regret in abandoning this research at its present stage is very great. I have discussed it as an investigation of the deformation of the Bonneville basin, but it has a broader meaning. The condition of the interior of the earth is one of the great problems of our generation. Those who have approached it from the geologic side have based a broad induction on the structural phenomena of the visible portion of the earth's crust, and have reached the conclusion that the nucleus is mobile. Those who have approached it from the physical and astronomic side have reached the conclusion that the nucleus is rigid. Here seems an opportunity for a crucial observation. If the crust of the earth floats upon a fluid nucleus, the evaporation of Lake Bonneville, by lifting from it a great weight, must have produced an uplift of determinate form. If the whole earth is solid, such a result could not have been wrought. The decisive phenomena are known to exist, and to be accessible, but they are scattered over a broad desert, and they can be gathered in only at the cost of much money and great labor.

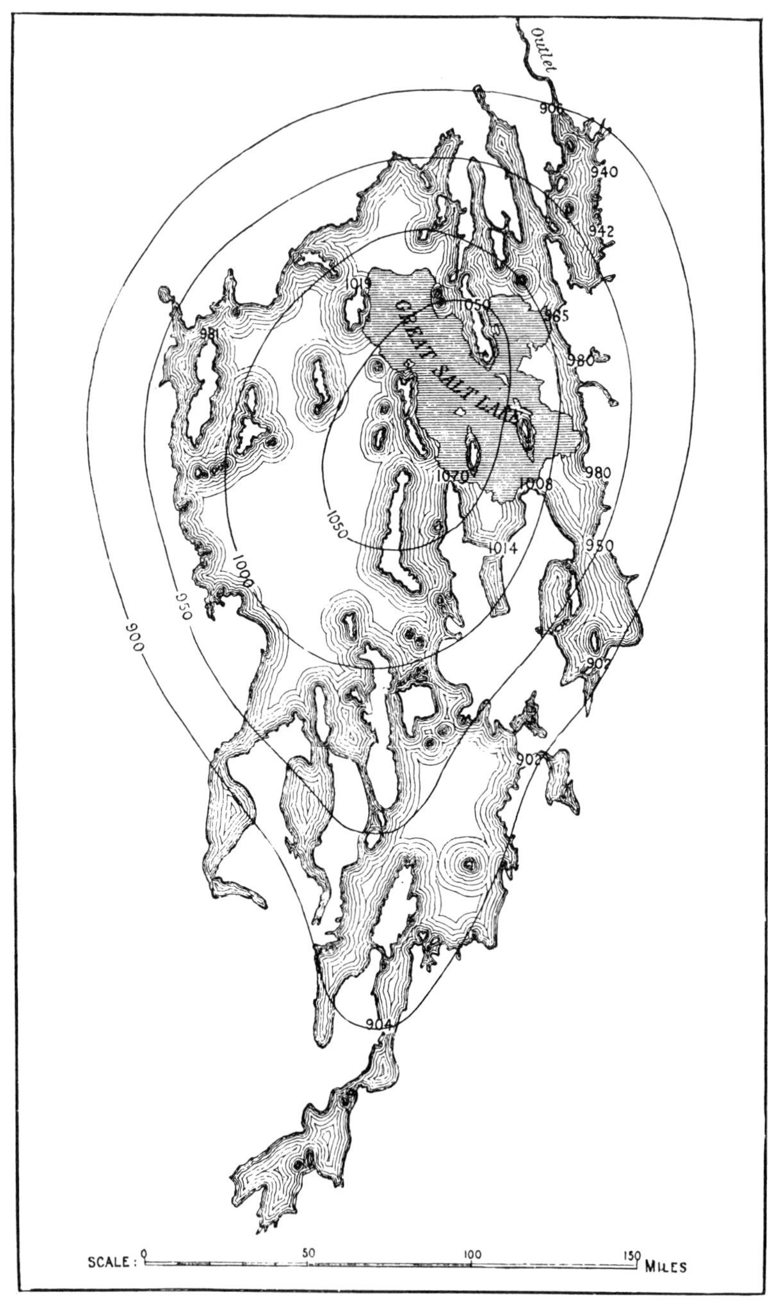

Map of the extinct Lake Bonneville, with hypothetic contours to illustrate the
subsequent deformation of the earth's crust, by G. K. Gilbert. The horizontal
figures mark determinations of the height of the old shore-line above Great Salt
Lake.

The Value of Outrageous Geological Hypotheses[1]

W. M. Davis
Harvard University

Meetings of geological societies in these modern days are often somewhat prosaic as compared to those of an earlier time when the limits as well as the methods of geological speculation were less defined than now, and when contradictory differences of opinion were commonly expressed even with regard to fundamental ideas concerning the conditions and processes of earth history. That was a time when the scientific imagination, not so much hampered as it is now by standardized principles, was accustomed to roam with little restraint over the unexplored fields of geological investigation; a time when the facts regarding the earth's crust had been gathered from a relatively small part of its surface, when a theory was thought to be established if it explained nothing more than the facts which it had been invented to explain, and when lively discussion as to the merits of rival theories too often degenerated into polemical diatribes between rival theorists.

In those earlier days, attendance at the meetings of Section E of the American Association of Science—the only meetings in which geologists from different parts of the country were then brought together—was likely to be rewarded by a vigorous, not to say vituperative dispute between Marsh and Cope, not merely as to the completed structures and systematic relationships of the fossil vertebrates that they were finding in the fresh-water Tertiary deposits of the west, but also as to mere priority in finding and naming the fossils; and so eager was each of those eminent worthies to secure his prior claim for a new find before the other came upon it that, according to stories then current, one or both of them sometimes, while still in the western field, resorted to the telegraphic announcement of a name for a newly discovered fossil to be published in the eastern newspapers. In the years, half a century

ago, when I first attended the meetings of the Boston Society of Natural History, they were occasionally the scene of emphatic contradictions between T. Sterry Hunt and M.E. Wadsworth on matters petrographic, for that recondite branch of geological science was then just taking form among us. Hunt knew exactly how rocks ought, in accordance with his theoretical views of terrestrial chemistry, to be constituted; while Wadsworth, in view of his observational study of thin sections, knew exactly how rocks are constituted; and each of these convinced positivists maintained his view with earnest vehemence.

It is as a result of many verbal battles then fought without asking or giving quarter that geology has come in these modern days to be a relatively well-restrained and orderly science. How much more carefully are facts scrutinized, and how much larger and safer is the inductive base of our generalizations now than formerly. How narrowly limited is the special field, either in subject or locality, upon which a member of the Geological Society of America now ventures to address his colleagues; so narrow that he often has it pretty much all to himself, and so thoroughly does he cover it that when his statement is completed there is little or nothing left for any one else to say. How much more rigorously logical is the guidance of the train of thought by which advance is made from the facts of observation to the conclusions of theory; and if by good fortune a hearer differs from a speaker as to the track along which the train of thought should be directed, how seldom does he intimate his difference of judgement in any but the most courteous manner! How utterly extinct is rudely polemical dissension; so extinct indeed that the younger geologists of to-day must be surprised to learn that it ever flourished. I wonder sometimes if those younger men do not find our meetings rather demure, not to say a trifle dull; and whether they would not enjoy a return to the livelier manners of earlier times.

Yes, our meetings are certainly prosaic to-day as compared to those of the earlier formative period when speculation was freer and when differences of

---

[1]An address delivered before the Leconte Club of the University of California at its annual meeting at Berkeley, February 21, 1925.

opinion on major principles were almost the rule rather than the exception. Our younger members may perhaps experience a feeling of disappointment, or even of discouragement at the unanimity with which the conclusions of an elder are received by a geological audience; for it must dampen the enthusiasm of beginners if they gain the impression that all the larger generalizations of our science have been established, thus leaving for them to discover only items of localized fact. And a like feeling of discouragement must often be shared by the chairman of a meeting when, after his encouraging invitation, "This interesting paper is now open for discussion," only silence follows. Are we not in danger of reaching a stage of theoretical stagnation, similar to that of physics a generation ago, when its whole realm appeared to have been explored? We shall be indeed fortunate if geology is so marvelously enlarged in the next thirty years as physics has been in the last thirty. But to make such progress, violence must be done to many of our accepted principles; and it is here that the value of outrageous hypotheses, of which I wish to speak, appears. For inasmuch as the great advances of physics in recent years and as the great advances of geology in the past have been made by outraging in one way or another a body of preconceived opinions, we may be pretty sure that the advances yet to be made in geology will be at first regarded as outrage upon the accumulated convictions of to-day, which we are too prone to regard as geologically sacred.

It was outrageous, two centuries ago, to interpret fossils as records of ancient life; for that interpretation did violence to the view then accepted as to the manner in which the earth had been formed and as to the date at which life had come to exist upon it. It was outrageous, little more than a century ago, to discover fossils of marine organisms in the disordered strata of lofty mountains high above sea level; for that discovery did violence to the ideas then obtaining as to the stability of the earth's crust. And it was equally outrageous, half a century ago, to be told that after mountains had been lifted up, they might in time be worn down to lowlands again, for that idea did violence to the views that had then come to be held regarding the instability of the earth's crust. It was an outrage upon the tacitly accepted principles of geological climatology, based on the postulate of a cooling earth, that there already should have been a glacial period in the past; and for that matter, the form in which the glacial theory was first promulgated was truly enough outrageous; nevertheless it now, in a much modified form, holds good as a standardized geological verity.

It was altogether outrageous to think that man had long been an inhabitant of the earth, instead of looking upon him as a new comer; and it was equally outrageous to discover that the sequence of fossils preserved in successive stratified formations indicated such a progression of life as would result from the evolution of later forms from earlier forms, instead of simply an arbitrary succession of independent creations. It is still rather outrageous to think that the earth has long been and possibly is still heating itself up by the slow compression of an originally uncompacted interior under the weight of a heavy exterior, instead of thinking that it has long been and still is cooling by the slow loss of a great original store of heat. And in view of the many evidences of crowding in the outer crust, it may be thought wantonly outrageous to look upon the earth as possessing an expanding interior which, like the caged starling, "wants to get out." Yet I believe it the part of wisdom to view even that outrage, as well as the Wegener outrage of wandering continents and the Joly outrage of periodical subcrustal heating-up and breaking out, calmly, as if they were all possibilities; and it may also come to be the part of wisdom to ask ourselves in what way and how far our present conception of the earth must be modified in order to transform such outraging possibilities into reasonable actualities; for that is precisely the way in which the above-listed outrages and many others have gained an established place in our science. Of course, if we do not approve of the necessary modifications we may reject them, and with them the outrages that they countenance.

Let it be noted in passing that the omission of the original L from the leading word of the preceding paragraph unfortunately results in its being pronounced as if it were derived from "out" and "rage"; its true meaning would be better indicated if its form were ultrageous, as it might well have been had not the L been lost on the way from Latin through French into English; for with the L preserved, the T and R would be joined in the second syllable and properly separated from the first. A word of opposite meaning would then be, not in-rageous, but in-trageous; and our language would be much more symmetrically developed if that and many similar opposites were added to it. However, if we are not allowed to say ultrageous, we might—or at least those of us who pronounce the French-English word, "route," like the English word "root" might—say oo-trageous, and thus reasonably avoid the implication of an erroneous popular etymology. But this is an irrelevant digression.

All that was necessary to make the outrageous occurrence of fossils reasonable and believable was to remodel our conception of the earth from that of a recently-and-ready-made planet into that of a very ancient and slowly changing planet, on which life had existed for ages and ages, always under the influence of environing conditions and in the presence of slow-acting processes very much like those of to-day; and when the ideal counterpart of the actual earth was once conceived in this fashion, the earth was still found to be just as comfortable a planet to live on as it had been in association with the earlier concept of a ready-made earth. All that was needed to explain the occurrence of marine fossils in the disordered strata of mountain tops was to replace the concept of an immovable earth's crust by that of a deformable crust; and although the rate of deformation was at first thought to have been violently rapid, the need of such hurry was later seen to be no need at all, but only a fancy; and thereafter the deformation was conceived to be a slow process. And so it has been with one of these outrages after another: their accommodation is easily accomplished by merely replacing one concept of the earth, under which they are unacceptable, by another under which they are acceptable; and the replacement once made, we are just as happy as we were before. To be sure, the process of replacement may be mentally uncomfortable, even distressing, while it is going on; but the moral of that is that we must not allow our concepts of the earth, in so far as they transcend the reach of observation, to root themselves so deeply and so firmly in our minds that the process of uprooting them causes mental discomfort; and one of the best aids toward the realization of this moral may be found in frequently making explicit announcement of all the unproved postulates on which our favorite concepts are based; for then we shall not be so likely to forget that they are all preceded by a great big IF.

We shall be aided in following this counsel if we strive to recognize how far most of our concepts of the earth really do transcend the short reach of observation. It is usual for a field observer to record that he has seen, for example, a ridge of sandstone; yet all that he has actually seen is a series of small and disconnected sandstone outcrops, perhaps not occupying more than a twentieth or a hundredth of the ridge surface; and the composition of the rest of the surface and of all the interior of the ridge is only a matter of inference; truly, a good and justifiable inference, but not the less an inference for being good and justifiable. Similarly, it is customary for a field geologist to record the presence of a fault when he detects the repetition of a given sequence of strata, and indeed to believe in the displacement that the term, fault, implies, as if it as well as the recurrence of the sequence of strata were a fact of observation; yet not only are the underground extensions of the strata and their long-past displacement merely matters of inference, but even the fault-fracture itself is usually inferred instead of being seen; or if seen at all, it is only in small linear extent, thus leaving all the rest of its superficial trace as well as all of its surface, either lost in the air or buried underground, to the imagination. In thus making distinction between the few facts of actual observation and their large extension in a superstructure of inference, it is not intended to impugn for a moment the validity of well-reasoned superstructures, but only to emphasize the inevitable disproportion that must exist between them and their observed basis; and thus to

make clearer the enormously speculative nature of geological science. For let it be noted that, in the case of a fault, we have to do with a double inference; first, the inference as to underground structures from surface outcrops; second, the inference of displacement because of the repetition of the inferred underground structures. Nevertheless, we believe that faults actually exist.

The very foundation of our science is only an inference; for the whole of it rests on the unprovable assumption that, all through the inferred lapse of time which the inferred performance of inferred geological processes involves, they have been going on in a manner consistent with the laws of nature as we know them now. We seldom realize the magnitude of that assumption. A philosopher of the would-be absolute school once said to me, in effect: "You geologists have an easy way of solving difficult questions: you account for the structures of the earth's crust by assuming that time and processes have been going on for millions and millions of years in the past as they go on to-day; but how do you know that time did not begin only a few hundred thousand years ago after the earth had been suddenly created in imitation of what it would have been if it had been slowly constructed in the manner that you assume?" The answer is as easy as the question: We do not *know*; we merely make a pragmatic choice between the concept of such an imitative creation which seems to us absurd, and a long and orderly evolution which seems to us reasonable. We might, to be sure, were we disposed to be disputatious, turn upon the would-be absolutist and ask him what he is going to do about it; but we have better use for our time than that.

The more clearly the immensely speculative nature of geological science is recognized, the easier it becomes to remodel our concepts of any inferred terrestrial conditions and processes in order to make outrages upon them not outrageous. The more definitely it is understood that the concept of a shrinking earth is based upon certain anterior concepts as to the status of its unobservable interior; the more readily can we entertain the concept of an expanding earth, based upon certain other concepts as to the status of its interior; and it is that particular outrage upon our standardized beliefs that I propose we should contemplate, calmly if possible, and patiently at any rate. To encourage our patience, let me recall another outrageous idea of recent introduction, which in itself is only a sort of reaction from an outrage of somewhat earlier invention and a return toward a more primitive view; namely, the recent idea that those topographical features which we call mountains owe their leading feature, namely their height, not as has been until lately supposed to a vertical movement of escape from the horizontal thrust by which their rocks have been crowded together, but to an uplifting force which acted long after the rocks were crowded together, and in which, as was thought when the view of a mobile earth crust was first promulgated, no component of horizontal thrusting is necessarily involved. A chief difference between that primitive view and its revival in the recent outrage is that the first view took little account of erosion and implied that each individual ridge and peak was the result of an individual or localized uplift; while the second view takes great account of erosion, not only in ascribing the present intermont valleys to the long and slow action of that patient process during and after recent uplift, but still more in ascribing the destruction of the surface inequalities, that must have been earlier produced when horizontal thrusting forces crowded the mountain rocks together, to a vastly longer action of erosion before the recent uplift of the worn-down mass was begun; for where in the whole world can we find mountains that to-day owe their height to an upward escape from horizontal thrusting; in other words, where in the world can we find any existing mountains that are still in the cycle of erosion which was introduced by an upward escape from the horizontal thrusting that deformed their rocks, and not in a later cycle of erosion which was introduced by uplift alone after the inequality of surface form due to earlier thrusting had been greatly reduced, if not practically obliterated!

The conventional phrase, horizontal compression, has been avoided in the preceding paragraph and the alternative phrase, horizontal thrusting, has been used in its stead, in order to prepare the way for the rather mild idea that the same terrestrial forces which produce great overthrusts may also, if somewhat differently applied, produce rock folds, slaty cleavage, and various other phenomena ordinarily explained under the earlier phrase; and thus to prepare the further way for the altogether outrageous idea that overthrusts do not result from the effort of the outer crust to adjust itself to a cooling and shrinking interior, but from the effort of an in-any-way warming and expanding interior to rearrange the outer crust. Of course, this is "impossible"; that is, it is impossible in an earth of the kind that we ordinarily imagine the earth to be; but it not at all impossible in an earth of the kind in which it would be possible. Our task therefore is to try to discover, as judicially and as complacently as we may, what sort of an earth that sort of an earth would be; and then to entertain the concept of that sort of an earth as hospitably as we can and to examine the behavior of such an earth at our leisure.

If an earth with an expanding interior had nothing more to do than to stretch its crust, there would be little trouble in our endeavor; but the concept before us

compels the expanding earth to do various other things also; and especially to produce great crustal overthrusts, the cross-country advance of which is measured in tens or twenties of miles. Hence the outward radial push of the expanding interior must somehow be turned into an almost tangential thrust, and how that is to be done it is difficult to imagine. However, there is no reason for immediate discouragement on that account; it very natural that our imagination to-day should fall short of conceiving all the possible behaviors of a warming and expanding earth, because we are not practised in imagining— that is, is making an image of—that sort of an earth, although a good beginning in that direction has been made in such an essay as that by Boucher on "The pattern of the earth's mobile belts."[2] But we surely have yet much to learn as to what may be all the various reactions of an expanding earth-interior on the shell that encloses it, even though many possible reactions may be now conceived.

For example, let the enclosing shell be defined as that part of the whole sphere which is exterior to the depth at which the next inner shell is warming more rapidly than any other. If that depth be great, the chief thrust of the expanding interior will be exerted on a thick shell; if small, on a thin shell; and the effects of interior expansion visible on the surface will surely be different in the two cases. It seems conceivable that the total thrust of expansion may, in so far as it produces batholithic movement, be slowly concentrated at the weakest part of the shell, and there permit the interior movement to be locally increased by the conversion of cubic expansion into linear expansion or intrusion; this being the converse of the process by which an unduly heavy and therefore isostatically subsiding part of the crust may slowly distribute its local movement through the whole of the interior and there produce a diminished spherical extension, as Lawson has suggested. It seems also conceivable that the movement of a localized batholithic introduction may find advantage in making an outward escape from compelling interior pressures, by changing the direction of its ascent from vertical to oblique, and thus diminishing the rate at which it has to raise the overhead crust. Whether an obliquely ascending mass of this kind could eventually, as it approaches the surface, drive along a slice of crust ahead of it and thus produce what we call an overthrust, is evidently problematical; but if an overthrust could be produced in that way it would be gratifying in certain respects.

If an obliquely ascending batholithic intrusion works its way through a heavy shell toward the surface and there drives ahead of it a crustal slice which we recognize as an overthrust, the oblique emergence of such an overthrust slice will cause an underdrag of the covering rocks in the rear of thrusting advance, and thus displace them with more or less extensional jostling so that they will cover a greater breadth of

---

[2]*Journ. Geol.*, XXXII, 1924, 265–290.

surface than that which they occupied before being underdragged. Surely the need of some such underdragging ought to have been recognized long ago when the prevalence of so-called normal faults which indicate superficial extension, over other faults which indicate compression, was inductively established; and the need is still greater to-day when great faults, such as those of the Basin Ranges, have been found to dip at moderate angles, such as 50° or 40° to the downthrow; for such was the conclusion reached by Gilbert in his latest season of field work in the Great Basin a little over ten years ago.[3] It is true that some geologists maintain the possibility of producing so-called normal faults as an indirect effect of horizontal compressional forces; but even if so contradictory an effect may thus be possibly produced, it by no means follows that such faults can not also be produced much more directly by extensional forces; and the possible cause and working of extensional forces should therefore be investigated; for there is no generally accepted mechanism, like an underdrag, adequate for the strong extensional dislocation of crustal blocks with diverse displacements, to be found in the usual schemes of dynamical geology; and in the lack of such a mechanism, any process, even a fantastic process, that will cause a strong underdrag seems worthy of at least an hour's consideration.

But let no one imagine that I here put forth the idea of an expanding earth interior, with its imagined consequences of an obliquely out-and-over-thrust mass

exerting an underdrag on the superficial crust in its rear, as an idea to be believed. I do not believe it myself, and am therefore doubly far from asking any one else to believe it. The idea is set forth simply as an outrage, to do violence to certain generally established views about the earth's behavior that perhaps do not deserve to be regarded as established; and it is set forth chiefly as a means of encouraging the contemplation of other possible behaviors; not, however, merely a brief contemplation followed by an off-hand verdict of "impossible" or "absurd," but a contemplation deliberate enough to seek out just what conditions would make the outrage seem permissible and reasonable.

Let me close this address by explaining to this hospitable and sympathetic conclave why it seems peculiarly appropriate for me, an easterner, to set before the westerners here gathered the particular outrage with which I have detained them. It is because my contacts with the geology of the Pacific slope during the winter of 1924-25—very unconformable contacts, because of my preconceptions—have been outraging the views that I have more or less unconsciously gained on the Atlantic slope as to the demure quietude of the later geological periods. In the east, the Miocene, Pliocene and Pleistocene have witnessed only leisurely processes of degradation, deposition and deformation, all of small relatively measure; but here on the Pacific slope those periods have been characterized by an extraordinary activity; deposits of enormous thickness have been laid down, and those deposits have been deformed and eroded on a scale that is really rather disconcerting. Is it not fair, therefore, that in return for the incredible stories that have been told me here as to what has happened lately in Californian geology, I should take a turn at telling some outrageously impossible stories myself? In any case, there stand the Basin Range fault blocks, just beyond the eastern skyline of California, displaced in such a manner as to extend over a greater breadth of country than that which they previously occupied; and if it is not possible to explain their extension by underdrag, as an indirect reaction of a passive exterior crust on an expanding earth interior, then we must ask by what other outrageous process it is proposed to explain them.

---

[3]Since giving the address on which is article is based I have opportunity of seeing several Basin Ranges, some in southeastern California, in company with that most competent of guides, Dr. L. F. Noble, of the U. S. Geological Survey, and some in Utah in the helpful company of Professors Schneider and Mathew, of the State University at Salt Lake City, and of Professor M. O. Hayes, of Brigham Young University at Provo; and the evidence then found for the occurrence of slanting fault surfaces seems to me indisputable, not only in the Basin Ranges themselves but also in the much longer bounding ranges of the Wasatch mountains on the east and the Sierra Nevada on the west. Far from the Range blocks being vertically uplifted without compression, as Gilbert first proposed in his report on the Wheeler Survey fifty years ago, still farther from their being the crowded blocks of a collapsed arch, as others have supposed, the Basin Range blocks seem to be the irregularly uplifted and diversely tilted blocks of a former lowland of erosion which has suffered a pronounced extension of its former east-west breadth, as I have briefly stated in the Proceedings of the National Academy of Sciences for July, 1925.

BULLETIN OF THE GEOLOGICAL SOCIETY OF AMERICA

VOL. 44, PP. 461-494, 1 FIG.   JUNE 30, 1933

RÔLE OF ANALYSIS IN SCIENTIFIC INVESTIGATION [1] *

BY DOUGLAS JOHNSON

(Read before Section E of the American Association for the Advancement of Science, December 27, 1932)

CONTENTS

INTRODUCTION

Custom decrees that the Chairman of your Section should, at the interval of a year following his presidency, deliver before you an appropriate address. It has seemed to me that I could best command your interest in some field of discussion where every one of us, geographer and geologist alike, has had experience. So I have selected the broad field connoted by the highly inclusive term, "scientific investigation"; and I would direct your attention, not to any particular results of such investigation, but to a concrete problem of method which I suppose must concern every scientific worker. This problem can briefly be stated as follows: What is the

---

[1] Manuscript received by the Secretary of the Society April 20, 1933.

* Expanded form of address as retiring Vice-President and Chairman of Section E, American Association for the Advancement of Science.

(461)

precise rôle of analysis in a properly conceived and successfully executed scientific investigation?

To answer such a question it is obvious that we must first determine, as best we may, what constitutes "a properly conceived and successfully executed scientific investigation." The early part of my address will be directed to this end. The latter part may then the more effectively discuss the value of analysis as an instrument of precision in research.

For a preface to my remarks I can do no better than quote from the opening paragraphs of Gilbert's presidential address before the American Society of Naturalists, delivered almost half a century ago: "In the statement of these considerations it is impossible to avoid that which is familiar, and even much that is trite. Indeed all expectation of entertaining or edifying you with the original or the new may as well be disclaimed at the outset. I shall merely attempt to outline certain familiar principles, the common property of scientific men, with such accentuations of light and shade as belong to my individual point of view." [2]

Obviously, my particular point of view, like Gilbert's, must be that of the student of earth science. It goes without saying that I am not competent to speak of methods of research in chemistry and physics, where experiment plays a far larger rôle than in geology and geography. So, also, the biologist, the astronomer, and investigators in other fields must speak for themselves. For this reason the title of my address may seem unduly ambitious. Yet I prefer the broader, more inclusive term "scientific investigation" to the more restricted, if more accurate, "geologic and geographic investigation," because it seems to me probable that some of the principles here discussed may find application beyond the limits of my particular field.

One characteristic of earth science deserves special emphasis in this connection. Most problems in geology, including geomorphology, involve the invisible past. We study phenomena produced long ago; and even where the responsible processes continue to operate, it is usually at a rate so slow as to give limited and imperfect information concerning their past performance. Furthermore, we can utilize experiment in solving our problems only in restricted measure. For us, the scientific method involves extended and deeply penetrating mental processes to which special attention must be given in our discussion.

It is a pleasure to express here my indebtedness to several of my Colum-

---

[2] G. K. Gilbert: The Inculcation of Scientific Method by Example, with an Illustration drawn from the Quaternary Geology of Utah. Amer. Jour. Sci., 3rd ser., vol. 31, p. 284, 1886.

bia colleagues, Robert S. Woodworth, Professor of Psychology; Adam Leroy Jones, Associate Professor of Philosophy; Sam F. Trelease, Professor of Botany; and George B. Pegram, Professor of Physics; who were good enough to read the address in manuscript form and give me the benefit of helpful criticisms. To Professor William Morris Davis I am indebted not only for very full comments on the manuscript which have enabled me greatly to improve its form and contents, but above all for penetrating and constructive discussions of method during association with him for thirty years as student, colleague, and friend. But neither Professor Davis nor others have any responsibility for the views expressed in these pages.[3]

THE SCIENTIFIC METHOD

"Nothing has such power to broaden the mind as the ability to investigate systematically and truly all that comes under thy observation in life."—Meditations of Marcus Aurelius.

BACKGROUND AND DEFINITION

The quest of Truth remains today, as it has remained throughout the ages, the supreme manifestation of the human intellect. And the orderly procedure of the mind by which this quest is prosecuted has long provided a fascinating subject for study. From the days of Zeno, Socrates, Plato, and Aristotle the "intellectual processes operative in the acquisition and in the creation of knowledge" have been intensively investigated. Out of these studies came the deductive, or Aristotelian, logic of the ancient Greeks. Perverted by the scholastic philosophers of the Middle Ages, the deductive process had fallen into disrepute when the scientific era opened with the great discoveries of the fifteenth and sixteenth centuries. William Gilbert in his "Treatise on Magnetism" and Francis Bacon in his "Novum Organum" condemned the scholastic method of reasoning, and pointed the way to a process in which the mind should start with the more common and obvious facts of nature, then move forward to things more complex until generalizations could be formulated and hidden causes discovered. Thus was born modern inductive logic, sometimes called "the scientific method."[4] But the scientific method of today is as far removed from that

---

[3] Professor Davis, in particular, dissents from my use of "inductive inference"; from the treatment of "Verification and Elimination" as a single stage in investigation; from the use of "Interpretation" rather than "Explanation" as the name of the final stage; and from various other phrases and passages in the text. Nor does he recognize that analysis is a process properly involved in the several stages of an investigation.

[4] Cf. D. S. Robinson: The Principles of Reasoning, pp. 333-343, 1930.

advocated by Bacon as is modern deductive logic from the legalistic quibbles of the schoolmen.

For purposes of this discussion I shall define "scientific method" as any method which effectively utilizes the several powers of the mind in a systematic, impersonal, non-emotional, unprejudiced effort to discover the origin and relationships of phenomena, and the laws which control their manifestation. The powers of the mind must be *effectively* utilized, for incorrect habits of thought may lead to the obscuring, rather than to the discovering, of truth. Not one capacity of the mind, but its *several capacities,* must be utilized. Observation, memory, comparison, classification, generalization, analysis and synthesis, induction and deduction, invention, experimentation, prediction, verification, revision, confirmation and interpretation—all these may, and a number of them must, be present in any truly scientific investigation. The procedure must be *systematic,*[5] for unguided, haphazard inquiry leads to omission of essentials, and consequent falsifying of results. It must be *impersonal,* since the discovery of things as they are is impossible to him who seeks to find them as he or others think they should be. It must be *non-emotional,* for when emotion claims the throne, reason abdicates. And finally, the procedure must be *unprejudiced,* for biased judgment is a poor instrument with which to seek the truth.

PLACE OF DEDUCTION IN THE SCIENTIFIC METHOD

Of all the factors involved in the scientific method as above defined, only one is likely to be challenged by any considerable number of men. Over three hundred years ago Francis Bacon was declaiming against the employment of hypothesis and deduction in scientific investigation. Today there are those who echo his criticism. You may call to mind the names of eminent geographers and geologists who view with disfavor Davis's extended employment of the deductive method in geomorphology. Their criticism is born, not so much of failure to recognize that Davis has made skillful and fruitful use of the method, as of a deep conviction that the method itself is dangerous, and apt to give results of artificial, rather than intrinsic, value.

What is this "deduction" which the ancient Greeks developed, the scholastic philosophers perverted, Francis Bacon condemned, and some of our colleagues hesitate to employ? Since we are to hold it an essential of correct scientific method, it behooves us to understand each other when

---

[5] "Systematic" does not necessarily imply unvarying order in the sequence of procedure. If the system exists, one may utilize its different parts in variable order, providing all parts are ultimately utilized.

we use the term, deduction, and to make sure that we all draw the same distinctions between it and the inductive process. Unfortunately, authorities are far from agreement in defining logic and the nature and scope of its parts. This is true of induction to a greater degree than it is of deduction. When the elect disagree, the layman who must use their terms can do little more than specify the sense in which he employs them. This I do, with apologies to the psychologist and logician who find that in these pages "mind," "mental faculties," "induction," "deduction," and "analysis" are employed with less skill and precision than they could wish. To the student of natural science the terms as here employed will, I believe, convey the general conceptions essential to our discussion.

Both induction and deduction start with something, do something with it, and get something else. In inductive reasoning we start with observed facts, subject them to certain logical processes, and seek thereby to discover general principles, or laws. In deductive reasoning we start with some generalization, subject it to certain other logical processes, and seek thereby to derive specific consequences which should correspond to observed facts. Induction proceeds from the particular to the general: deduction proceeds from the general to the particular.

When employing the inductive method the investigator is deeply concerned with the validity of the observations which serve as the starting point of his mental processes. If the supposed facts are not facts, it matters little how correct may be his mental processes; the results obtained are not trustworthy. But when employing the deductive method, the investigator is not primarily concerned with the truth of the generalization which serves as the starting point of his deductive reasoning. For the time being he takes the general proposition for granted, and concerns himself with the consequences which may logically be deduced from it. If the general proposition be erroneous, he will discover its invalidity when he has derived from it specific consequences, compared these with known or newly observed facts, and found an obvious lack of accordance between deduced expectations and observed realities.

In both types of reasoning, appeal is made to observed facts, although not necessarily to the same facts. But in one case, facts come into the picture first; in the other, last.

It is easy to see why logicians should hold that induction is the reverse of deduction. But it is a fallacy for the scientific investigator to suppose that the two methods of reasoning are antagonistic, or that one of them is good, the other, bad. The wise investigator will use one to supplement the other; for he secures great advantage if he first employs inductive reasoning to derive from observed facts certain general conclusions, then reverses the process and, using the conclusions as working hypotheses, deduces their reasonable consequences, checking these last against observed facts as the best proof of the correctness of his reasoning.

NATURE OF THE SCIENTIFIC METHOD

With deduction accorded its rightful place in the scheme of scientific reasoning, we are ready to inquire more closely into the precise nature of the scientific method. Fortunately, distinguished guidance is here afforded us. Gilbert's paper on "The Inculcation of Scientific Method by Example" [6] placed in clear relief the outstanding differences between the theorist and the investigator, defined the rôle of hypothesis and deduction in research, and emphasized the superiority of multiple working hypotheses as an instrument for discovering scientific truth. Chamberlin's essay on "The Method of Multiple Working Hypotheses" [7] later demonstrated the dangerous defects of what he terms the method of the ruling theory and the method of the single working hypothesis, and convincingly advocated the employment of a variety of working hypotheses as the most effective means of discovering all facts pertinent to a problem and the best guarantee of arriving at their correct interpretation. Davis has in most striking manner illuminated the deductive processes essential to this last method,[8] applying them to a wide range of geomorphic problems. An experience of more than thirty years devoted to scientific study, in the course of which I have given conscious attention to the relative value of work accomplished by different methods of investigation, convinces me that I shall make no mistake if I follow the lead of the masters just cited, and accept the method of multiple working hypotheses as the best procedure in research. And I hope it will not seem presumptuous if I venture to lay before you certain considerations not fully discussed by Gilbert, Chamberlin, or Davis, which seem to me essential to the best understanding and use of the method they regarded as superior to all others.

---

[6] G. K. Gilbert: The Inculcation of Scientific Method by Example, with an Illustration drawn from the Quaternary Geology of Utah. Amer. Jour. Sci., 3rd ser., vol. 31, p. 284, 1886.

[7] T. C. Chamberlin: The Method of Multiple Working Hypotheses. Jour. Geol., vol. 5, pp. 837-848, 1897.

[8] See especially his paper on the "Disciplinary Value of Geography, Part I, The Science of Geographical Investigation." Pop. Sci. Mo., vol. 78, pp. 105-119, 1911.

STAGES OF INVESTIGATION

In actual use the method of multiple working hypotheses naturally resolves itself into fairly distinct steps, or stages. It is desirable to summarize these, and to state the essential nature of each stage, before proceeding to our inquiry as to the rôle analysis should play in each.

The first stage is that of *Observation,* during which the investigator takes cognizance of certain facts or things, and of the existence of a problem concerning their interpretation. Here, and repeatedly in succeeding stages, memory plays an important rôle, the investigator recalling from past experience, acquired either personally or through the work of others, additional things of similar character, already named and catalogued in his mind.

In the second stage, *Classification* occurs. The observed facts are compared, both among themselves and with others called up from memory; fundamental, as opposed to superficial, resemblances are noted; and like facts are grouped together.

Conditions are then ripe for the third stage, in which *Generalization* takes place. Here, the inductive process comes most prominently into play, the mind inducing from the classified facts broad generalizations respecting them. These generalizations may take the form of a law or principle expressing some significant relation between the observed facts. Such law or principle may, in turn, constitute a partial or complete explanation of the facts, as did Harvey's famous induction regarding circulation of the blood. More often, perhaps, the generalization merely paves the way for an explanation, which comes fully into being in the next, or fourth, stage.

Here, *Invention* further extends the inductive process. If the inductive inference of stage three involves a complete explanation of the facts, the investigator deliberately proceeds to invent additional possible explanations. If the earlier induction provided but a partial explanation, or merely paved the way toward an explanation, invention completes the process, and then presses on to the formulation of additional explanations for the same facts. It is this deliberate effort to invent the greatest possible number of reasonable explanations for a given phenomenon or set of phenomena which distinguishes the method of multiple working hypotheses from other methods of research. For it is these tentative explanations which become the working hypotheses of the next, or fifth, stage. Where others have previously invented explanations of the phenomena in question, these are, by the impartial investigator, tentatively taken into his family of working hypotheses, and accorded the same hospitable consideration given to those of his own creation. The precise nature of the inventive process is a debatable question which need not here concern us.[9] It is sufficient for our purposes to know that it is a real and valuable faculty of the human intellect, and that it is capable of deliberate cultivation.

In the fifth stage, *Verification* and *Elimination* require extended use of deductive reasoning. The investigator now takes each working hypothesis in turn, and, reversing the mental processes previously employed, first deduces its reasonable consequences as fully and as precisely as possible, then confronts the deduced expectations with facts observed in the field or established by experimentation, thus verifying the competence of the hypothesis to explain the facts, or eliminating it as incompetent and hence unworthy of further consideration. As will later appear, the verification of this stage is limited, and requires some measure of independent confirmation.

Confirmation and *Revision* is secured in the sixth stage. Each hypothesis which survives stage four requires further study. Its deduced consequences guide the investigator in predicting the existence of facts not yet discovered, or in devising experiments which will establish facts not previously known. The establishing of these facts, or the failure to establish them, will confirm the validity of the hypothesis, or require its revision, possibly even its ultimate rejection. From this severer test there should emerge one or more hypotheses which alone or jointly offer satisfactory explanation of the observed facts.

The investigator now passes to the seventh and last stage, fully formulating his ultimate *Interpretation* of the phenomena investigated.

No one would pretend that each of the seven stages described above is a sealed compartment, in which the mind performs no function appropriate to the other stages. The human intellect is an unruly member, leaping forward and backward in a most wilful way. One could not, even if he would, carry through an investigation strictly in the orderly sequence just enumerated. Nor is it desirable that he should do so. The development of any stage may throw valuable light on the procedure appropriate to some earlier or some later stage; and the intellect instinctively, inevitably turns backward or rushes forward to profit by this better illumination.

Furthermore, the solution of many problems is not the work of a single

---

[9] "No one has ever given any explanation how the hypotheses arise in the mind." Encyclopedia Britannica, Article on *Induction*, 11th Edition, 1910. For a hypothesis as to the origin of hypotheses, see G. K. Gilbert: "The Origin of Hypotheses," Science, n. s., vol. 3, pp. 1-13, 1896.

mind, but of the collective mind of those who, sometimes over long periods, contribute to the advancement of knowledge in a given field. The individual thus enters upon an investigation when others have more or less completely traversed some or all of the stages enumerated above. If wise, he will retrace their steps sufficiently to verify or revise such of their results as he wishes to utilize in his own study. But in view of the work already done, his mind is likely to push forward here, or turn back there, as expediency rather than system may dictate. There is thus set up a plexus of paths by which the mind traverses every part of the problem under investigation.

The plexus will vary for different problems, and must vary widely for problems in such widely different fields as physics and botany, chemistry and geomorphology. I have merely set forth what seems to me the normal, logical sequence of scientific research, viewed from the standpoint of a student of geomorphology. In this field certainly, possibly in other fields also, any well-ordered investigation will be found to present the seven enumerated stages in more or less distinct form, however complex may be the paths by which the mind traverses these stages.

RÔLE OF ANALYSIS IN SCIENTIFIC RESEARCH

"Look beneath the surface; let not the several quality of a thing nor its worth escape thee."—Meditations of Marcus Aurelius.

DEFINITION OF SCIENTIFIC ANALYSIS

What is analysis? The dictionaries tell us that it is the process of separating a thing or a concept into its constituent parts, in order to arrive at the essential or ultimate elements, causes, or principles; that it is the tracing of things back to their sources; and that it is designed to clarify and test knowledge. The chemist analyzes a complex substance to determine its precise composition. For the purposes of our discussion I would define scientific analysis as "the process of separating observations, arguments, and conclusions into their constituent parts, tracing each part back to its source and testing its validity, for the purpose of clarifying and perfecting knowledge."

What rôle should analysis play in the seven stages of a scientific investigation as previously defined? In proceeding to this inquiry I shall again fall back upon a precedent established by Gilbert. He realized the disadvantage of discussing abstract propositions, when the human mind grasps more readily things presented to it in concrete terms. Accordingly, he illustrated his conceptions of correct scientific method by experience gained in his study of the shores of Lake Bonneville. I am persuaded we can pursue our discussion most easily if I illustrate my observations, so far as feasible, by reference to some concrete example. For sake of simplicity I shall select a common phenomenon found on the rocky shores of many lands, and in tracing the part played by analysis at each stage of a comprehensive scientific investigation, shall illustrate my points freely by referring to the study of this phenomenon, the origin of which is still a subject of dispute. Those who follow this discussion should not attach undue importance to the particular example selected for illustrative purposes, nor to the statements made concerning it. I have deliberately selected a debatable matter, the study of which is not yet completed. Our concern here is not with the origin of the phenomenon, but with the method of investigating its origin.

Along the rocky shores bordering the sea are found nearly horizontal platforms or benches (figure 1), from a few feet to a few hundred feet in width, cut in solid rock. At their seaward margins these benches terminate in relatively abrupt slopes which descend toward or even into the water. From their inner, or landward, margins rise steep slopes or rocky cliffs. In elevation the benches range from one or two feet above ordinary high tide indefinitely upward, often several hundred feet above the same datum plane.

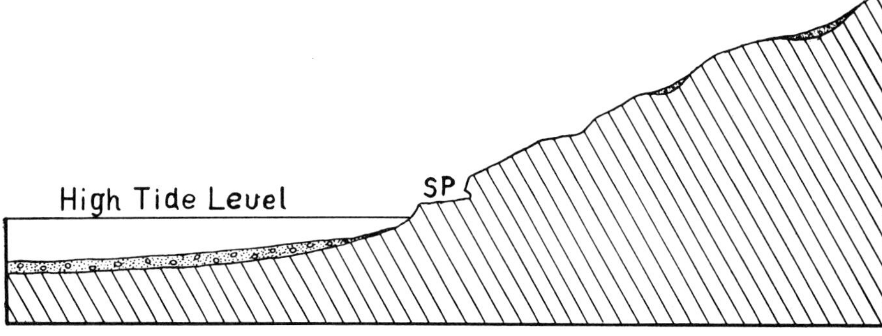

FIGURE 1.—*Marine Benches*

The combination of bench and cliff bordering the sea might arouse no significant reaction in the mind of the ordinary observer. But such a combination instantly excites the brain of the geomorphologist to activity. Memory recalls textbook diagrams, explanatory titles under photographs of similar features, discussions of the origin of such features in geologic and geographic treatises, perhaps earlier field examination of such forms. Rapidly the mind compares the forms, the situations, the mutual relations of the assembled features in the various cases called before it by observation and memory, and before the observer is conscious of the process he has leaped to an inductive inference: that both bench and cliff were cut into the margins of the land by wave erosion. In this case the inductive inference involves an explanation; but it is only a partial explanation. If the bench is wave-carved, why is it exposed to view above the level of the sea? The mind rushes on: Perhaps the land has been raised since the bench was carved; perhaps the sealevel has dropped.

The observer must now be on his guard against a dangerous tendency of the human intellect: the tendency to accept as valid a plausible explanation, and then to look for facts in support of that explanation. He must deliberately repress the tendency toward premature conclusions, and begin the task of gathering all the facts upon which alone can a satisfactory explanation be based. As already stated, we are not primarily concerned with the course of his particular investigation of these interest'ng forms, nor with the conclusions he may reach respecting their origin; but we shall draw freely upon this imaginary hypothetical study, to illustrate the uses of analysis in scientific research.

A final word of caution before we enter upon our task of tracing the rôle analysis plays in the several stages of an investigation. One should be careful not to confound the analytical process with the various processes which constitute the successive stages of research. It is our thesis that analysis is not one of these stages; rather, it is something which may be employed in many, indeed in all, of them. Analysis is not Observation; it is not Classification. It is an instrument of precision which may be so employed as to render Observation more complete and more accurate, Classification more effective and significant; and so on through each succeeding stage of a properly conceived and successfully executed scientific investigation.

I. ANALYSIS IN THE STAGE OF OBSERVATION

The initial stage in scientific investigation is normally that of *Observation*. The first employment of this mental process may be made incidentally, perhaps almost passively. But once the investigator realizes that the observed facts present a problem, for the solution of which additional facts are desirable, he becomes an active inquirer, seeking to discover all facts bearing on the problem. He observes as widely and as accurately as possible; the observed facts being automatically recorded in his memory, but also, since the memory is notoriously fallible, deliberately recorded in his notebook if he be a prudent investigator.

Since we are here concerned merely with the recording of external facts observed by the eye, it might appear that analysis has no rôle to play in this initial stage of an investigation. But let us dissect the matter and scrutinize its parts, first turning our attention to the material facts which are the object of observation. Can the analytical powers of the mind be brought to bear upon the facts themselves in such manner as to aid the investigator?

Let our hypothetical student of coastal benches answer. Having observed one such bench, and had his curiosity aroused to investigate, he begins his search for other examples. Very soon he is confronted by the necessity of discriminating coastal benches of the type described above from other forms similar in some respects yet different in others. He needs a name by which to designate the type forming the particular object of inquiry, and tentatively calls them "elevated marine benches" or simply "marine benches," because of their resemblance to submarine benches formed by wave abrasion and because of their apparent marine origin.

The question then arises: What constitutes a marine bench in the sense in which he is using the term? Is it any nearly level surface in the vicinity of the shore, terminated seaward by an abrupt descent, and landward by cliffs which rise steeply to higher levels? This combination of slopes appears repeatedly in Nature, as the product of a variety of causes. The non-critical investigator, who welcomes every shelf and scarp facing the sea as one more link in the chain of facts he seeks to explain, is almost sure to accumulate, without knowing it, an assortment of unrelated observations, many of them irrelevant to his problem. His investigation is seriously compromised from its very beginning. How can he explain the origin and history of elevated marine benches if he unwittingly bases his reasoning on a confused mixture of wave-carved surfaces, benches due to differential weathering, rock terraces cut by a river before drowning brought in the sea, notches cut by a glacial stream flowing between waning continental ice and the sloping hillside, and terraces due to landslides, faulting, or monoclinal warping? As all these features are found in the immediate vicinity of the shore in different places, and as all of them have repeatedly been ascribed to wave erosion, it is clear that the field data must, in the first instance, be subjected to rigid scrutiny.

The facts can not be lumped. They must be analyzed—separated into their constituent parts, and each part tested as to its relevancy to the problem under investigation. Here, the analytical process imposes a very uncomfortable but highly essential rule: that every fact the relevancy of which is in any degree doubtful be excluded from the group of facts for which explanation is being sought. A certain shelf, or terrace, closely resembles other marine benches, but its surface corresponds to the surface of a resistant rock layer. Obviously, this particular form need not necessarily have the same origin as those cut indifferently across various rock structures. Although the observer may feel certain that this feature is identical in origin with the other forms he is investigating, the analytical process demands that it be excluded from the study. And for a thoroughly sufficient reason. If this one terrace be really a product of differential weathering, and yet be included among the facts later to be explained, the explanation then advanced must be competent to account for it just as fully as for the others. This will inevitably exclude from consideration any hypothesis competent to explain the other benches only, and may thus prevent discovery of the true explanation of the forms constituting the main object of inquiry.

I have said that the rule requiring exclusion of all data of doubtful relevancy is an uncomfortable one. The investigator does not willingly push aside some of his most spectacular facts. He is disturbed when analysis points a questioning finger at one after another of his bits of observational data. He may even find that if he rigidly applies the exclusion rule, the quota of competent data ultimately admitted to consideration is pitifully small. It may not even be sufficient to establish any theory on a sound basis. And the investigator craves a theory which will explain the facts.

Let us frankly admit the discomfort rigid analysis imposes; but let us, at the same time, acknowledge that there is no guarantee of safety in scientific investigation save that offered by such analysis. If the facts are all of equivocal character, lending themselves to a variety of interpretations, the formulation of a definite theory must await discovery of new and less equivocal facts. Postponement of judgment may bring discomfort; it does not bring disaster.

Thus far we have discussed the necessity of analyzing the things observed. But the investigator must probe more deeply than that. He must analyze the observational process itself. Does he really see what he thinks he sees? Not unless he is constantly on guard against the well-known dangers to which observation is subject. Chief among these, perhaps, is the tendency to include inference with fact, to confuse theory with observation. The observer thinks he sees a bench cut in granite, when in fact all he really observes is a few scattered outcrops of granite protruding through the soil; from these he infers, perhaps erroneously, that the vastly larger invisible mass is likewise granite. In the case imagined a moment ago the investigator may report that he saw one wave-carved bench which did not bevel the rock layers. All he really saw was a shelf, or bench, parallel to the rock layers. He inferred it was wave-carved, when in fact it may have been produced by differential weathering of weaker beds overlying a more resistant layer.

Another danger to which observation is subject is that of ocular deception. This danger looms large when one is looking for certain specific forms in the landscape. The power of suggestion is greater than many realize, and one looking for marine benches is in danger of finding them in faint undulations or irregularities of the terrain, which under other circumstances would not impress him as significant. If he is looking for benches which he expects to be horizontal, the inclined position of something he mistakes for such a bench may quite escape his eye, even when the inclination is so marked as to be quite obvious to another who is without preconception as to what should be observed.

Incompleteness of observation is another common danger. The eye tends to pick from the landscape that which seems to it for the moment significant, and fails to note much that later stages of the study may show to be vitally important.

Even where the initial observation is fairly complete, the benefits which should be derived from it may be lost through failure to record permanently in memory and in notebook all the facts observed. Many studies of shore terraces have been vitiated because only part of the relevant facts were recorded by the investigator.

No one is wholly free from the dangers of defective observation. But one who acquires the habit of analyzing his observational powers and processes is less exposed to these dangers than are those who give no conscious attention to this initial step of a scientific investigation. In research, as in other things, to be forewarned is to be forearmed; and the investigator who is conscious of the dangers inevitably associated with the observational process will be on his guard against those dangers. He will deliberately exclude all inferences from his group of supposedly observed facts, will avoid the possibility of ocular deception by pushing each observation far enough to establish its validity, and will make a persistent effort to see all things, record all things, and remember all things bearing upon his particular problem.

In this last effort he may derive one great practical advantage in respect to field and laboratory methods. Having analyzed the observational process, he will realize that it is almost always doubly defective so long as observations are recorded by memory alone; defective, first, because the mind does not retain all it initially records; but also, and far worse, defective because the exposure of the mind to the facts is normally too rapid for all of them to make an impression upon even that sensitive recording instrument. Once aware of this source of danger, the investigator can, by a simple procedure, remove the double difficulty. By fully recording his observations in writing, in the presence of the facts, he not only insures against the fallibility of memory, but at the same time insures far more complete and effective observation. Both the time required for a written record, and the stimulation and direction of the observing faculties inevitably consequent upon systematic recording, enormously increase the completeness of the record.

II. ANALYSIS IN THE STAGE OF CLASSIFICATION

Is there any need for employing analysis in the second stage of an investigation, the stage of *Classification*? Let us note first that the analysis involved in stage one led to a sort of rudimentary classification, the separation of observed facts into those which were relevant and those of doubtful relevancy. We at once suspect that analysis must likewise be involved in any further effective classification of the facts of observation surviving the analysis of stage one.

The investigator is not content merely to exclude from consideration every doubtful bit of data. For example, our hypothetical student does not rest when he has excluded from his study every topographic form which does not certainly belong in the group of elevated marine benches. He raises the question as to whether all these benches need necessarily have had the same history. He takes cognizance of the fact that if the benches were carved at different times, and some of them were affected by events, such as continental uplift, which did not affect the others, a serious difficulty may be introduced into the attempt to explain their history. The investigation, to be successful, must deal with a group of facts which are comparable. A simple hypothesis competent to explain benches five feet above high tide, may conceivably be quite incompetent to explain benches fifteen feet above high tide.

The investigator therefore seeks to segregate the relevant facts into groups having like characteristics. Nor is he willing to base his classification on superficial resemblances and obvious differences. Just as the biologist classifies the whale with mammals rather than with fishes, so the student in any other field must make a deliberate effort to discover fundamental similarities and differences which escape the ordinary observer, but which may have great significance for the scientific inquirer. To accomplish this task most effectively he brings his analytical faculties to bear upon the data to be classified. He separates his observations into their constituent parts, compares them one with another, and strives to discover their essential elements. Only thus can he hope to see revealed those similarities and differences not immediately obvious.

In the case of our hypothetical study of marine benches, it may happen that one group of benches falls systematically within a limited altitude range, while another group has a different and more variable range. If so, the investigator will discover that fact, and classify the groups separately. He may also find that some of the benches are more weathered than others; that some are mantled with debris while others are not. If so, appropriate groupings will be established.

III. ANALYSIS IN THE STAGE OF GENERALIZATION

Classification paves the way to the third stage of investigation, in which *Generalization* occurs. Thus, in our hypothetical study of marine benches, it is only when Classification has been effected that the investigator can make significant generalizations; such, for example, as that in one group the benches have a narrow range in altitude, are always close to the sea, present no evidence of weathering of the rock surfaces, and are free from debris; that in a second group the benches have higher altitudes but equally narrow range in altitude, present evidence of moderate weathering, and have small quantities of debris near their inner margins; while in a third group the benches are still higher, vary extensively in altitude, and have surfaces which are deeply weathered and prevailingly covered with debris.

As thus stated, the foregoing generalizations involve no explanation of the classified facts, although it is obvious that they represent a step forward toward possible explanations. We may, however, easily imagine an inductive inference, based on the same mass of observed facts, which does involve partial or complete explanation of them. The inference that the benches were carved by wave erosion involves a partial explanation. The inference that the benches represent former shorelines of the sea raised by some upheaval of the land involves a more nearly complete explanation. These are all generalizations based on a number of particular facts, and are essentially inductive in nature; although they include, as do all inductions, an element of deduction in that they assume uniformity in Nature's laws and the repetition of certain shore forms wherever the waves of the sea have operated.

We have now to inquire whether the analytical faculties of the investigator can serve any useful purpose in the stage of generalization. Let us first note that the empirical generalizations stated above, free from any explanatory element, had their roots in the careful analysis that made effective classification possible. But it is when inducing generalizations which involve more or less explanation that the analytical process can render greatest service. Such generalizations, like those free from explanation, must be rooted in facts and represent a normal outgrowth of them. They must be legitimate inferences induced from the facts themselves. As we shall see in a later section, the investigator may formulate conceptions involving explanations which are not an immediate normal outgrowth of the facts, but the products of invention. As these two types of explanation, the induced and the invented, have somewhat different standing before the court of the intellect, it is important that they be not confused. It is in making this discrimination that the investigator must again employ his analytical powers. The nature of the facts, the nature of their distribution, their relationships, and other pertinent elements must critically be examined, to the end that only legitimate inferences may be induced from them.

For example, the first two groups of classified facts described above, in both of which uniformity of altitude above sealevel is a characteristic of the benches, may properly give rise to the generalization that after the benches were carved the sealevel dropped. Uniformity of bench altitude suggests this as a legitimate inference involving explanation. But no such generalization may be based on the third group of facts, in which great variation in elevation is an outstanding characteristic of the benches. Such heterogeneity in elevation does not suggest either uniform drop of sealevel or systematic differential uplift of the land. Perhaps the facts can be explained on the basis of one or the other of these explanations, or on both combined; but if so, the explanations must be deliberately invented and applied to the facts. They are not normal outgrowths of the facts.

If the investigator analyzes the process of generalization a little further he will realize that it is not sufficient that generalizations grow out of the facts. They must grow out of a sufficient number of facts to be truly general. One of the commonest errors in research is to generalize prematurely, on the basis of a wholly inadequate foundation of facts. The investigator who consciously analyzes both processes and their results at each stage of his research, is less likely to fall into this and other errors than is one who thinks without thinking about his thinking.

When one analyzes the grouping of classified facts respecting elevated marine benches, another important consideration becomes clear; generalizations involving explanation, to be effective, must be made separately with respect to each class of facts. Further study may show that a single explanation applies equally to all the benches. But it is quite as likely to show the reverse. Hence the necessity of treating the different groups of facts independently, until a common origin and a common history for all have been fully demonstrated.

IV. ANALYSIS IN THE STAGE OF INVENTION

In the fourth stage of our hypothetical investigation the inquirer takes the classified facts and the generalizations concerning their nature, and uses them as a basis for *Invention* of as many explanations of the facts as may be possible. It is here that "The Scientific Use of the Imagination," as Tyndall has happily phrased it, comes most prominently into play. The invention may be deliberate, the result of conscious effort. Often it springs unexpectedly into consciousness, the result of a mind well equipped with pertinent knowledge repeatedly "mulling over" the facts in variable combinations. If others have anticipated our inquirer in the inventive process, as is usually true in greater or less measure, he welcomes every idea of alien origin which offers a possible explanation of the facts, and accords it just as hospitable treatment as those born of his own intellect. If the generalizations of stage three involved explanation of the facts, the investigator seeks additional and independent explanations. Where generalization involved but partial explanation, or merely paved the way for an explanation, the investigator employs his inventive powers to complete the unfinished task; and then moves forward to the invention of alternative explanations.

As has previously been pointed out, the process of invention is not fully understood. It resembles induction to the extent that the mind starts with concrete facts, and from them passes to conceptions of broad application which take the form of tentative explanations of the facts. But the mind has far greater liberty in the stage of Invention than in the preceding stage of Generalization. The relation to facts is here less close. A proper generalization, being an outgrowth of the facts, must have its roots well grounded in them. An invention may spring from "thin air," the rarefied atmosphere of more abstract reasoning. The stimulus to invention comes from the facts, and the thing invented must not palpably be contradicted by the facts; but it need not be a normal outgrowth of the facts. Thus, the conception of a uniform drop of sealevel is not a normal outgrowth of the third group of facts mentioned above, because scattered remnants of marine benches occurring at a great variety of altitudes do not, of themselves, suggest the idea that the sea surface has been lowered rather than the land surface uplifted. But one may invent the conception of uniform but intermittent drops of sealevel, and offer it as a possible explanation of benches occurring at many different altitudes, since none of the observed facts necessarily contradicts such an explanation.

We have noted the greater liberty enjoyed by the mind in the stage of Invention. But it is not wholly free. Since the explanations induced in stage three, and those invented in stage four, are to serve as working hypotheses in the deductive operations of stage five, they must conform to certain fundamental requirements governing the formulations of hypotheses. In the first place, it is essential that the tentative explanations should be so precisely molded that specific deductions may be derived from them. If an explanation is so vague in its inherent nature, or so unskillfully molded in its formulation, that specific deductions subject to empirical verification or refutation can not be based upon it, then it can never serve as a working hypothesis. A hypothesis with which one can not work is not a working hypothesis.

Again, the explanation as formulated must be possible, for explanations clearly contradicted by well-established natural laws waste time and energy without advancing the investigator toward the true goal of his researches. On the other hand, it is manifestly dangerous to exclude explanations which merely appear incredible or absurd because they run counter to established opinions. Many a door to truth has thus prematurely been closed, and long remained closed. The hypothesis of continental glaciation seemed incredible in 1840, as did the hypothesis of evolution in 1860.

How then shall the inventive mind properly mold its hypothesis, and distinguish between explanations which appear plausible but are really unsound, and those which immediately provoke distrust yet merit hospitable consideration? It seems to me that here the analytical process offers us an instrument of incalculable value. Let us dissect each tentative explanation into its component elements, trace each element back to the assumptions and inferences which lie hidden behind it, and test the reasonableness of the whole by testing the validity of each part. Once more we turn to the shore for concrete illustrations of abstract principles.

Our investigator of marine benches, having first observed and classified his facts, and then derived certain generalizations from these facts, decides that hereafter he will concentrate his study upon the first group of benches earlier described—those lying close to the present shore, ranging from 1 or 2 feet to 5 or 6 feet only above ordinary high tides, showing fresh rock surfaces usually free from debris. To distinguish this group of benches from others, he calls them "shore platforms," thus recognizing their restriction to the immediate vicinity of the shore (SP, figure 1).

He now proceeds to the invention of as many explanations as possible for the facts he has collected concerning these shore platforms. Among some eight or ten explanations which occur to him, or which he appropriates from others, we consider but two. One, already briefly mentioned, is to the effect that the shore platforms, originally carved below sealevel, have been raised by regional uplift of the coast. The second supposes that the platforms were carved in their present position wholly above high tides, by storm waves of the present sealevel. The first explanation is in accordance with current geologic opinions, both regarding the conditions under which waves carve platforms and regarding crustal movements. The mind welcomes it hospitably as a hopeful direction in which to pursue the study. The second runs counter to long established views that the level of platform cutting lies distinctly below sealevel. The mind immediately recoils from this explanation as being inherently improbable. But our investigator, distrusting the mind's intuitive reactions, determines neither to accept the first as a tentative working hypothesis, nor to reject the second, until he has first analyzed them both. Note that his analysis is here directed merely to determining the acceptability of each as a member of his family of working hypotheses. Testing the competence of the hypotheses to explain all facts observed comes later, and involves quite different processes.

As to the first suggested explanation, he asks a number of pertinent questions. What is the fundamental assumption on which the explanation rests? Obviously, that such platforms *must* be cut below sealevel, and that the invention of another factor, such as regional uplift, *must* be invoked to explain their position above the sea. But what is back of these "musts," of these implications of compulsion? First, there is the teaching of standard treatises on geology, which commonly show both wave-cut and wave-built platforms below the surface of the water. Second, there are the reports of various observers who state that they find rock platforms under the sea, and the bases of rocky sea cliffs a number of feet below the level of low tide.

But do the treatises present convincing evidence that text and diagram correctly represent conditions of wave-cutting? They do not. An occasional text represents wave-cutting above sealevel, so that the testimony of the treatises is, in some measure, contradictory. And do the observers submit conclusive evidence that the submerged platforms and cliff bases described by them are in process of normal development at the present time? It must be admitted that they do not. They may believe such to be the case; but the facts they present do not seem to exclude the possibility that platforms and cliffs were carved at a higher level, and recently submerged by subsidence of the land or by rise of sealevel. Thus the investigator pushes his inquiries, and as he does so the tentative explanation which first appealed to him most strongly takes on an increasingly doubtful aspect. He may ultimately be impelled to reject it as lacking the necessary degree of reasonableness. Or caution may dictate that he preserve it as a possible working hypothesis, to be later subjected to the more searching test of deductive reasoning. In any case, critical analysis has assigned the explanation (of wave erosion below sealevel followed by coastal uplift) a position far less dominating than it originally occupied.

The investigator next critically examines that tentative explanation which he first was tempted to reject as being unworthy of serious consideration. What lies back of the suggestion that storm waves may have carved the platforms during the present relations of land and sealevel? For one thing, an assumption that there may be factors involved in wave attack which permit effective erosion above ordinary high tide level. What then are the factors involved in wave attack, and do any of them favor activity at levels relatively high? Observation and memory present certain pertinent facts bearing on this point. The momentum of masses of water hurled up the sloping shore zone by advancing waves insures some erosive activity above still-water level. Furthermore, it is well established that in both waves of oscillation and waves of translation the crests rise higher above still-water level than the troughs sink below it. For this reason the average level of wave attack is above still-water level. Onshore winds are known to pile up the water to levels often a number of feet above normal. Thus, at least three factors in wave attack tend to produce high-level cutting, and it seems not unreasonable to suppose that their combined result might be wave erosion at the level where the low lying platforms are actually observed.

Along these and other lines the investigator pushes the analysis, until he is satisfied that the tentative explanation of shore platforms as a product of storm-wave attack under existing conditions of land and sealevel, while presenting obvious difficulties, is sufficiently reasonable to deserve a place among the tentative hypotheses later to be tested. So in turn he dissects and scrutinizes each tentative explanation which his inventive powers suggest, or which he adopts from the work of others. Throughout this stage of the investigation it is evident that analysis is an invaluable aid to his studies, and the best guarantee of their eventual success.

V. ANALYSIS IN THE STAGE OF VERIFICATION AND ELIMINATION

On entering the fifth stage of the investigation our hypothetical inquirer possesses, let us suppose, some half dozen tentative explanations which have survived the analytical processes applied in the preceding stage. He now treats each of these in turn as a working hypothesis, and, reversing the mental processes employed in the third stage, endeavors by *deductive* reasoning to determine as precisely as possible just what features should characterize the shore platforms in case the particular hypothesis for the moment under scrutiny be the true explanation of these forms. His object is to verify, so far as possible, the competence of some hypotheses, and to eliminate others which critical tests show to be manifestly incompetent.

He secures the required tests by an appeal to facts. After deducing the reasonable consequences of a working hypothesis, he confronts these by such facts as are already in his possession, drawing upon memory, upon his field notes, and upon the published observations of others. If the consequences logically deduced from the hypothesis are in accord with the facts, he accepts this as verification of the competence of the hypothesis to explain the facts already in hand. Where deduced consequences find no counterpart in observed facts, he feels justified in eliminating the hypothesis from further consideration.

Here it should be noted that the success of the deductive operations involved in this fifth stage of an investigation depends primarily upon two things: first, the fullness and accuracy of the logical operations performed by the mind; second, adequate testing. The careful analysis employed in each of the preceding stages has done all that properly can be done to assure that only competent hypotheses shall serve as the starting point for logical deduction. The process of deduction itself, while not directly concerned with the validity of these hypotheses, is vitally concerned with the laws of thought governing a peculiar mental operation to which each hypothesis is now subjected. Obviously, the mind must operate with completeness and precision according to the laws of logic if the results are to be trustworthy.

The second condition of success is adequate testing of the products of deduction. Let us recall that the very essence of the method of multiple hypotheses is hospitable consideration of every possible explanation of a given group of facts, in the full knowledge that not all these explanations can correspond to reality. Let us remember further that the deductive process itself neither adds anything to, nor takes anything from, the original content of any hypothesis, but merely develops and elaborates its inherent possibilities. With these considerations in mind, we can understand how critically important it becomes to test the greatest possible number of deduced consequences for each hypothesis by confronting them with the greatest possible number of facts competent to either verify or contradict them. The whole method stands or falls with the success or failure of this operation.

When all possible consequences of each working hypothesis have been deduced, it will often appear that some of the competing hypotheses have certain consequences in common. These have relatively low critical or discriminative value, since if facts are found corresponding to the deduced expectations, they merely show that any one of several hypotheses may be true. There will usually be found, however, some one or more consequences peculiar to hypothesis A, while certain others are peculiar to hypothesis B, and so on. It is these unlike consequences which have the highest critical value in discriminating between valid and invalid hypotheses, and it is on these that the investigator will most depend when drawing conclusions.

To what extent, if any, can analysis serve the investigator in this stage of *Verification and Elimination*? Since, as we have just seen, effective use of the deductive process depends upon one's ability to make logically correct deductions, it is expedient that the investigator should subject his own mental processes to careful scrutiny. There are fallacies in reasoning which must be avoided. These are treated in works on logic and need not be discussed here. But some study of logic will direct conscious attention to one's mental habits, and help to secure the precision and completeness requisite for careful deduction.

The investigator must realize that if his deductions are false, it matters little whether or not observed facts correspond to them. The conclusions are, in any case, invalidated, and he may either reject a perfectly valid hypothesis or embrace one which is invalid, on the basis of false reasoning. Safety lies only in analysis. Deductions, like inductions, must be separated into their component parts, and each part scrutinized and its validity tested. Let us illustrate what this involves by turning once again to the problem of the shore platforms.

Our investigator has accepted as one working hypothesis the tentative explanation of storm-wave erosion under present conditions of land and sealevel. From this hypothesis he deduces certain expectable consequences, among which let us consider only one. If the platforms are the product of storm waves operating at the present time, he reasons, then there should be found upon their surfaces rock debris constituting the cutting tools with which waves accomplish their erosive work. Turning to his record of field observations, he finds that many of the platforms are remarkably free from debris. Indeed, their outer edges commonly drop abruptly into moderately deep water, so that it is difficult to see how much debris could be cast upon the platforms or maintained there. The facts in hand conspicuously fail to match the expectations deduced from the hypothesis, so our investigator rejects the hypothesis of storm-wave erosion as incompetent. In doing so, he rejects a perfectly valid hypothesis solely because he failed to examine his deductions critically before using them as tests of the competence of the hypothesis.

Let us suppose, however, that our investigator was so fully convinced of the value of analysis as an instrument of research that he would neither accept nor reject a hypothesis until he had subjected each deduction derived from it, and each test of its correspondence with the facts, to such rigid scrutiny as he was able to make. Under these circumstances he would first have examined the deduction relating to the expected presence of wave-cutting tools on the platforms. He would have asked: Upon what does the deduction rest? Evidently it rests upon the assumption that waves of water unarmed with debris cannot effectively erode the coast. And on what, in turn, does this assumption rest? Current geologic opinion may run this way, and geologic textbooks may assert that waves without cutting tools are feeble. But current opinion and the textbooks have often proved fallible guides. So our investigator would have gone back to the sources on which both opinion and textbook should be based; the actual recorded evidence as to the nature and causes of damage accomplished by waves. Here he would have found much information concerning havoc wrought by waves under conditions which seem clearly to preclude the effective intervention of debris. The terrific impact of the water itself, the force of the currents it generates, the direct pressures exerted upon air and water imprisoned in crevices in the rock, the sudden expansion of air in crevices and pore spaces when the rapid retreat of a wave creates a partial vacuum outside, all these are described as effective causes of damage by waves, on the basis of concrete and seemingly reliable evidence. Our investigator must then have concluded that while shore debris is presumably a highly important factor in wave erosion, there is no reason to exclude the possibility that waves relatively free from debris, and striking with a force varying from hundreds to thousands of pounds per square foot, may accomplish much erosive work. He would therefore have rejected, not the hypothesis of storm-wave erosion, but his tentative deduction respecting the necessity of finding debris on the platforms; and would have looked for other deductions from the same hypothesis which he could use as a more reliable basis for testing its validity.

It is not alone in making deductions from hypotheses that one's critical powers should render aid in this fifth stage of an investigation. When confronting deduced consequences with observed facts, the investigator must be equally on the alert. If harmony is apparent, he will be on his guard against too ready acceptance of the pleasing result. He will probe the situation to make sure that harmony is real and not merely apparent. If discordance is revealed, he will dissect and scrutinize the facts to determine whether the discord may not be the result of accidental introduction into the scene of some irrelevant factor not germane to the problem.

The investigator who fully analyzes the process of testing hypotheses by deducing their consequences and confronting these with observed facts, will gain a keen appreciation of one fundamental consideration: the effectiveness of the test will vary with the number and character of the deductions made from any given hypothesis, and the variety and nature of the facts available to match against them. If the deductions are great in number, and some of them at least are peculiar in character; and if they are nicely matched by facts which are equally numerous and peculiar, then the test of the hypothesis is particularly good.

We have now to note certain important limitations of the "verification" involved in this fifth stage of a scientific investigation. "To verify" is to prove to be true. But what is it that we prove to be true in this stage of Verification and Elimination? Not the hypothesis as the true and only explanation of the facts, for seldom, if ever, is a hypothesis directly verified. The thing we prove is the existence of a certain measure of harmony between the deduced consequences of a hypothesis and the observed facts. The "verification" is merely verification of one or more hypotheses as *competent* to explain certain facts, not verification of them as the *true* explanation of those facts. Several hypotheses may be competent to explain a given set of facts, where only one of them can be the true explanation. Obviously, there is need of further study before any hypothesis can be accepted as fully established. This further study is secured in the next following stage, Confirmation and Revision, discussed on subsequent pages. The verification of the present stage is partial and limited.

It is in the present stage that one is especially apt to commit an error in the deductive process which has wrecked many a train of reasoning. I have characterized this as the error of "limiting Nature." It consists in setting narrow and rigid limits to what natural forces can accomplish under hypothetical conditions. "So long as a stream is ungraded through a hard rock barrier, weak rock areas above the barrier cannot be reduced to a peneplane." "A waterfall in homogenous unconsolidated material cannot retreat upstream for any great distance, but must be rapidly flattened out to a gradual slope." So the human intellect rashly sets metes and bounds to Nature. And Nature proceeds calmly to perform the impossible. The deductive process inevitably involves judgment as to what Nature can and can not accomplish. One must see to it, then, that the limits set are sufficiently generous and flexible, and that they are established only after careful analysis of all factors involved. Only thus will deductions be reasonable, and dependence upon them be safe.

One further observation should find place in this part of our discussion. Some who oppose the use of hypothesis and deduction in scientific investigation have asked: "What is the use of setting up men of straw, only to knock them over?" The answer is that men of straw have no place in a scientific investigation. We have already seen that in the earlier stage of Invention only possible explanations are accepted into the family of working hypotheses. Analysis is there deliberately employed to exclude absurd or incredible hypotheses. The men of straw are not set up. They are early cast aside. Deduction has therefore, in the stage of Verification and Elimination, to deal only with such hypotheses as have some valid title to credence. To determine how far they are entitled to credence is the province of this fifth stage. Some hypotheses, when beaten on the anvil of analysis, will break. But this measures the success, not the futility, of the method. Let any who question the efficacy of a method that invents many hypotheses only to destroy some of them, ponder the words penned by Gilbert: "The demolition of hypotheses, instead of testifying to the futility of research, is the method and condition of progress."

We conclude, therefore, that during the fifth stage of the investigation, in the deduction of consequences from the several working hypotheses, and in the confrontation of deduced consequences with observed facts, analysis is both the anvil where falsehood is shattered and the fire where truth is forged.

VI. ANALYSIS IN THE STAGE OF CONFIRMATION AND REVISION

From the fifth stage of the investigation the inquirer should emerge with a much depleted stock of working hypotheses. In some cases only one will remain as apparently competent to explain all the recorded facts. In other cases two or more may survive thus far. If but one, the cautious investigator will seek confirmation of its validity before utilizing it in his ultimate interpretation of the facts. If more than one, he pushes his studies further to discover which one represents the true explanation of the facts, or whether they are jointly responsible for a complex result.

Whether one or several hypotheses survive the tests applied in the preceding stage, the survival is almost always in an "unfinished" state. The tests remain incomplete, and therefore inconclusive, because not all the desired facts are available, and perhaps not all the desirable deductions have been made. The investigator, let us imagine, has deduced five reasonable expectations from a certain hypothesis, of which three only are matched by facts in his possession. On the other hand, he has five categories of facts, of which three only correspond to expectations thus far deduced. He has two deductions having no counterpart in observed facts, and two groups of facts which are disturbing because not related to anything thus far deduced from the hypothesis. Just what these discrepancies mean, he does not know. They may mean that observation was incomplete, and that new facts remain to be discovered. They may mean that deduction was incomplete or imperfect. In either of these two cases the hypothesis may be valid as it stands. But there are other possibilities. The discrepancies may mean that the hypothesis is invalid and must ultimately be rejected; or they may mean that it only requires a certain amount of revision to bring it into harmony with all the facts.

It is the task of the investigator in this sixth stage to find out just what the discrepancies do mean. He searches for the missing facts which, if found, will constitute logical proof of the validity of the generalization involved in the hypothesis. Here, the extraordinary value of the deductive processes of stage five become apparent. The observation of stage one was a wandering, unguided observation. But the renewed observation of stage six is skillfully directed observation. He now knows what to look for, and hence how and where to look. If, like the geologist or geographer, he is dealing with facts to be found in the field, directed observation will materially shorten his field work and greatly increase its fruitfulness. He goes directly to the places where the new facts should be found. If, like the physicist, chemist, or modern student of biology, he must discover new facts through experimentation, skilfully directed experiment will replace the vague gropings of a less advanced stage of research. The deductive process has added new interest to the quest for truth, and has enormously increased the probabilities of finding it through improved induction.

The probative value of facts discovered in this new quest is much higher than that of facts earlier observed. Heretofore facts were observed, and hypotheses adapted to them. Now the investigator has, by a process of deduction, predicted the finding of facts the existence of which is as yet unknown. The prediction is provisional, of course. "If hypothesis A be correct, then I should find, etc." If he finds the facts provisionally predicted for hypothesis A, and does not find those provisionally predicted for hypotheses B, C and D, his confidence in the validity of hypothesis A is doubly strengthened: first, positively by finding the facts appropriate to A; and second, negatively by not finding those which should occur were the other hypotheses valid.

Directed observation has another great value. It is apt to lead to the discovery of facts wholly unanticipated by the deductions which direct the search. As Davis has well said, the investigator's outsight on the facts is sharpened by his insight into the nature of his problem. Things which formerly made no impression on eye or mind now instantly arrest his attention, some because they are expected, but others for the very reason that they are quite unexpected. The discovery of facts that are wholly unanticipated, but which, none the less, find full explanation in a hypothesis under investigation, has exceptionally high value as *Confirmation* of the hypothesis. If many investigators discover many facts, all fully accounted for by the hypothesis, and this is repeated over a long period of time, the confirmation becomes of such high quality that the hypothesis is transformed into a well attested theory.

Where none of the deductions of stage five fully accounts for certain of the facts already in hand; or where newly discovered facts fail to find any counterpart in previous deductions; or, again, where some of the facts which should occur, according to a hypothesis otherwise well substantiated, cannot be found, *Revision* of the hypothesis is strongly indicated. This revision carries the investigator back over all the stages he has previously traversed. Beginning with Observation, he again scrutinizes the facts, especially those which fail to match any deduced consequences. Perhaps they are irrelevant to his problem and should have been excluded; but he dare not exclude them merely because they are inconvenient. They may be critically important for his problem, and show the necessity of revising or rejecting some hypothesis. He must determine whether exclusion of the facts, revision of the hypothesis, or rejection of the hypothesis is called for. Next, the investigator painstakingly reviews his Classification of the facts, to discover if revision is there called for; after which he turns his attention to the

Generalizations based on the classified facts. He then scans his Inventions with care, revising explanations and adding to them where either gives promise of bringing expectations into closer harmony with observed facts. He is particularly careful in his re-examination of the deductions which serve as the basis for Verification and Elimination in stage five, for he realizes the many opportunities for error in deductive reasoning, and hence the great probability that revision may there be required. Lack of completeness in deduction is especially apt to cause difficulty, and proper revision, to include some new consequences not previously foreseen, may bring complete harmony out of partial discord.

The process of securing the requisite confirmation, revision, or rejection of hypotheses makes new and heavy demands upon the critical faculties. Especially is this true of the process of revision and rejection, which carries the investigator back over previous stages of the investigation in a search for some weak link in his chain of reasoning, some factor overlooked, or some item improperly included. If, as we have seen, analysis had a rôle to play in each of these preceding stages, how much more important is it that the analytical faculties should be on the alert in a review of one's reasoning in search for error. If previous work has been carefully performed, the error is not apt to be obvious. More likely it is obscurely involved in some apparently sound procedure, in which case it will continue to escape detection unless each bit of evidence, each argument, and each conclusion are separated into their component parts and subjected to critical scrutiny. It is such analysis as this which must, in the circumstances detailed above, determine whether certain apparently incongruous facts should be rejected as irrelevant to the problem, or included as fully pertinent; and, in the latter case, whether a given hypothesis must be rejected or merely revised, before deduction and observation are brought into harmony.

To illustrate the essential nature of the confirmation and revision envisaged in stage six, let us take one more illustration from the study of shore platforms. Suppose, for example, that after the verification and elimination of stage five, our investigator of shore platforms finds that only one hypothesis offers a reasonable explanation of the facts thus far in his possession. This, let us say, is the hypothesis of storm-wave erosion under present conditions of land and sealevel.

He has previously deduced from this hypothesis certain expectable consequences which are not matched by any facts he has thus far observed. Thus, he has reasoned that if the platforms be the product of storm waves they should be found at higher elevations on exposed headlands where waves are large, and at lower elevations in bays or behind barriers where waves are small. Furthermore, the platforms should be broadest where the most violent storm waves, breaking on a coast, wear it back most rapidly. Here are two perfectly concrete deductions from the hypothesis; but he has never observed any such relation between intensity of wave attack and height and breadth of platform as the deductions call for. Perhaps this is partly because he never gave conscious attention to this relation, and partly because the shores he has studied did not present sufficiently strong contrasts in exposure to produce differences in height and breadth of platforms so great as to force themselves upon his attention.

In any case, the investigator now plans field work on more remote coasts, carefully analyzing the requirements of his problem, and seeking shores where conditions are especially favorable for testing the validity of these two deductions. His powers of observation are now greatly enhanced. He sees many things about shore benches which quite escaped attention before. Among others, he notes the remarkable freshness of the cliff face back of the platforms, the frequent occurrence of a groove or notch eroded in the base of the cliffs, occasional fresh landslides on the platforms, overhanging sod at the cliff top, and marine organisms living in pools of salt water on the platform surface. These are all details he had failed to predict, yet these unanticipated observations find such reasonable explanation in the hypothesis of continued active development of the platforms by present storm waves that he accepts them as particularly valuable confirmation of that hypothesis. When he finds, in addition, that the benches do range higher on the headlands and lower in the bays, he regards the confirmation as doubly strengthened.

There remains the prediction that platforms of maximum breadth should, other things being equal, be found on those coasts subjected to maximum violence of wave attack. Once more our investigator analyzes the variable factors presented by a number of coasts, in the effort to select wisely several localities which may offer him favorable conditions for testing the validity of this last deduction. He chooses his hunting-grounds, visits them in turn, and in each case finds no shore platforms at all. Instead, angry seas beat themselves to foam on a chaos of tumbled rocks.

Much disconcerted by this glaring discordance between deduced expectations and observed facts, our investigator turns to analysis for an answer to the enigma. Again he dissects and scrutinizes in detail the several stages of his work: the reasoning which led to invention of the parent hypothesis, the deduction which caused him to expect the broadest benches where wave attack is most violent, and the facts which fail so conspicuously to harmonize with expectations. Each in turn is probed and tested with greatest care and the most complete detachment and impartiality.

Perhaps the outcome of this analysis is discovery that the hypothesis is valid in its essentials, but requires revision of details. He may conclude that where storm waves attack a coast most violently and undercut it most rapidly, falling debris breaks the waves into confused and surging waters which cannot etch a neatly carved platform at a specific level. If such be the result of his analysis, he will recast the hypothesis in terms of the erosive behavior of storm waves of moderate intensity, realizing that this may well differ widely from the behavior of maximum storm waves. If some other explanation better resolves the difficulty, he will recast the hypothesis accordingly. He will then proceed to test the hypothesis in its new form by inspection of a wide variety of shores where waves of every type and degree of intensity are engaged in consuming the land.

VII. ANALYSIS IN THE STAGE OF INTERPRETATION

Let us suppose that the hypothesis of wave erosion under present conditions of land and sealevel survives, in its modified form, every test applied to it. The investigator has reached the seventh, and final, stage of his labors, in which he proceeds to formulate for publication his *Interpretation* of the origin of shore platforms. Can he at last dispense with analysis, the instrument which has served him so well at every previous step in his labors? Not wholly, I think. In any event, there must remain with him a caution, inspired by his full understanding of the nature of the proofs with which he is dealing, an understanding which comes only from critical analysis of both evidence and arguments. Rarely can geologist or geographer offer mathematical demonstration of the truth of his conclusions. He can, as a rule, go no further than to show that a given proposition has a high degree of probability. Certain facts, carefully observed and properly classified, prompt the invention of various hypotheses. The deduced consequences of one of these hypotheses alone is matched by all the facts, including some which are newly discovered and highly peculiar. The investigator, therefore, has much confidence that an interpretation based on this hypothesis will be valid, but he realizes there is still opportunity for error. He may not have thought of all possible hypotheses. He may not have deduced all the reasonable consequences of those hypotheses considered. He may not have discovered all pertinent facts. Some future contribution to knowledge may throw unexpected light on his problem, and radically affect his conclusions. He therefore wisely regards his present interpretation as highly probably theory, rather than as demonstrated fact.

In previous stages an error led only one investigator astray, and that one could retrace his steps and repair damage by embarking on some new line of reasoning. But in this final stage an error committed to the printed page may lead many astray, and may show wholly unexpected vitality and power for evil. The printed statements must go far enough; but they must not go farther than the evidence and fully tested reasoning warrant. A given interpretation must not be applied to all shore platforms before it has completely been demonstrated that there are no such platforms for which it is not the most satisfactory explanation. It must not even be asserted as the sole cause of any platform, unless and until it has clearly been proven that no contributing cause has played a part in their formation.

In these and other ways the investigator must be on his guard against overstatement, understatement, erroneous statement of his results. Analysis is the weapon with which he defends himself against too broad generalization and other errors which are wont to intrude themselves into the final stage of an investigation. Each sentence he scrutinizes; each conclusion he dissects; checking here and verifying there, to the end that the formulation of his interpretation shall contain no less, and particularly no more, than critically observed facts and carefully tested reasoning may justify. An interpretation, thus cautiously reached and conservatively formulated, will surely command the serious consideration of scientific men.

Conclusion

Have we, in the method of multiple working hypotheses, applied with the aid of rigorous analysis, something which will guide us unfailingly to the discovery of truth? We are compelled to answer this question in the negative. No device, however perfect, can wholly deprive the human intellect of its capacity for making mistakes. De Leon searched in vain for the fountain of youth. Can we hope for a magical fountain of truth?

The most for which we may reasonably hope is by correct methods of research to reduce the chances of error to a minimum, and to raise to its maximum the probability of discovering the real causes and relations of things. This we have done, so far as lies within our power, when we are accurate in observing facts, careful in classifying them, cautious in generalizing from them, fertile in inventing hypotheses, ingenious and impartial in testing their validity, skillful in securing their confirmation or revision, and judicial in formulating ultimate interpretations.

Multiple working hypotheses as a method, employed in connection with critical analysis as an instrument of precision, offer us, in my opinion, the best guarantee of success in scientific research.

Reprinted from the *Journal of Geology*, v. 5 (1897), p. 837-848.

Studies for Students.

THE METHOD OF MULTIPLE WORKING HYPOTHESES.[1]

THERE are two fundamental modes of study. The one is an attempt to follow by close imitation the processes of previous thinkers and to acquire the results of their investigations by memorizing. It is study of a merely secondary, imitative, or acquisitive nature. In the other mode the effort is to think independently, or at least individually. It is primary or creative study. The endeavor is to discover new truth or to make a new combination of truth or at least to develop by one's own effort an individualized assemblage of truth. The endeavor is to think for one's self, whether the thinking lies wholly in the fields of previous thought or not. It is not necessary to this mode of study that the subject-matter should be new. Old material may be reworked. But it is essential that the process of thought and its results be individual and independent, not the mere following of previous lines of thought ending in predetermined results. The demonstration of a problem in Euclid precisely as laid down is an illustration of the former; the demonstration of the same proposition by a method of one's own or in a manner distinctively individual is an illustration of the latter, both lying entirely within the realm of the known and old.

Creative study however finds its largest application in those subjects in which, while much is known, more remains to be learned. The geological field is preëminently full of such sub-

---

[1] A paper on this subject was read before the Society of Western Naturalists in 1892, and was published in a scientific periodical. Inquiries for the article have recently been such as to lead to the belief that a revision and republication are desirable. The article has been freely altered and abbreviated so as to limit it to aspects related to geological study.

837

jects, indeed it presents few of any other class. There is probably no field of thought which is not sufficiently rich in such subjects to give full play to investigative modes of study.

Three phases of mental procedure have been prominent in the history of intellectual evolution thus far. What additional phases may be in store for us in the evolutions of the future it may not be prudent to attempt to forecast. These three phases may be styled the method of the ruling theory, the method of the working hypothesis, and the method of multiple working hypotheses.

In the earlier days of intellectual development the sphere of knowledge was limited and could be brought much more nearly than now within the compass of a single individual. As a natural result those who then assumed to be wise men, or aspired to be thought so, felt the need of knowing, or at least seeming to know, all that was known, as a justification of their claims. So also as a natural counterpart there grew up an expectancy on the part of the multitude that the wise and the learned would explain whatever new thing presented itself. Thus pride and ambition on the one side and expectancy on the other joined hands in developing the putative all-wise man whose knowledge boxed the compass and whose acumen found an explanation for every new puzzle which presented itself. Although the pretended compassing of the entire horizon of knowledge has long since become an abandoned affectation, it has left its representatives in certain intellectual predilections. As in the earlier days, so still, it is a too frequent habit to hastily conjure up an explanation for every new phenomenon that presents itself. Interpretation leaves its proper place at the end of the intellectual procession and rushes to the forefront. Too often a theory is promptly born and evidence hunted up to fit in afterward. Laudable as the effort at explanation is in its proper place, it is an almost certain source of confusion and error when it runs before a serious inquiry into the phenomenon itself. A strenuous endeavor to find out precisely what the phenomenon really is should take the lead and crowd back the question, commendable at a later stage, "How came this so?" First the full facts, then the interpretation thereof, is the normal order.

The habit of precipitate explanation leads rapidly on to the birth of general theories.[1] When once an explanation or special theory has been offered for a given phenomenon, self-consistency prompts to the offering of the same explanation or theory for like phenomena when they present themselves and there is soon developed a general theory explanatory of a large class of phenomena similar to the original one. In support of the general theory there may not be any further evidence or investigation than was involved in the first hasty conclusion. But the repetition of its application to new phenomena, though of the same kind, leads the mind insidiously into the delusion that the theory has been strengthened by additional facts. A thousand applications of the supposed principle of levity to the explanation of ascending bodies brought no increase of evidence that it was the true theory of the phenomena, but it doubtless created the impression in the minds of ancient physical philosophers that it did, for so many additional facts seemed to harmonize with it.

For a time these hastily born theories are likely to be held in a tentative way with some measure of candor or at least some self-illusion of candor. With this tentative spirit and measurable candor, the mind satisfies its moral sense and deceives itself with the thought that it is proceeding cautiously and impartially toward the goal of ultimate truth. It fails to recognize that no amount of provisional holding of a theory, no amount of application of the theory, so long as the study lacks in incisiveness and exhaustiveness, justifies an ultimate conviction. It is not the slowness with which conclusions are arrived at that should give satisfaction to the moral sense, but the precision, the completeness and the impartiality of the investigation.

[1] I use the term theory here instead of hypothesis because the latter is associated with a better controlled and more circumspect habit of the mind. This restrained habit leads to the use of the less assertive term hypothesis, while the mind in the habit here sketched more often believes itself to have reached the higher ground of a theory and more often employs the term theory. Historically also I believe the word theory was the term commonly used at the time this method was predominant.

It is in this tentative stage that the affections enter with their blinding influence. Love was long since discerned to be blind and what is true in the personal realm is measurably true in the intellectual realm. Important as the intellectual affections are as stimuli and as rewards, they are nevertheless dangerous factors in research. All too often they put under strain the integrity of the intellectual processes. The moment one has offered an original explanation for a phenomenon which seems satisfactory, that moment affection for his intellectual child springs into existence, and as the explanation grows into a definite theory his parental affections cluster about his offspring and it grows more and more dear to him. While he persuades himself that he holds it still as tentative, it is none the less lovingly tentative and not impartially and indifferently tentative. So soon as this parental affection takes possession of the mind, there is apt to be a rapid passage to the unreserved adoption of the theory. There is then imminent danger of an unconscious selection and of a magnifying of phenomena that fall into harmony with the theory and support it and an unconscious neglect of phenomena that fail of coincidence. The mind lingers with pleasure upon the facts that fall happily into the embrace of the theory, and feels a natural coldness toward those that assume a refractory attitude. Instinctively there is a special searching-out of phenomena that support it, for the mind is led by its desires. There springs up also unwittingly a pressing of the theory to make it fit the facts and a pressing of the facts to make them fit the theory. When these biasing tendencies set in, the mind rapidly degenerates into the partiality of paternalism. The search for facts, the observation of phenomena and their interpretation are all dominated by affection for the favored theory until it appears to its author or its advocate to have been overwhelmingly established. The theory then rapidly rises to a position of control in the processes of the mind and observation, induction and interpretation are guided by it. From an unduly favored child it readily grows to be a master and leads its author whithersoever it will. The subsequent history of that mind in respect to that theme is but the progressive dominance of a ruling idea. Briefly summed up, the evolution is this : a premature explanation passes first into a tentative theory, then into an adopted theory, and lastly into a ruling theory.

When this last stage has been reached, unless the theory happens perchance to be the true one, all hope of the best results is gone.   To be sure truth may be brought forth by an investigator dominated by a false ruling idea.   His very errors may indeed stimulate investigation on the part of others.   But the condition is scarcely the less unfortunate.

As previously implied, the method of the ruling theory occupied a chief place during the infancy of investigation.   It is an expression of a more or less infantile condition of the mind.   I believe it is an accepted generalization that in the earlier stages of development the feelings and impulses are relatively stronger than in later stages.

Unfortunately the method did not wholly pass away with the infancy of investigation.   It has lingered on, and reappears in not a few individual instances at the present time.   It finds illustration in quarters where its dominance is quite unsuspected by those most concerned.

The defects of the method are obvious and its errors grave If one were to name the central psychological fault, it might be stated as the admission of intellectual affection to the place that should be dominated by impartial, intellectual rectitude alone.

So long as intellectual interest dealt chiefly with the intangible, so long it was possible for this habit of thought to survive and to maintain its dominance, because the phenomena themselves, being largely subjective, were plastic in the hands of the ruling idea; but so soon as investigation turned itself earnestly to an inquiry into natural phenomena whose manifestations are tangible, whose properties are inflexible, and whose laws are rigorous, the defects of the method became manifest and an effort at reformation ensued.   The first great endeavor was repressive.   The advocates of reform insisted that theorizing should be restrained and the simple determination of facts should take its place. The effort was to make scientific study statistical instead of causal. Because theorizing in narrow lines had led to manifest evils theorizing was to be condemned. The reformation urged was not the proper control and utilization of theoretical effort but its suppression. We do not need to go backward more than a very few decades to find ourselves in the midst of this attempted reformation. Its weakness lay in its narrowness and its restrictiveness. There is no nobler aspiration of the human intellect than the desire to compass the causes of things. The disposition to find explanations and to develop theories is laudable in itself. It is only its ill-placed use and its abuse that are reprehensible. The vitality of study quickly disappears when the object sought is a mere collocation of unmeaning facts.

The inefficiency of this simply repressive reformation becoming apparent, improvement was sought in the method of the working hypothesis. This has been affirmed to be *the* scientific method. But it is rash to assume that any method is *the* method, at least that it is the ultimate method. The working hypothesis differs from the ruling theory in that it is used as a means of determining facts rather than as a proposition to be established. It has for its chief function the suggestion and guidance of lines of inquiry; the inquiry being made, not for the sake of the hypothesis, but for the sake of the facts and their elucidation. The hypothesis is a mode rather than an end. Under the ruling theory, the stimulus is directed to the finding of facts for the support of the theory. Under the working hypothesis, the facts are sought for the purpose of ultimate induction and demonstration, the hypothesis being but a means for the more ready development of facts and their relations.

It will be observed that the distinction is not such as to prevent a working hypothesis from gliding with the utmost ease into a ruling theory. Affection may as easily cling about a beloved intellectual child when named an hypothesis as if named a theory, and its establishment in the one guise may become a ruling passion very much as in the other. The historical antecedents and the moral atmosphere associated with the working hypothesis lend some good influence however toward the preservation of its integrity.

Conscientiously followed, the method of the working hypothesis is an incalculable advance upon the method of the ruling theory; but it has some serious defects. One of these takes concrete form, as just noted, in the ease with which the hypothesis becomes a controlling idea. To avoid this grave danger, the method of multiple working hypotheses is urged. It differs from the simple working hypothesis in that it distributes the effort and divides the affections. It is thus in some measure protected against the radical defect of the two other methods. In developing the multiple hypotheses, the effort is to bring up into view every rational explanation of the phenomenon in hand and to develop every tenable hypothesis relative to its nature, cause or origin, and to give to all of these as impartially as possible a working form and a due place in the investigation. The investigator thus becomes the parent of a family of hypotheses; and by his parental relations to all is morally forbidden to fasten his affections unduly upon any one. In the very nature of the case, the chief danger that springs from affection is counteracted. Where some of the hypotheses have been already proposed and used, while others are the investigator's own creation. a natural difficulty arises, but the right use of the method requires the impartial adoption of all alike into the working family. The investigator thus at the outset puts himself in cordial sympathy and in parental relations (of adoption, if not of authorship,) with every hypothesis that is at all applicable to the case under investigation. Having thus neutralized so far as may be the partialities of his emotional nature, he proceeds with a certain natural and enforced erectness of mental attitude to the inquiry, knowing well that some of his intellectual children (by birth or adoption) must needs perish before maturity, but yet with the hope that several of them may survive the ordeal of crucial research, since it often proves in the end that several agencies were conjoined in the production of the phenomena.   Honors must often be divided between hypotheses.   One of the superiorities of multiple hypotheses as a working mode lies just here.   In following a single hypothesis the mind is biased by the presumptions of its method toward a single explanatory conception. But an adequate explanation often involves the coördination of several causes.   This is especially true when the research deals with a class of complicated phenomena naturally associated, but not necessarily of the same origin and nature, as for example the Basement Complex or the Pleistocene drift.   Several agencies may participate not only but their proportions and importance may vary from instance to instance in the same field.   The true explanation is therefore necessarily complex, and the elements of the complex are constantly varying.   Such distributive explanations of phenomena are especially contemplated and encouraged by the method of multiple hypotheses and constitute one of its chief merits.   For many reasons we are prone to refer phenomena to a single cause.   It naturally follows that when we find an effective agency present, we are predisposed to be satisfied therewith.   We are thus easily led to stop short of full results, sometimes short of the chief factors.   The factor we find may not even be the dominant one, much less the full complement of agencies engaged in the accomplishment of the total phenomena under inquiry.   The mooted question of the origin of the Great Lake basins may serve as an illustration. Several hypotheses have been urged by as many different students of the problem as the cause of these great excavations. All of these have been pressed with great force and with an admirable array of facts.   Up to a certain point we are compelled to go with each advocate.   It is practically demonstrable that these basins were river valleys antecedent to the glacial incursion.   It is equally demonstrable that there was a blocking up of outlets.   We must conclude then that the present basins owe their origin in part to the preëxistence of river valleys and to the blocking up of their outlets by drift.   That there is a temptation to rest here, the history of the question shows.   But on the other hand it is demonstrable that these basins were occupied by great lobes of ice and were important channels of glacial movement. The leeward drift shows much material derived from their bottoms. We cannot therefore refuse assent to the doctrine that the basins owe something to glacial excavation. Still again it has been urged that the earth's crust beneath these basins was flexed downward by the weight of the ice load and contracted by its low temperature and that the basins owe something to crustal deformation. This third cause tallies with certain features not readily explained by the others. And still it is doubtful whether all these combined constitute an adequate explanation of the phenomena. Certain it is, at least, that the measure of participation of each must be determined before a satisfactory elucidation can be reached. The full solution therefore involves not only the recognition of multiple participation but an estimate of the measure and mode of each participation. For this the simultaneous use of a full staff of working hypotheses is demanded. The method of the single working hypothesis or the predominant working hypothesis is incompetent.

In practice it is not always possible to give all hypotheses like places nor does the method contemplate precisely equable treatment. In forming specific plans for field, office or laboratory work it may often be necessary to follow the lines of inquiry suggested by some one hypothesis, rather than those of another. The favored hypothesis may derive some advantage therefrom or go to an earlier death as the case may be, but this is rather a matter of executive detail than of principle.

A special merit of the use of a full staff of hypotheses coördinately is that in the very nature of the case it invites thoroughness. The value of a working hypothesis lies largely in the significance it gives to phenomena which might otherwise be meaningless and in the new lines of inquiry which spring from the suggestions called forth by the significance thus disclosed. Facts that are trivial in themselves are brought forth into importance by the revelation of their bearings upon the hypothesis and the elucidation sought through the hypothesis. The phenomenal influence which the Darwinian hypothesis has exerted upon the investigations of the past two decades is a monumental illustration. But while a single working hypothesis may lead investigation very effectively along a given line, it may in that very fact invite the neglect of other lines equally important. Very many biologists would doubtless be disposed today to cite the hypothesis of natural selection, extraordinary as its influence for good has been, as an illustration of this. While inquiry is thus promoted in certain quarters, the lack of balance and completeness gives unsymmetrical and imperfect results. But if on the contrary all rational hypotheses bearing on a subject are worked coördinately, thoroughness, equipoise, and symmetry are the presumptive results in the very nature of the case.

In the use of the multiple method, the reaction of one hypothesis upon another tends to amplify the recognized scope of each. Every hypothesis is quite sure to call forth into clear recognition new or neglected aspects of the phenomena in its own interests, but ofttimes these are found to be important contributions to the full deployment of other hypotheses. The eloquent expositions of "prophetic" characters at the hands of Agassiz were profoundly suggestive and helpful in the explication of "undifferentiated" types in the hand of the evolutionary theory.

So also the mutual conflicts of hypotheses whet the discriminative edge of each. The keenness of the analytic process advocates the closeness of differentiating criteria, and the sharpness of discrimination is promoted by the coördinate working of several competitive hypotheses.

Fertility in processes is also a natural sequence. Each hypothesis suggests its own criteria, its own means of proof, its own method of developing the truth; and if a group of hypotheses encompass the subject on all sides, the total outcome of means and of methods is full and rich.

The loyal pursuit of the method for a period of years leads to certain distinctive habits of mind which deserve more than the passing notice which alone can be given them here. As a factor in education the disciplinary value of the method is one of prime importance. When faithfully followed for a sufficient time, it develops a mode of thought of its own kind which may be designated the habit of parallel thought, or of complex thought. It is contra-distinguished from the linear order of thought which is necessarily cultivated in language and mathe-matics because their modes are linear and successive. The procedure is complex and largely simultaneously complex. The mind appears to become possessed of the power of simultaneous vision from different points of view. The power of viewing phenomena analytically and synthetically at the same time appears to be gained. It is not altogether unlike the intellectual procedure in the study of a landscape. From every quarter of the broad area of the landscape there come into the mind myriads of lines of potential intelligence which are received and coördinated simultaneously producing a complex impression which is recorded and studied directly in its complexity. If the landscape is to be delineated in language it must be taken part by part in linear succession.

Over against the great value of this power of thinking in complexes there is an unavoidable disadvantage. No good thing is without its drawbacks. It is obvious upon studious consideration that a complex or parallel method of thought cannot be rendered into verbal expression directly and immediately as it takes place. We cannot put into words more than a single line of thought at the same time, and even in that the order of expression must be conformed to the idiosyncrasies of the language. Moreover the rate must be incalculably slower than the mental process. When the habit of complex or parallel thought is not highly developed there is usually a leading line of thought to which the others are subordinate. Following this leading line the difficulty of expression does not rise to serious proportions. But when the method of simultaneous mental action along different lines is so highly developed that the thoughts running in different channels are nearly equivalent, there is an obvious embarrassment in making a selection for verbal expression and there arises a disinclination to make the attempt. Furthermore the impossibility of expressing the mental operation in words leads to their disuse in the silent processes of thought and hence words and thoughts lose that close association which they are accustomed to maintain with those whose silent as well as spoken thoughts predominantly run in linear verbal courses. There is therefore a certain predisposition on the part of the practitioner of this method to taciturnity. The remedy obviously lies in coördinate literary work.

An infelicity also seems to attend the use of the method with young students. It is far easier, and apparently in general more interesting, for those of limited training and maturity to accept a simple interpretation or a single theory and to give it wide application, than to recognize several concurrent factors and to evaluate these as the true elucidation often requires. Recalling again for illustration the problem of the Great Lake basins, it is more to the immature taste to be taught that these were scooped out by the mighty power of the great glaciers than to be urged to conceive of three or more great agencies working successively in part and simultaneously in part and to endeavor to estimate the fraction of the total results which was accomplished by each of these agencies. The complex and the quantitative do not fascinate the young student as they do the veteran investigator.

The studies of the geologist are peculiarly complex. It is rare that his problem is a simple unitary phenomenon explicable by a single simple cause. Even when it happens to be so in a given instance, or at a given stage of work, the subject is quite sure, if pursued broadly, to grade into some complication or undergo some transition. He must therefore ever be on the alert for mutations and for the insidious entrance of new factors. If therefore there are any advantages in any field in being armed with a full panoply of working hypotheses and in habitually employing them, it is doubtless the field of the geologist.

T. C. CHAMBERLIN.

BULLETIN OF THE AMERICAN ASSOCIATION OF PETROLEUM GEOLOGISTS
VOL. 43, NO. 7 (JULY, 1959), PP. 1500-1502

THERE IS A REASON[1]

GORDON RITTENHOUSE[2]
Houston, Texas

Mr. Chairman, fellow members of S.E.P.M. and A.A.P.G., and guests, if I were a preacher instead of a geologist, I might take for my text today the words "There is a reason." I am not a preacher, of course, but as some of you know, I have been somewhat of a missionary within our profession for a way of thinking about rocks—or if you wish, a philosophy about how to look at rocks. This philosophy or way of thinking is not new— certainly it is not original with me but antedates by far my entrance into the field of geology some 30 years ago.

What is this way of thinking—this philosophy? It is based on the assumption that rocks as they occur today did not just happen, but that there is a reason for their being what they are and where they are. What rocks are and where they are are due to a combination of physical, chemical, and biological factors that have acted through time according to definite laws. These factors have left their imprints on the rock in varying degrees and, from these imprints or clues, the history of the rock may be reconstructed with varying degrees of success. This reconstruction of the total history of the rock is to me of paramount importance. To do it one needs not only to observe and ask "why"; one must believe "There is a reason."

This may seem quite simple—something that we all do—and therefore you may be wondering why I am talking about it today. It is simple—but it certainly is *not* something we all do.

Let me illustrate this with two examples. I keep on my desk a piece of core that I call my "Separator of Men from Boys." I have given many budding geologists fresh from more than a score of universities a hand lens and knife and asked them to study this core for a few minutes and then to describe it, to give the rock a name, and to indicate what geological history can be deduced from it. I have found few "Men" and many "Boys." Less than half recognized that information about the relief, the climate, and the kind and age of rocks in the source area can be deduced from the kind and size of pebbles, or that size, sorting, and rounding of the pebbles suggest a relatively short distance of transport and deposition under fluvial conditions, probably in an alluvial fan.

Fewer than one in ten recognized that the orientation of the pebbles indicates that the beds are almost vertical, or that low porosity, close packing, and the kinds and number of contacts between pebbles indicate extensive post-depositional modification. Most of those young geologists were not asking themselves "What properties does this rock have that will tell where it came from, how and under what environmental conditions was it formed, and how has it been changed since it was deposited?" Nor were they saying "There is a reason" why it is like it is today.

As a second example, I might point to descriptions of measurèd sections. In publications these commonly read "sandstone, brown, fine-grained, cross-bedded, 14 feet," and I have no reason to believe the descriptions are more extensive in many field notebooks. Here we have major lithology, color, size of grain, and the presence of a primary structure. For distinguishing this bed from those above and below, or for comparing it or the beds in a sequence, with beds in another measured section this may appear to be adequate, in fact, may appear to be a first very necessary step. But is it adequate? From the terms "sandstone" and "fine-grained" it may be inferred that the agent of deposition was a current of

[1] Presidential address, 33d annual meeting of the Society of Economic Paleontologists and Mineralogists, Dallas, March 17, 1959.

[2] Shell Development Company.

moderate velocity. What kind of current and what was its direction of flow? I submit that these are important questions to answer, since the answers may bear on selection of where to measure the next section or how the lithologic variations between sections are to be explained. In our example, cross-bedding may be a useful clue. But what do we know from just the word "cross-bedded?" Our answer may come from the direction of dip of the cross-beds, their inclination, their thickness and lateral extent, and their relation to beds above and below.

I might follow this example further, using the lack of information on composition conveyed by the broad term "sandstone" and the lack of information on how much induration has occurred and what caused the induration. I hope, however, that I have made my point; namely, that it is common to overlook many clues that could be used in reconstructing the history of rocks because we are not wondering what they mean or assuming that "There is a reason."

In looking at any rock we have some purpose—and a limited time in which to achieve that purpose. Therefore we can not describe everything—we must be selective, observing or measuring those features of the rock that will permit us to accomplish our purpose. In making this selection we are guided by our experience—and by our way of thinking about rocks.

When our objective is limited, such as in identifying a subsurface structural datum, our choice may be easy, because our experience or that of others may indicate what to look for. Generally little thought is required to identify the datum—although much thought may be needed to interpret what the position of the datum means.

When our objective is broader, the choice may be more difficult, requiring a combination of experience and a way of thinking about rocks. Thus in our "sandstone, brown, fine-grained, cross-bedded, 14 feet" example, reliance is being placed on experience but this time with the *hope* of accomplishing the purpose, experience having shown that useful geological interpretations may be made in some cases if vertical and lateral variations in lithology are known. Outlining lithologic geometries may be a powerful tool, particularly in subsurface work—provided enough control points are available to define the geometry and provided we know what the geometry means when we have it. Commonly our control is not adequate, and how we interpolate between control points to outline the geometry will depend on our way of thinking about rocks.

All too often, experience is substituted for thought—not made its partner. Because certain types of observations or measurements have contributed to the successful solution of problems in the past, they are selected and applied indiscriminately to new problems. Here we have one type of "shot gun" approach, based on a philosophy which, in effect, says "If we can make enough observations on enough rocks, put them in a machine and turn the crank, something useful may come out." To some, this is the "modern statistical approach" to geological problems. This maligns statistical methods—which can be valuable in geology—though not as a substitute for thinking. Since this approach does not reach the objective in a minimum of time, if at all, it is wasteful of time, manpower, and money. I prefer the "There is a reason" philosophy.

In selection of features that may be significant in rocks, we are faced with ever increasing knowledge in geology and related sciences, and ever increasing specialization. In looking at rocks and at problems involving rocks there is an increasing tendency to look at specialized aspects of the rock, rather than at the rock as a whole. The sands "belong" to one specialist; the clays, the carbonates, the isotopes, the trace elements and various fossil groups, to others. Specialization has great potentialities for our science, since it can lead to a better understanding of the physical, chemical, and biological factors that have made rocks what they are today. But it has great hazards as well. Let's not forget how the elephant appeared to the three blind men—like a wall, like a post, like a snake. Just as we need to consider the entire elephant, we need to consider the entire rock—putting each type of observation and deduction in its proper perspective.

With an increasing number of things to observe or measure, it becomes increasingly important to make the proper choice—the choice that will permit us to reach our objective in the time available. To me the obvious procedure is to go from the simple to the complex, from the rapid to the slow, from the less expensive to the more expensive. In general, but not always, this will mean going from the large to the small, from the visible to the invisible, from the direct to the indirect. There is no magic formula, no "gimmick," that will indicate what to look for or when. This will depend on the problem, the purpose, the time, the material available, and the way one thinks about rocks. Because I am firmly convinced that what rocks are and where they are has resulted from a combination of physical, chemical, and biologic factors that have acted through time according to definite laws, and that those factors have left their imprints on the rock in varying degrees, I believe that as we study rocks we should observe, ask "why," and remember "There is a reason." Thank you.

Models Versus Data—An Imbalance

Gerald P. Salisbury
Unocal Corporation
Los Angeles, California

It is increasingly obvious that the volume of information on world geological patterns generated by industry and government organizations has outstripped the ability of the geological community to assimilate and integrate it. Nevertheless, the impulse to generalize remains, and regional syntheses continue to be proposed by individuals without access to the entire existent data file. Although the problem of making all the existing data available to concerned investigators is probably unsolvable, the likely existence of relevant information should at least be recognized and an effort made to obtain access.

Since the late 1940s, geological endeavors have followed two principal interrelated but curiously contradictory directions: one, the most obvious, the development of comprehensive integrations of sedimentation, tectonic, and alteration processes; the other, the commonly unremarked accumulation of vast volumes of detailed information about virtually every sector and aspect of the world and its history.

In the 1930s, the geological profession generally felt that significant conclusions about worldwide processes must wait on more information. Responsible scientists should await further data before preparing speculative generalizations. The attitude was a long held one as expressed by Archibald Geikie in 1905:

In the third and last place, it seems to me that one important lesson to be learnt from a review of the successive stages in the foundation and development of geology is the absolute necessity of avoiding dogmatism. Let us remember how often geological theory has altered. The Catastrophists had it all their own way until the Uniformitarians got the upper hand, only to be in turn displaced by the Evolutionists. The Wernerians were as certain of the origin and sequence of rocks as if they had been present at the formation of the earth's crust. Yet in a few years their notions and overweening confidence became a laughing stock. From the very nature of its subject, as I have already remarked, geology does not generally admit of the mathematical demonstration of its conclusions. They rest upon a balance of probabilities. But this balance is liable to alteration, as facts accumulate or are better

understood. Hence what seems to be a well-established deduction in one age may be seen to be more or less erroneous in the next. Every year, however, the data on which these inferences are based are more thoroughly comprehended and more rigidly tested. Geology now possesses a large and ever-growing body of well-ascertained fact, which will be destroyed by no discovery of the future, though it will doubtless be vastly augmented, while new light may be cast on many parts of it now supposed to be thoroughly known.

. . .

Each of us has it in his power to add to this accumulation of knowledge. Careful and accurate observation is always welcome, and may eventually prove of signal importance. While availing ourselves freely of the use of hypothesis as an aid in ascertaining the connection and significance of facts, we must be ever on our guard against premature speculation and theory, clearly distinguishing between what is fact and what may be our own gloss or interpretation of it.
—Archibald Geikie, 1905, *The Founders of Geology*, second edition: London, Macmillan, p. 472, 473.

Geology became widely regarded—and derided—as simply a "descriptive" science with no orienting conceptual framework, certainly no mathematical foundation. The geosynclinal concept, biological evolution, a model of igneous emplacement and metamorphism, a simple model of cyclic sedimentation, a systematic concept of surface morphology, and (for the oil industry) the anticlinal theory of oil accumulation served in the 1930s for working frames. Continental drift had been agreed—at least in the U.S.—to be physically improbable.

Quite unexpectedly the impetus of military and post-war technology in the late 1940s and 1950s brought into being a swelling flood of information that resulted by the 1960s in the framework of a comprehensive world model, plate tectonics and its associated vision of world dynamism, by the late 1960s and 1970s in a comprehensive concept of thermal molecular alteration of hydrocarbons, and by the 1970s and 1980s in a concept of cyclic sea-level variation integrating worldwide erosional and depositional patterns. Largely unnoticed in their scope, exploration surveys for hydrocarbons and minerals using increasingly subtle and complex data accumulated detailed structural and stratigraphic information on all parts of the world. The influx of multiple facts led, on the one hand, to efforts to generalize and, on the other, to attempts to document the data in all its specific complexity.

Beginning in the 1950s at the same time that global models began to be developed, the urge somehow to place geology in a mathematical frame yielded a growing application of statistical methods to the burgeoning mass of data.

Application of statistical methods to geology was accompanied by a radically different philosophy: an emphasis upon the random nature of most geological data and a rejection of strictly deterministic causal relations. This philosophy did not merely attempt to accommodate observational error, but rather regarded most processes as inherently nondeterministic. The practical result is that accuracy in observation is no longer particularly critical. Ranges of error can be easily tolerated. Only a small sample of data is necessary. In geomorphology, concern for deterministic explanations for particular land forms was largely replaced by mathematical analyses of essentially random drainage systems. Davis' anthropomorphic analogies of topographic evolution were set aside. In the journals, concern for publication of large-scale, accurate, detailed maps was forgotten and replaced with demands for easily read page-size draftsman's illustrations.

The two opposite approaches—model-making generalization and voluminous data collection—which are now the two faces of our knowledge of the world, are worth considering.

The world concepts—remarkably successful generalizations—fill the objective of all science: to somehow embody a morass of facts in a simple frame that can be easily understood. In short, they provide explanations (if possible, ones that will allow prediction). Characteristically the concepts have been based on bold, remarkably insightful, plausible speculations which, by extensive application to a variety of geological situations, have been expanded over the years into widely accepted theories. They are conspicuously flexible and adaptable and are essentially undisprovable. Their "truth" rests upon consensus of plausibility—a consensus of a small group of specialists—and upon their practical utility to a very large group of uncritical users. Oddly, their utility lies largely in providing ad hoc explanations for selected information, but very little in the way of testable prediction. The activity of a considerable sector of the academic community seems to be that of generating local models of the world theory: a "story" with a plate-tectonics, thermal-maturation, or sea-level–cycle plot. Perhaps like the original insights, the current models, however extensive their descriptions, are nearly always based on limited data and are represented by cartoons or by very small-scale maps casually ignoring unsightly specifics. The academic community, largely ignorant of the vast volume of data in industry files, appears unaware of the contrast between models and data. Industry personnel on the other hand are little concerned by their inconsistencies.

As the product of the past 50 years of exploration of North America, there are in the files the logs of several hundred thousand wells integrated in detailed (commonly 1 inch to the mile or greater) structural and stratigraphic maps. All of the outcropping sedimentary structure has been mapped on similar or larger scales. The basins of the continent are covered by magnetometer, gravity meter, and seismic surveys. The continental shelves and deep Gulf of Mexico have been mapped in detail with seismic surveys. On close inspection, the details of these maps can be seen to be quite different from the generalizations of published regional maps and of the very small-scale "maps" and sketches of the journals. Critical tectonic analysis, of necessity based upon accurate structural descriptions (location, orientation, scale), cannot be accurately done using published maps.

The large but seemingly unrecognized disparity between existent data and the selective information upon which published models are based should at least be recognized. Those who undertake to synthesize the world's geology accept a responsibility to gather and honor all the data. The geologist without access to that information is in the same position as an earthbound astronomer who wishes to investigate the planets without access to the latest results of space exploration. Neither can reasonably proceed without the whole database.

Much basic data is in the hands of the mineral and oil and gas industry, an industry now being partially dismantled and quickly losing its memory of filed material. There is a need for:

(1) Some kind of national archive to store all kinds of surface and subsurface geological information and geophysical information.

(2) A renewed effort to develop accurate, comprehensive descriptions of the geology of the U.S., and for that matter the world, upon which to base meaningful syntheses rather than merely imaginative speculations.

The American Association of Petroleum Geologists Bulletin
V. 55, No. 11 (November 1971), P. 1927-1938, 16 Figs.

The Geological Attitude[1]

JOHN G. C. M. FULLER[2]

Calgary 2, Alberta

Abstract Geological activity takes place mainly in response to industrial and social pressures. Past geological reaction to these pressures profoundly altered popular conceptions of time, the Church, man, and the balance of nature. The present-day circumstances of geology are not essentially different from those of the past.

About 180 years ago when the history of the earth was enlarged into a geological time scale, the actions of geologists came into conflict with established beliefs on the creation and destiny of the world. Social concern for these matters was evident in the public attention given to geology during the early 1800s, and also later, when evolutionary concepts of man and his place in nature began to permeate society.

Petroleum geology in North America illustrates the role of technology in determining the style and scope of geological work. Peaks of activity cluster obviously on the introduction from time to time of new instrumental capabilities (geophysical apparatus, for example), although not infrequently such activity is testing concepts or relationships perceived long before. Organic metamorphism and continental drift provide two examples.

The petroleum industry now faces the dilemma of satisfying predicted demands for fuel, without doing irreparable injury to its environment of operation. Awareness of man's place in nature, which is a fundamental perception of geology, governs the geological attitude.

The Past Is the Key to the Present

By "the geological attitude," I mean that particular point of view which has been a characteristic of geologists since they first tracked mud and heresy into the schools of learning some 200 years ago. To explain this attitude we must know something about the nature of geological activity itself, and something about the social environment in which it is found. I begin with two statements to cover these points.

First, geological activity takes place mainly in response to industrial and social pressures.

Second, past geological reaction to these pressures

[1] Manuscript received, June 11, 1971; accepted, July 16, 1971. Keynote address: 56th Annual Meeting of The American Association of Petroleum Geologists, Houston, Texas, March 29–31, 1971.

[2] Amoco Canada Petroleum Company Ltd.

The manuscript of this address was read by Richard L. Evans, Maurice Kamen-Kaye, Arthur A. Meyerhoff, and Edgar W. Owen, to all of whom I am indebted for advice.

profoundly altered popular conceptions of time, the Church, man, and the balance of nature.

Examples both of pressure and reaction range through the whole of geology; and although the emphasis is always shifting, according to the particular demands of the time, present-day circumstances are not essentially different from those of the past, as I hope to show.

A third point about geological activity is that for at least 100 years its practical style has depended on technology. Peaks of activity cluster obviously on the introduction from time to time of new instrumental capabilities, and not infrequently such activity is testing concepts or relationships which had been perceived as long as a century beforehand.

Scotland, 1785

Modern geology began in style. In 1785, James Hutton, a 58-year-old naturalist and farmer, presented to the Royal Society of Edinburgh his thoughts "concerning the system of the earth, its duration and stability." The paper dealt with the lands and the seas in a succession of former worlds, and the power of water to wear away rocks. It painted a scene of immeasurable antiquity, constant in change, and evidently endless. "With respect to human observations," he wrote, "this world has neither a beginning nor an end." Hutton's words, as later events were to prove, struck a mortal blow to the ruling notion of secular time.

And of what importance was that? When Isaac Newton and the astronomers a century before had expanded the universe in space, beyond any constraint other than natural law, the established and popular view of Creation suffered no harm. When Hutton and the geologists, however, expanded the created world in time, they seemed to be challenging the Divine word. The reason was this: the English-speaking peoples had for two centuries resisted the authority of their original Church, preferring to substitute the authority of the English Bible (Fig. 1). Divine authority rested in the Scriptures, and to question them seemed to question Providence itself. The creation of the world, said the Bible, was accomplished in six days:

1927

FIG. 1.—Rowlandson, *Syntax Preaching*, 1815. "But, beloved, be not ignorant of this one thing, that one day is with the Lord as a thousand years, and a thousand years as one day" (Second Epistle General of Peter, ch. 3, v. 8).

Hutton and the geologists were saying that even six millennia could not meet the facts. Suddenly, the geological world view became a religious issue, and a conflict of appropriate ferocity commenced.

INDUSTRIAL BASIS OF STRATIGRAPHY

The battle of time was fought mainly on the evidence of stratigraphy, which was then new to academic geology, although it had been in use for 100 years or more as traditional working rules and empirical observations among colliers and quarrymen. In its primitive state stratigraphy had been of little or no philosophical interest to scholars, whose chief geological concerns were focused on cosmogony.

The change came with the Industrial Revolution in the second half of the 18th century, when the demand for coal and minerals, and the resulting acceleration of mining and quarrying operations, brought together practical geology and industrial interests. By this means, a working knowledge of the strata—their order, regularity, and distinguishing features—came to the notice of naturalists and scientists. One can argue, in a sense, that organized stratigraphy as a branch of geology was a product of 18th-century industrial pressure. By the beginning of the 19th century stratigraphy was reacting on popular beliefs.

FLOOD AND MILLENNIUM

Hutton's world view, and much of the geological doctrine of uniformity that followed from it (Hooykaas, 1963) can be summarized in a single phrase: the inevitability of gradualness. But the stratal succession was not gradual; it was manifestly serial and punctuated by relics of past turmoil. At many places the uppermost layer of sand and clay was so clearly a deposit such as might have resulted from a sudden inundation of the landscape, that stratigraphy seemed to prove with certitude the fact of the Flood; and thus proved, only the foolhardy and tough-minded would doubt Moses' account of it. These glacial deposits were mapped for years as "Diluvium" or deluge-stuff, long after the Flood peak had passed.

The apparently catastrophic nature of the event that left the diluvium, and the nearly universal extinction of life which it was believed to have caused, together with the relics of upheaval in the stratigraphic succession, became the chief elements in a new doctrine of successive creations which held the field for 50 years. Decayed fragments of it can still be found in sensational literature (Fig. 2).

It was a popular theory, for it incorporated an old belief that the major divisions of secular time, the six celestial days of Creation, were indeed millennia. Shakespeare knew in 1599 that the poor world was almost 6,000 years old, and said so; and at Cambridge in 1659 John Lightfoot fixed the exact date.

PUBLIC ATTENTION

Geology so stirred up public imagination during the middle years of the last century that people responded in thousands. The story of the rocks was news. In 1835, at Boston, 1,200 people turned out day after day for six weeks to hear Benjamin Silliman (Fig. 3), Professor at Yale College, lecturing on geology. "Clergymen thank me warmly," he wrote, "for the manner in which they say that delicate points are treated" (Fisher, 1866). No less than 4,500 ticket applications were made for Charles Lyell's lectures in the same city in 1841 (Fig. 4). Henry D. Rogers, whose name reappears in connection with petroleum geology, drew an audience of 1,000 in the comparatively small town of Portsmouth, New Hampshire, in 1845 (Struik, 1957). In England, Adam Sedgwick (Fig. 5), founder of the Cambrian System, made a public sensation in 1838 with a speech to some 3,000–4,000 people on the seashore at Tynemouth. Hugh Miller (Fig. 6) in 1854 gave a geological lecture in London at which there was an attendance of 5,000.

FIG. 3.—Benjamin Silliman. "I gave **full** scope to time precedent to man, but omitted any discussion of the days. At the conclusion many persons pressed forward towards me, and almost overwhelmed me." The "days" were the six celestial days. Portrait (detail) by Samuel Finley Breese Morse, Yale University Art Gallery. Gift of Bartlett Arkell, B.A., 1866.

A

Judging by the length of the seventh day of the creative week, which the Bible indicates is still in progress with a little more than 1,000 years to go, each creative day was at least 7,000 years in length. Man was created at the very end of the "sixth [creative] day"; and according to chronology of the Bible, humans have now been on earth almost six thousand years.

B

Yes, the facts unearthed by geologists agree with what the Bible says about special creation.

Man About 6,000 Years Old
According to the Bible, God created man about 6,000 years ago, and in recent years this has many times been confirmed.

FIG. 2.—The doctrine of successive creations is permanently embedded in folklore. Four and a half million copies of these versions of geological opinion circulated in 1967 (*Awake,* v. 48, no. 8, 1967, p. 7, 13).

FIG. 4.—Charles Lyell. At Boston he **had** to give his lectures in duplicate to keep audience down to about 1,500 at a sitting (Bailey, 1962). Lithograph after portrait by J. E. Mayall; Geological Society of London.

FIG. 5.—Adam Sedgwick. "Nearly all the population of Tynemouth turned out to join us. I addressed them at the utmost stretch of my voice six different times." Portrait by T. H. Maguire; Ipswich Museum.

Another spectacular display of popular enthusiasm took place near Birmingham in 1849, when Sir Roderick Murchison (Fig. 7), celebrated author of the Silurian System, led a field meeting on the Silurian outcrops at Dudley (Fig. 8). H. B. Woodward (1907) recorded that the Bishop of Birmingham, taking a gigantic speaking-trumpet, called on those present to offer their salutations to Murchison, repeating after him the words "Hail, King of Siluria." This they did three times; and 25,000 people visited the place that day. This kind of thing was not new to Murchison. When he was appointed Director of the Geological Survey of Great Britain, members of the House of Commons rose to their feet and cheered on hearing the Prime Minister's announcement.

THE EXPLOSION OF 1859

But during all this time, while stratigraphy piled system upon system, and the fossils multiplied, the conviction was growing among some geologists that a continuous process governed the evolution of living forms, a process in which the hand of Providence was scarcely to be seen, and which had produced man almost incidentally. When Charles Darwin published these convictions in 1859 he detonated an explosion that scorched the very floorboards of

FIG. 6.—Hugh Miller. "Five thousand persons, and I carried them with me throughout." Lithograph after a portrait by W. Bonnar.

FIG. 7.—Sir Roderick Murchison. "I gave them a regular half-hour's harangue." Portrait engraving *in* Geikie (1875).

society; and the burning fallout can still be seen (Fig. 9).

This confrontation with the established order affected everybody. Darwin's book, *On the Origin of Species,* quickly went through three editions. All the scientific doubts of the preceding years seemed suddenly to be public issues. "The rational cat was among the mystical pigeons," wrote Nicolas Bentley (1968), "but with beak and claw and with celestial vision dimmed by tears of outrage, the pigeons fought back." The Bishop of Oxford enquired of Thomas Huxley (Fig. 10) whether he claimed descent from an ape on his father's or his mother's side, and Huxley retorted that he would rather have an ape for a grandfather than a man who misused his gifts to bring ridicule and religious prejudice into a grave scientific discussion.

Huxley (1863) followed Darwin's book with *Evidence as to Man's Place in Nature,* and this subject, after lying dormant for a long time, has become the central issue of our day. It is, in fact, the key to the social impact of geological activity past and present.

Going back to Darwin for a moment, one finds that he was actually following a well-trodden path, which his grandfather, Erasmus Darwin, had traveled before him. And among other well-known precursors it is something of a curiosity—a quite astonishing one in view of later events—to find John Wesley, founder of the Methodist Church. Nearly 100 years before the *Origin of Species,* Wesley (1770) was putting his name to questions which asked: "By what degrees does nature raise herself up to man? How change these paws into flexible arms?

FIG. 8.—At the Wren's Nest, Dudley, Worcestershire. "About 25,000 persons visited the Dudley caverns on that day in 1849" (Woodward, 1907). Photograph shows part of old workings. Silurian limestone was particularly valuable to local iron-smelting industry.

<table>
<tr><td>

A

"Scientists who go about teaching that evolution is a fact of life are great con men, and the story they are telling may be the greatest hoax ever.

B

Belief in evolution requires *faith.* But which is harder to believe — the *theory* of evolution or the Genesis account of creation?

Scientists must "infer" that evolution occurred. They cannot see it.

</td></tr>
</table>

FIG. 9.—Partisan opposition to Darwin's evolutionary concept still circulates widely in sensational literature (**A.** *Awake,* v. 48, no. 8, 1967, p. 7; **B.** *The Plain Truth,* v. 35, no. 6–7, 1970, p. 42).

What method will she make use of to transform these crooked feet into supple and skilful hands?" (Fig. 11).

This only goes to show that a search for the originator of an idea or the first time it was recorded really does not shed much light on the social pressures and industrial demands which

FIG. 10.—Thomas Huxley. "I unhesitatingly affirm my preference for the ape." Portrait, Geological Society of London.

Fig. 11.—"The ape bears affinity to the quadruped and the man." "The whole progress of nature is so gradual, that the entire chasm from plant to man, is filled up with divers kinds of creatures, rising one above another, by so gentle an ascent that the transitions from one species to another, are almost insensible" (Wesley, 1823).

These creatures, the *Anthropomorpha* of Linnaeus, were illustrated in scientific literature of Wesley's time. They reappeared in Huxley's *Man's Place in Nature* (Collected Essays, 1894, Fig. 6).

bring that idea into prominence, and cause it to function as the focal point of new activity. Geological ideas grow and spread when social conditions are right for them; and the social pressures, working on the technology at hand, determine the style and scope of the geological activity which results.

Petroleum Geology

Consider, for example, the history of American petroleum geology. Stated simply, drilling for petroleum in Pennsylvania during the 1860s was an outcome of overhunting the whales whose oil served as lamp oil during the first half of the century. Kerosene provided a substitute for whale oil. And although petroleum at that time was already known to yield kerosene, a second competitive answer to whale-oil shortage was found in oil distilled from cannel coal and bituminous shale. The technology of oil-well drilling and casing moreover was also in existence during the first half of the century, having been developed in response to the demand for brine suitable for salt production in the territory west of the Alleghenies.

Petroleum geology thus presents a peculiar feature: the industry possessed a working technology before the geological facts of its commodity were known, and as a consequence its operational and theoretical conceptions diverged wildly.

Forty or fifty years passed in Pennsylvania and the Mid-Continent area before geological principles relating to the origin and accumulation of oil—in particular the "anticlinal" theory

—found their way into industrial practice. The reason was quite simply that science was not needed to make a profit from sweet oil lying in huge quantity near the surface of the ground. Drilling mania and greed sustained the momentum until the early 1900s when, by coincidence, wildcat prospecting which had spread southward into Oklahoma encountered proof of mappable geological control over the sites of the oil fields. From that point, between 1913 and 1915, the formal organization of geology in the major American oil companies began (Dickey, 1959; Owen, 1959). That is one reason why the headquarters of this Association of petroleum geologists, founded in 1917, is in Oklahoma, at Tulsa, and not in Pennsylvania, at Titusville, where the oil industry originated.

Yet it is a fact that an integrated geological conception of the origin of petroleum in black shales and its concentration into oil pools, and of the link between coal rank and indigenous petroleum, went back 50 years to Henry D. Rogers, the first State Geologist of Pennsylvania. Rogers published his findings in 1863, only four years after Drake's discovery well was drilled at Titusville.

That the volatile hydrocarbons were distilled, as it were, from out the low-lying carbonaceous strata, into the pores and fissures of the over-resting ones, receives strong confirmation from the fact that the elsewhere bituminous shales of the Silurian and Devonian ages, deep under the coal, are altogether as much desiccated and debituminized everywhere in the districts contiguous to the anthracites as the coal beds themselves.

As he wrote this passage did Rogers see in his mind's eye the workings of a refiner's retort, distilling lamp oil from a charge of heated shale? Within a few years such models for metamorphism of organic matter in sedimentary rocks had disappeared from America.

EYES IN THE GROUND

Geology in the petroleum industry serves one function only: to provide eyes in the ground. And all geological activity that is of any use whatever to the industry is directed toward this function. Consequently, petroleum geology clusters round the inventions and innovations which, quite literally, see oil.

The man of action in organized petroleum geology in the United States after the First World War undoubtedly was Everette De-Golyer (Fig. 12). What the plane table could do on the surface, a remote sensor could do underground. He succeeded first in Gulf Coast salt-dome terrane, in 1922, with Eötvös torsion balances (Fig. 13). Three years later, by

FIG. 12.—Everette DeGolyer. "The oil industry is not particularly interested in geology or geologists, physics, or geophysicists as such; it is interested in all of these various efforts in the degree that they can help to find oil" (DeGolyer, 1940). Photograph *in* Deegan and MacNaughton (1945).

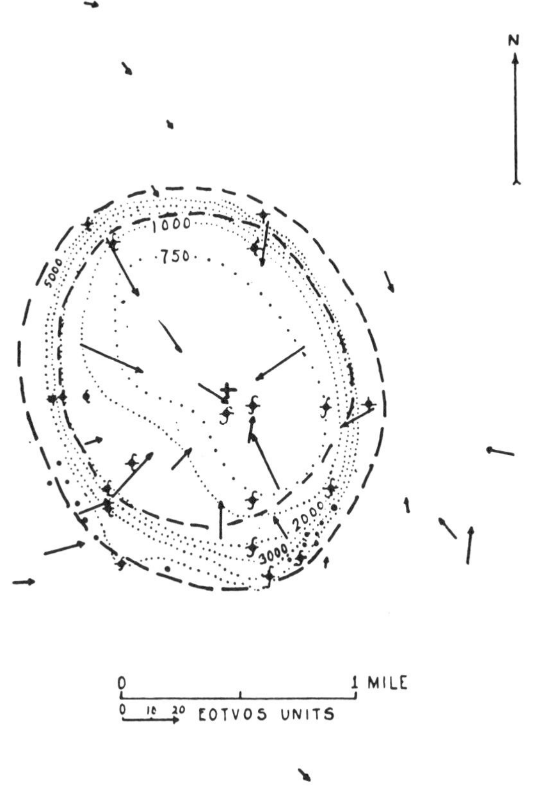

FIG. 13.—Torsion balance survey of Nash salt dome, site of first oil field to be discovered by geophysical methods in United States (March, 1924).

Length of each arrow represents magnitude of the horizontal gradient of gravity and its direction gives the direction of maximum gradient. Dashed lines show predicted position of salt at 500–900 ft and at 4,000–5,000 ft. Dotted contours are structure contours based on subsequent drilling. Survey by Rycade Oil Corporation and D. C. Barton (from Barton, 1930).

converting the principles of German acoustical-seismic apparatus to electrical instrumentation, he assembled what was to be the most effective physical sensing device ever used in geology. Its effect on structural prospecting was immediate and explosive.

The Schlumbergers' invention of "electrical coring" and its introduction to the United States in 1929, later provided the means of detecting by downhole physical attributes alone the existence of unsuspected closure to potential reservoirs. From this sprang a vogue for "stratigraphic traps."

But the main problem in exploration is now no longer one of simply locating favorable structure or favorable reservoirs, for the fact that there exists only a finite number of pools in a finite amount of continental rock increasingly involves statistical conceptions of hydrocarbon potential and sedimentary volume in the exploratory process.

NEW TECHNOLOGY

What about recent technological developments? The major concentrations of activity are related to new instrumentation. Consider for example the scanning electron microscope, which is essentially a microtopographic plotting machine. In petroleum geology it is transforming rotary-drill and core chips into storehouses of submicroscopic fossil remains, and to judge by its performance since it was introduced about 7 years ago, it promises to be truly a magic lantern in the dusty passageways of stratigraphy. Its capability of revealing microtopography is being applied also to inorganic forms and it is already a leading instrument in the study of mineral diagenesis and reservoir porosity. Another leader, also operating on the same scale, is the electron microprobe, a microanalytical instrument of unparalleled discrimination.

For petroleum itself, the long-awaited realization now bursts out that organic matter is a nearly ubiquitous ingredient of aqueous sediments, and that in conditions of burial that scarcely reach the rank of inorganic or mineral metamorphism the organic matter undergoes profound diagenetic and metamorphic change. Gas chromatography, absorption spectrography, and isotope spectrometry have enabled geologists to perceive that petroleum does indeed form by a process of geothermally activated organic metamorphism in the way that H. D. Rogers visualized it in Pennsylvania 108 years ago.

The whole diagenetic scene in sedimentary rocks, both organic and inorganic, is being illuminated more brightly, and the expanding view will eventually be seen, from the petroleum standpoint, to be of wider scope than either the sedimentary or stratigraphic perspectives alone. As before, the geology is dependent on technology.

In the seismic field, what may turn out to be a literal realization of seeing in the ground is now theoretically attainable by means of holography, or wavefront reconstruction as it was called by its inventor, Dennis Gabor, in 1948 (Gabor and Stroke, 1969). In geological prospecting the aim of holography is to transform received acoustic signals into three-dimensional optical images that may be viewed in the round.

As another example of delayed explosion,

consider the concept of continental drift, which might turn out to be a classic. The idea itself, as promulgated by Wegener (1912), is now about 60 years old. It combined from the start both academic standing and popular appeal, and was regularly served up with Lost Atlantis in newspaper articles of the 1920s and 1930s.

On the whole, European and British geologists accepted it in preference to the alternative of a spidery apparatus of land bridges that seemed to be required to explain some paleontological problems. They did this on the ground that it was self evident, rather like evolution and granitization, though the mechanism was wholly unknown.

United States geologists on the other hand tended not to accept it, and there the matter lay until the end of the 1950s when new instruments began returning information from the ocean floors that galvanized the imagination. I think there is no doubt that the burst of activity which now clusters on paleomagnetic vectors and reversals, radiometrics, sonars, and deep-sea core-drilling devices, had its origins in the International Geophysical Year of 1960–1961 and the unlucky Mohole project, and ultimately in the vision of the geologists and geophysicists who launched these enterprises. Going a step farther to the popular reaction, I think we can perceive already that the one-world concept which is at the heart of the new tectonics reinforces and feeds back to a general concern for the total world environment.

GEOLOGICAL OPTIONS, 1971

Concepts generally thought of as belonging to "conservation" are now finding their way into exploration, particularly on the continental shelves, for geologists and the public alike are beginning to visualize the total inventory of prospective rock that extends under the sea to the edge of the continents as final sources of supply, and questions about the meaning of "ownership" are in the newspapers. This is to be expected because social pressure on the industry exceeds anything that it has experienced before.

Public concern for the environment is now linked tenaciously to wrecked oil tankers and automobile fumes, and however unjust it may be, that is the fact of the situation. It affects the whole petroleum industry. What can the geologist do? In my view he must speak his mind: the knowledge possessed by this Association must be used to make it abundantly plain to the public that we are as deeply concerned about the environment as anyone else. In fact the Association is already on the move. The published statements and constitutional acts of the AAPG Presidents, K. H. Crandall (1970) and W. H. Curry (1970), on this problem and on the fundamental purposes of AAPG indicate the line that we should be taking, and there is tangible expression of it in the work of the Association's Environmental Geology Committee and its new Advisory Council.

But the fact remains that the industry faces an immediate necessity of finding and producing unprecedented amounts of petroleum to meet the foreseeable demand; and it is at this point that the petroleum geologist can be confused by the options presented in a pollution-conservation dilemma, of which the essential issues are not at all clear. One can treat the two main options separately.

First, petroleum accounts for something close to three quarters of world energy demand, most of which is concentrated in specific areas of very high consumption. Waste products of this high consumption visibly pollute the environment. Each one of us, for example, pro-

duces about 20 times as much pollution by consumption of nonrenewable resources as a so-called underdeveloped person. Problems of this kind involve the petroleum geologist to the extent that he is part of a community which is consuming more energy in its lifetime than the sum total consumed by all preceding generations. Pressure is on the geologist to find oil and gas to satisfy the predicted rates of future consumption (Figs. 14, 15), but at the same time he must express his concern for the maintenance of the quality of life in his surroundings, and act on his concerns if he sees any deterioration.

The second part of the dilemma is this: petroleum exploration has now reached areas of the continental shelf and the remote Arctic that were formerly closed through lack of a technology that could operate in those places. The marine and Arctic environments, particularly the latter, are in their own ways peculiarly prone to damage by operations which elsewhere might seem quite normal and inoffensive. The petroleum geologist, if he has any mind for conservation principles at all, is involved to the degree that the natural environment of the ground in

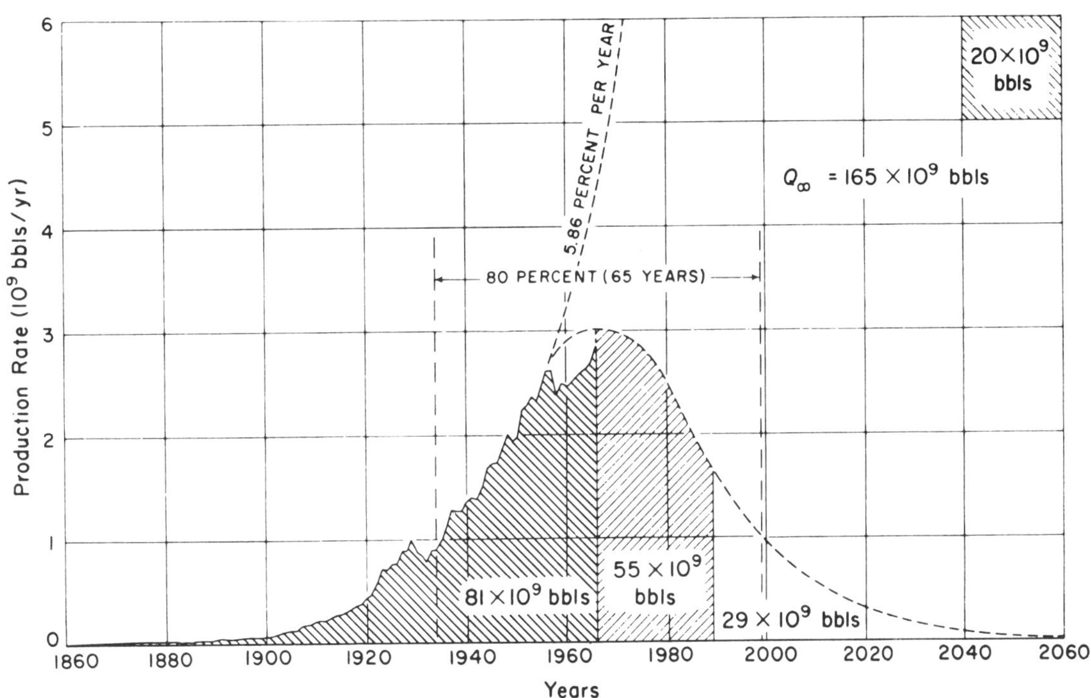

Fig. 14.—Complete cycle of crude-oil production in United States and adjacent continental shelves, exclusive of Alaska (Hubbert, 1969, Fig. 8.17).
The time that will be required to produce and consume the middle 80 percent of the ultimate amount of crude oil to be produced in the conterminous United States is only about 65 years, or less than a single life-time (Hubbert, 1969).
From *Resources and Man: a Study and Recommendations* by the Committee on Resources and Man of the Division of Earth Sciences, National Academy of Sciences–National Research Council, with the cooperation of the Division of Biology and Agriculture. W. H. Freeman and Company. Copyright © 1969.

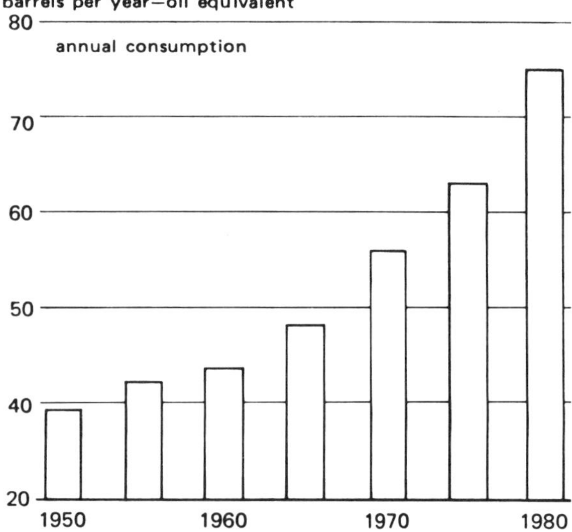

FIG. 15.—By 1980, some surveys show, the average American will be using an annual energy equivalent of about 75 bbl of petroleum. This compares with 48 bbl in 1965, and 39 bbl in 1950 (*Horizons,* v. 20, no. 5, 1970, p. 14; Amoco Production Company).

Petroleum now provides nearly three-quarters of all energy consumed in the United States. With this information, and assuming a population of 225 million, it is evident that for 1980 (neglecting gas production) the forecast production/consumption rates of petroleum given by Figures 13 and 14 fail to match by factor of 5 times.

which he hopes to find petroleum is itself a resource, the use of which cannot be simply regarded as a matter of capture, or even of prorated exhaustion (Fig. 16).

These alternatives, which we can summarize as consumption-pollution and exploitation-injury are to my mind the horns of the dilemma. The geologist can escape from the dilemma if he wishes by saying that the problems are mainly of recovery, carriage, and use, rather than problems of exploration. But if he does this without first analyzing the whole geological scene he can find himself confused in aim, and very likely suffering from what K. H. Crandall (1969) called "motivation mortality." I think many of us suffer bouts of it from time to time, and it is something, as I mentioned before, that the educational and informational resources of the Association can help to correct. Quite a lot depends on how far we align ourselves with the trend that Linn Hoover (1971) noted "toward a more humanistic view of the geosciences," because this trend, if it continues for the next few years, he wrote, "may have as much impact as a new scientific concept." That really is

the crux of the pollution-conservation problem. It is a lighted fuse.

CONCLUDING REMARKS

Summarizing briefly, I began by linking geological activity to industrial demand, and suggested that geology in turn had profoundly altered some popular concepts of man and nature. That is still true. I also claimed that geological activity was dependent on technology, and this I illustrated by the accelerated activity which takes place when new instrumentation is applied to old concepts.

Lastly, I endeavored to say that awareness of man's place in nature, which is a fundamental perception of geology, governs the geological attitude.

In this Association let us use the earth for the benefit of all. This world is one world. Perhaps the biggest achievement of geologists in this century will once again turn out to be a general principle, a worldwide concept, that reveals beyond its strictly geological meaning the essential unity of nature.

A

Cousteau said between **30** and 40 per cent of all sea life has disappeared in the last **50** years.

B

The first year's efforts will concentrate in these areas:

● Identifying factors that will help man predict modifications in the oceans, whether caused by nature or by man himself.

● Investigating specific ocean areas with special attention to food chains and pollutants.

FIG. 16.—Marine and arctic environments in which petroleum exploration is now proceeding are themselves resources, the use of which cannot be simply regarded as a matter of capture, or even of prorated exhaustion.
A. From 1969 newspaper review of television film by Commander Jacques Cousteau.
B. Two objectives from National Science Foundation program, International Decade of Ocean Exploration, for which $15 million was requested in fiscal 1971. The Foundation requested $2 million for its Arctic Research Program (*Geotimes,* v. 15, no. 4, 1970, p. 24).

REFERENCES CITED (ANNOTATED)

Bailey, Sir Edward, 1962, Charles Lyell: London, Nelson, 214 p.

Barton, D. C., 1930, Geophysical prospecting for oil: Am. Assoc. Petroleum Geologists Bull., v. 14, no. 2, p. 201–226.

Bentley, Nicolas, 1968, The Victorian scene, a picture book of the period 1837–1901: London, Weidenfeld and Nicolson, 296 p.

Crandall, K. H., 1969, Our profession: Am. Assoc. Petroleum Geologists Bull., v. 53, no. 8, p. 1561–1562.

——— 1970, Decisions for the demanding decade: five "E's" of survival: Am. Assoc. Petroleum Geologists Bull., v. 54, no. 11, p. 1980–1983.

Curry, W. H., 1970, A.A.P.G., environment, and the Federal Government: Am. Assoc. Petroleum Geologists Bull., v. 54, no. 10, p. 1805–1808.

Darwin, Charles, 1859, On the origin of species by means of natural selection or the preservation of favoured races in the struggle for life: London, Murray, 149 p. (repr., London, Penguin, 1968). (Reprint has useful historical introduction by J. W. Burrow, p. 11–48.)

Deegan, C. J., and L. W. MacNaughton, 1945, E. DeGolyer, Honorary Member: Am. Assoc. Petroleum Geologists Bull., v. 29, no. 3, p. 397–401.

DeGolyer, E. L., 1940, Future position of petroleum geology in the oil industry: Am. Assoc. Petroleum Geologists Bull., v. 24, no. 8, p. 1389–1399.

Dickey, P. A., 1959, 100 years of oil geology: Geotimes, v. 3, no. 6, p. 6–9, 24–25; no. 7, p. 6–7, 24–25. (A concise and accessible history of petroleum geology in North America. See also Owen [1959].)

Fisher, G. P., 1866, Life of Benjamin Silliman, M.D., LL.D.: New York, Scribner, v. 1, 407 p.; v. 2, 408 p.

Gabor, Dennis, and G. W. Stroke, 1969, Holography and its applications: Endeavour, v. 28, no. 103, p. 40–47.

Geikie, Archibald, 1875, Life of Sir Roderick I. Murchison, bart.; K.C.B., F.R.S.: London, Murray, 2 vols.

Hoover, Linn, 1971, Geology is also people: Geotimes, v. 16, no. 1, p. 9.

Hooykaas, R., 1963, The principle of uniformity in geology, biology and theology: Leiden, Brill, 237 p. (This valuable work examines the problems and attitudes of geological thought, chiefly, though not exclusively, in the first half of the 19th century.)

Howell, J. V., 1930, How old is petroleum geology?: Am. Assoc. Petroleum Geologists Bull., v. 14, no. 5, p. 607–616.

Hubbert, M. K., 1969, Energy resources, in Resources and man, a study and recommendations by National Academy of Sciences—National Research Council: San Francisco, Freeman, 259 p.

Hutton, James, 1785, System of the earth. (This paper was published in 1788 under the title *Theory of the Earth; or an investigation of the laws observable in the composition, dissolution, and restoration of land upon the globe* [Roy. Soc. Edinburgh Trans., v. 1, p. 209–304]. An abstract of *System of the Earth* [1785] was reprinted in Eyles, V. A., 1950, Roy. Soc. Edinburgh Proc., sec. B, v. 63, p. 380–382. Both the *System* and the *Theory* were reprinted in *Contributions to the history of geology*, v. 5, G. W. White, ed., New York, Hafner, 1970, 204 p.)

Huxley, Thomas, 1863, Evidence as to man's place in nature: Edinburgh; repr. in *Collected Essays,* London, v. 7, 1894. (The exchange of words between Huxley and the Bishop of Oxford [Fig. 10 and text] took place in 1860 at the meeting in Oxford of the British Association for the Advancement of Science. Accounts differ on the exact words spoken. A letter written by Huxley about two months afterward referred to the Bishop's remarks on "my personal predilections in the matter of ancestry," and if the question is put to me, he wrote, "I unhesitatingly affirm my preference for the ape.")

Lightfoot, John, 1659: (Quoted by White [1896]: "Heaven and earth, centre and circumference, were created together, in the same instant . . . on the twenty-third of October, 4004 B.C., at nine o'clock in the morning" [John Lightfoot, Vice-Chancellor of Cambridge University].)

Lyell, Charles, [1841] 1845, Travels in North America in the years 1841–42, with geological observations on the United States, Canada, and Nova Scotia: New York, Wiley and Putnam, v. 1, 251 p.; v. 2, 221 p. (Lyell's lectures were given at the Lowell Institute, where Benjamin Silliman and Louis Agassiz also drew huge audiences [Struik, 1957].)

Miller, Hugh, [1854] 1857, The two records, Mosaic and geological, in The testimony of the rocks; or geology in its bearings on the two theologies, natural and revealed: Edinburgh and London, 500 p. (In a letter to his wife, telling her about the success of his lecture in London, Miller wrote: "I had a noble audience of, as the *Morning Advertiser* says, five thousand persons, and I carried them with me throughout.")

Owen, E. W., 1959, Remarks on the history of American petroleum geology: Washington Acad. Sci. Jour., v. 49, no. 7, p. 256–260.

Rogers, H. D., 1863, Coal and petroleum [Anon.]: Harper's New Monthly Magazine, v. 27, p. 259–264. (H. V. Howell [1930] pointed out that a reference in this anonymous article to "my brother, Professor W. B. Rogers," made it plain that the writer was H. D. Rogers. When Lyell visited North America, 1841–1842, H. D. Rogers accompanied him in the field.)

Sedgwick, Adam, [1838]: (Sir John Herschel in 1838 described in a letter how Sedgwick, standing "on the point of a rock, a little raised," addressed "some 3,000 or 4,000 colliers and rabble (mixed with a sprinkling of their employers), which has produced a sensation such as is not likely to die away for years.")

Shakespeare, William, [1599], As you like it. ("The poor world is almost six thousand years old." Popular belief held that the six millennia, or "days" of creation, which preceded the first "Sabbath-day," foretold the passage of six millennia on earth before the last "Millennium.")

Struik, D. J., 1957, The origins of American science (New England): 2d ed., New York, Cameron, 430 p. ("Understanding the science, the learning and the technology of an age means not only a knowledge of the content of the individual professions and techniques, but also an insight into the ways in which they are related to the social structure, the cultural aspirations and the traditions of this age.")

Wegener, Alfred, 1912, Die Entstehung der Kontinente: Geol. Rundschau, v. 3, no. 4, p. 276–292.

Wesley, John, [1770] 1823, A survey of the wisdom of God in the creation: or, a compendium of natural philosophy: London, Wm. Pine (3d American ed., New York, Bangs and Mason, 1823, 2 vols. (Quotations in the text are from this edition.)

(Wesley stated that the *Compendium* was in great measure, translated from the Latin work of John Francis Buddaeus. "But I have found occasion to retrench, enlarge, or alter every chapter, and almost every section" (Preface, 4th ed., 1784).

Wesley commented particularly on the "degrees of nature," and the "transitions from one species to another" among the "immense variety of species."

"Wherefore then was this variety bestowed upon brutes? Are they the more happy, or more useful to one another for it? No. This variety then is doubtless intended for the sake of man, to prevent confusion, and decide and ascertain his property.")

White, A. D., 1896, A history of the warfare of science with theology in Christendom: London, Macmillan, 2 vols.

Woodward, H. B., 1907, The history of the Geological Society of London: London, Geological Society of London, 336 p.

The American Association of Petroleum Geologists Bulletin
V. 58, No. 12 (December 1974), P. 2490-2496

Continental Drift and Scientific Revolution[1]

DAVID B. KITTS[2]

Norman, Oklahoma 73069

Abstract Some authors have suggested that the earth sciences are in the midst of a scientific revolution of the kind characterized by Thomas Kuhn. The shift in opinion concerning continental drift during the past decade has many of the features identified by Kuhn as characteristic of scientific revolutions. There are other features, relating to the historical character of earth science, which clearly distinguish the recent events in geology and geophysics from the revolutions described by Kuhn.

INTRODUCTION

It was inevitable that the geologists and geophysicists who contributed to the formulation of the "new global tectonics" should have discovered Thomas Kuhn. In 1962 Kuhn published his *The Structure of Scientific Revolutions.* This is not far from the date that earth scientists have come to mark as the beginning of a particularly significant decade. It now is held generally that during this decade a "scientific revolution" occurred. Some authors, notably Wilson (1968), Cox (1973), and Hallam (1973), explicitly have invoked Kuhn's theory of the history of science in their treatment of recent events in geology. These works are valuable historical accounts of a remarkable period of intellectual ferment. I find Cox's treatment particularly exciting and illuminating. But the doctrines of Kuhn serve these authors as little more than a source of rationalization for the opinion that a scientific revolution has occurred in earth science. I am concerned that in uncritically labeling recent events in earth science as a revolution, and by implication a "Kuhnian revolution," we may miss something significant about the history of geology and, more importantly, something fundamental about the very nature of geologic knowledge. I shall, therefore, examine with some care the claim that geology is just now in the midst of a scientific revolution.

PARADIGMS AND NORMAL SCIENCE

A

Kuhn rejected the view that science develops by the piecemeal addition of items to the stockpile of scientific knowledge. He suggested rather that periods of "normal science," which are characterized by theoretical stability, are broken occasionally by episodes of rather sudden change in theoretical foundation. Kuhn presented the essentials of this view early in "The Structure of Scientific Revolutions." He wrote (p. 10-11),

> In this essay, "normal science" means research firmly based upon one or more past scientific achievements, achievements that some particular scientific community acknowledges for a time as supplying the foundation for its further practice. Today such achievements are recounted, though seldom in their original form, by science textbooks, elementary and advanced. These textbooks expound the body of accepted theory, illustrate many or all of its successful applications, and compare these applications with exemplary observations and experiments. Before such books became popular early in the nineteenth century (and until even more recently in the newly matured sciences), many of the famous classics of science fulfilled a similar function. Aristotle's *Physics,* Ptolemy's *Almagest,* Newton's *Principia* and *Opticks,* Franklin's *Electricity,* Lavoisier's *Chemistry,* and Lyell's *Geology*—these and many other works served for a time implicitly to define the legitimate problems and methods of a research field for succeeding generations of practitioners. They were able to do so because they shared two essential characteristics. Their achievement was sufficiently unprecedented to attract an enduring group of adherents away from competing modes of scientific activity. Simultaneously, it was sufficiently open-ended to leave all sorts of problems for the redefined group of practitioners to resolve.
>
> Achievements that share these two characteristics I shall henceforth refer to as "paradigms," a term that relates closely to "normal science." By choosing it, I mean to suggest that some accepted examples of actual scientific practice—examples which include law, theory, application, and instrumentation together—provide models from which spring particular coherent traditions of scientific research.

Wilson, Cox, and Hallam held that the hypotheses of continental drift and plate tectonics constitute a paradigm which recently has been accepted by the geologic community. This suggestion seems plausible enough but, before accepting it, let us take a careful look at the whole body of geologic knowledge with the paradigm concept in mind.

[1]Manuscript received, October 2, 1973; accepted, February 28, 1974.

[2]School of Geology and Geophysics and Department of the History of Science, University of Oklahoma.

A draft of this paper was written while I was on sabbatical leave at Stanford University in the spring of 1972 and it was presented to the geology colloquium there. I extend my thanks to Norman J. Silberling with whom I discussed many of the problems treated here. Betty Bellis and Deborah Merritt have given valuable assistance in the preparation of the manuscript. Clerical assistance has been provided by a grant from the faculty research fund of the University of Oklahoma.

2490

B

There is an immensely complicated body of knowledge shared by geologists, a significant part of which is not expressed explicitly. There is, however, an important part of that body of knowledge that is identified easily. It consists of what geologists, and all others in the scientific community, consider to be the most fundamental and comprehensive scientific principles of their time. Scientific principles are formulated explicitly in theories. A theory, in this technical sense, is a set of statements of general form which, together with statements about initial and boundary conditions, provide a basis for the logical derivation of a great number and variety of singular descriptive statements. General theories thus cover or comprehend particular events. Geologists do not set out to verify or falsify theories. The laws of physics are not questioned within the context of geologic inference. They are simply presupposed. Geologic events must be such as not to falsify fundamental theories. When an inferred event appears to be forbidden by the laws of physics, the resulting "paradox" is not resolved by altering the laws of physics to accommodate the event, but by altering the event to accommodate the laws of physics. A striking example of this inferential stratagem is to be found in the attempt to resolve the "paradox of major overthrusts" (for a discussion of this case see Kitts, 1974).

Physical theory serves as a conceptual framework which the geologist does not and, according to the well-understood rules of his game, cannot transcend. In accepting an unproblematic background of theoretical propositions geologists are prevented from assuming that any logically conceivable event whatever may occur. To impose some restrictions upon what we are permitted to say about the world is simply to engage in rational discourse. The acceptance of physical theory provides the necessary protection against chaos in geologic discourse and at the same time frees geology from the stranglehold of "pure" observation. Observations alone cannot tell us what is possible, they can only tell us what is. If geologists were to invoke only observations and the empirical generalizations based upon them in their historical inferences, then the past would turn out to be exactly like the present. Under the umbrella of comprehensive theories we may consider what is possible. This is all that makes the pursuit of history interesting, for it is all that permits us to imagine that the past was in any significant way different from the present.

I have in the above remarks propounded a version of that venerable geologic doctrine, "the uni-

formitarian principle." It is an explicit statement of a view that is held, I believe, by most geologists. What I have said therefore is descriptive rather than prescriptive.

C

The concept of the paradigm is designed by Kuhn to enhance our understanding of the growth of science. For scientific growth to occur, according to Kuhn, paradigms must be doubted, attacked, and ultimately repudiated. Geologists never doubt, attack, or repudiate physical theory. Among geologists there is unshakable consensus concerning what is to be regarded as inviolate within the context of geologic inference. For them the fundamental laws of physics do not function as empirical statements subject to observational test, but rather as rules not to be violated under any circumstances (for a discussion of Newton's second law as a rule see Hanson, 1958, p. 100). Consider Kuhn's remarks on normal theoretical work (1962, p. 30),

A part of normal theoretical work, though only a small part, consists simply in the use of existing theory to predict factual information of intrinsic value. The manufacture of astronomical ephemerides, the computation of lens characteristics, and the production of radio propagation curves are examples of problems of this sort. Scientists, however, generally regard them as hack work to be relegated to engineers or technicians. At no time do very many of them appear in significant scientific journals. But these journals do contain a great many theoretical discussions of problems that, to the nonscientist, must seem almost identical. These are the manipulations of theory undertaken, not because the predictions in which they result are intrinsically valuable, but because they can be confronted directly with experiment. Their purpose is to display a new application of the paradigm or to increase the precision of an application that has already been made.

There is a good deal of work carried on by geologists that may appear to be of the sort described by Kuhn. But the theoretical investigations of the geologist are not undertaken to confront the accepted theory with experiments and thus to add to its empirical support, but rather to use the theory in new roles in historical investigation. A dramatic example is provided by the application of the principles of isotope chemistry to paleoecology. This work was not undertaken to bolster any paradigm. It was undertaken to permit a new and significant kind of historical inference. In historical inferences events are regarded as ends in themselves rather than as a means of getting at theories. However deeply the geologist becomes involved in the theoretical aspects of a current physical paradigm, his goal is not to verify or falsify the paradigm. His work is not directed at the paradigm at all. For him, physical theory is accepted unquestioningly in order that the primary business of historical inference may proceed.

The body of shared knowledge and belief may be in large part identical for geologists and physicists. Most physicists, like geologists, are not involved in any attempt to alter the fundamental theory of their day. But if Kuhn's paradigm is to figure in explanations of scientific growth, there must be from time to time a handful of "physicists" who are willing to attack it. What may be a paradigm for physical science is a kind of inviolable "superparadigm" for geology.

Again, I must make sure the reader understands that I am attempting to describe the character of geology. I make no claim that geology must be as I have described it or that it should be. It is conceivable in principle that a geologist confronted with a paradox resulting from the incompatability of a geologic event and physical theory might respond by insisting that physical theory be altered to bring it into accord with the geologic event. Paleontologists sometimes fault biologic theory because it is incompatible with the record of biologic events. Geologists never fault physical theory because it is incompatible with the record of physical events. The reason for this intriguing difference between paleontologists and geologists is probably as much a historical and sociological question as it is a methodological one and is, in any case, beyond the scope of this paper to consider.

Geologists have had no role in the revolutions which have led to the overthrow of comprehensive theoretical paradigms. When a theoretical paradigm is abandoned in favor of another it may have revolutionary consequences for geology. The revolution that led to the overthrow of classical mechanics in the late 19th and early 20th centuries affected geology in an important way, but it was not accomplished with the participation of geologists or at their instigation.

CONTINENTAL DRIFT

A

All this is preliminary to a discussion of a current revolution in geology. Geologists are not talking about a change in theoretical paradigm taking place outside geology. They are concerned with a change going on inside geology which leaves the theoretical paradigm unscathed.

Hallam (1973, p. 106) wrote, "A paradigm is an accepted model or pattern of beliefs but it is easier to explain by example than define. Kuhn draws his examples from physics and chemistry." It is no mere accident of Kuhn's background and academic training that he chooses his examples of paradigms from chemistry and physics. It is clear that for Kuhn, paradigms exercise their pervasive influence by virtue of their being general knowl-

edge systems. He recognizes different degrees of comprehension, but he does not consider any hypothesis which is concerned wholly with particular events. The hypothesis of continental drift *is* concerned wholly with particular events. It does not consist of a body of general propositions of which events must be instances, but is itself an instance which must be consistent with some body of general propositions. It is not formulated in universal or even general terms, but in singular descriptive terms. It makes no assertions about an untimebound and unspacebound natural order, but about conditions prevailing at particular times and places. It is, in short, historical rather than theoretical. This important fact was recognized by Chamberlin who wrote (1928), "Wegener's hypothesis is no general theory of earth behavior or earth deformation. It describes simply one supposed breaking up of a consolidated land mass and the migration of the different fragments."

In its most highly developed state the hypothesis of continental drift would consist of an exhaustive description of a nonrepeatable succession of historical events. "Continent" hardly functions as a general term in the hypothesis. To put something in the class of continents permits us to predicate very little of that thing—little more, in fact, than that it is capable of motion with respect to others of its kind. What is significant is not what we can say of a continent by virtue of its being a member of the class of continents, but what we can say about its motion within some designated coordinate system.

A general law containing the notion of continental drift could, in principle, be formulated. A geologist might be willing to say, for example, "Wherever there are continents, be they on earth or some other planet, they will move with respect to one another." This is not what is meant, however, by "the theory of continental drift," and it is not in this form that the concept figures in geologic inferences.

The singular character of the hypothesis is clear in Wegener (1924, p. 1, 2), who wrote,

He who examines the opposite coasts of the South Atlantic Ocean must be somewhat struck by the similarity of the shapes of the coast-lines of Brazil and Africa. Not only does the great right-angled bend formed by the Brazilian coast at Cape San Roque find its exact counterpart in the re-entrant angle of the African coast line near the Cameroons, but also, south of these two corresponding points, every projection of the Brazilian side corresponds to a similarly shaped bay in the African, and conversely each indention in the Brazilian coast has a complementary protuberance on the African. Experiment with a compass on a globe shows that their dimensions agree accurately.

This phenomenon was the starting point of a new conception of the nature of the earth's crust and of the movements occurring there; this new idea is called the theory of displacement of continents, or, more shortly, the displacement theory, as its most prominent component is the assumption of great horizontal drifting movements which the continental block underwent in the course of geologic time and which presumably continue even today.

Wegener has explained the present configuration of the South American and African Atlantic coastlines by supposing that the continents had been joined at some time in the past and had since drifted apart. The hypothesis of continental drift does not serve the function of covering generalization in this explanation, but of initial and boundary conditions. Then what are the unstated generalizations involved? One of them must be a statement to the effect that if two complex irregular physical boundaries may, in principle, be fitted together, then it is more plausible to assume that they were initially coincident than to assume that they just happen to have the same configuration. One might employ this generalization in support of the conclusions that two pieces of paper had been torn from a single sheet.

A statement describing a long-range historical trend is sometimes called a "law." Popper (1957) and others have pointed out that these so-called "laws of historical succession" or "laws of evolution" are not natural laws in the usually accepted sense of this term, because they consist of descriptions of sequences of events. The processes, Popper maintains, may proceed in accordance with certain laws, but the description itself is not a law but a finite conjunction of singular statements.

To claim that the hypothesis of continental drift is historical rather than theoretical is not to claim that it is trivial nor to claim that it is insignificant in the recent history of geology. But whatever its significance may be, it cannot be identical to the significance of a theory. And however sudden the shift in opinion concerning it, that shift cannot be just like the changes in opinion concerning theories that figure so importantly in Kuhn's treatment of the history of science. Having emphasized and, I hope, established this point, I should like now to consider the way in which the hypothesis of continental drift has in fact exerted its profound influence on the earth sciences.

A HISTORICAL PARADIGM

The hypothesis of continental drift will be made to conform to the unproblematic background of physical theory or it will be abandoned. The objection repeatedly has been raised that to suppose the continents moved is a violation of physics. Some geologists rejected continental drift on these grounds, whereas others proceeded to alter initial and boundary conditions in such a way as to permit it to occur. No one cited the lack of agreement between the historical event and physical theory as grounds for rejecting or altering the theory. To quote again from the well-known AAPG volume of 1928, Longwell wrote (p. 152),

It is obvious that the results of geophysical examination, so far as they go, are generally unfavorable to the displacement hypothesis, but they are not conclusive. In fact the geophysicists turn the problem back to us, with the statement that geologists alone can determine whether the geophysical forces have had "in geologic history an appreciable influence on the position and configuration of our continents." Let us therefore re-examine some of our evidence, to see whether it is compelling. If it is, then physical geologists should be content to accept the fact of displacement, and leave the explanation to the future.

A fully satisfactory explanation of continental drift has not yet been formulated. The attempt to achieve such an explanation goes on because by now western European and American geologists are, by and large, convinced that the evidence for continental drift is indeed compelling. The task that geologists and geophysicists have set for themselves has many aspects of what Kuhn has called "puzzle solving." He wrote (1962, p. 36), "Bringing a normal research project to a conclusion is achieving the anticipated in a new way, and it requires the solution of all sorts of complex instrumental, conceptual, and mathematical puzzles. The man who succeeds proves himself an expert puzzle-solver, and the challenge of the puzzle is an important part of what usually drives him on."

Kuhn made clear that for him puzzle solving is not a routine undertaking of little significance in the history of science. It is an important activity requiring imagination and ingenuity on the part of the investigator. But I am reluctant to call the attempt to accommodate continental drift and physical theory puzzle solving because it might not do justice to the profound significance of the hypothesis in geology. That significance lies in the fact that the hypothesis of continental drift, despite its singular character, has a unifying effect. Theories unify by covering a great number and variety of phenomena. By invoking the theory of mechanics, planetary motion, falling bodies, and major overthrusts may all be explained. The hypothesis of continental drift, being particular in form, cannot cover phenomena. It must itself be covered. Its unifying function lies in the fact that it figures in many explanations as a particular antecedent condition. The distribution of animals, the remanent magnetism of rocks, and the shapes of the continents can all be explained, it is claimed, by supposing as an initial condition that certain continental masses formerly were joined and have broken apart and drifted. It is not simply a matter of explaining phenomena with which we are already familiar. Once a hypothesis has been proposed it directs the attention of scientists to certain kinds of phenomena, this being true whether the scientists are trying to verify or falsify the hypothesis. There is no question, for example, that paleomagnetic studies have received their main impetus from the hypothesis of continental drift.

Because the hypothesis of continental drift is itself something to be explained, it unifies at another level. Munyan wrote (1963, p. 2),

> The implications of this proposal [continental drift] seriously challenged many of the beliefs and theories of the constitution of the earth, its physical properties, tectonics, and biological developments. As a result a considerable furor of opposition arose on all counts, but, in particular, the geophysicists alleged that drift was out of the question because the crust could not endure such forces.

The physical properties of the earth are discovered by attempting to construct them in such a way as to permit all the phenomena which we observe and infer. Continental drift is one phenomenon which has been inferred and consequently presents an opportunity for geologists to extend their knowledge of the earth's physical properties. For one convinced that continental drift has occurred, the earth must have physical properties which permit it.

The hypothesis of continental drift unifies in two ways. It figures as a common antecedent condition in a great variety of explanations, and attempts to explain it significantly influence the direction of investigation. The effect of the hypothesis of continental drift upon geology has not been exaggerated. A substantial proportion of geologists in North America is engaged in an attempt to verify, falsify, or explain continental drift. The hypothesis of continental drift constitutes a kind of paradigm. It is not like the theoretical paradigms of Kuhn. It is the sort of paradigm particularly appropriate to the most scrupulously historical of all sciences in that it is itself historical. It imposes upon those who accept it a particular version of history rather than a general theory of the world.

Plate Tectonics

In the attempt to bridge the inferential gap between physical theory and the historical hypothesis of continental drift, certain hypotheses have been recognized as particularly significant and have themselves been accorded the status of "theories." Such a hypothesis is "plate tectonics."

Cox (1973) boldly formulated the hypothesis of plate tectonics in a series of definitions and postulates. His Postulate I states (p. 41), "The plates are internally rigid but are uncoupled from each other. At their boundaries two plates may pull apart or slip one beneath the other, but within the plates

there is no deformation." But, Cox continued (p. 43),

> Of course, every geologist knows that these postulates do not provide a complete explanation for all geologic information—if the continents were rigid blocks, there would be no need to study structural geology. Yet the conclusions drawn from plate tectonics offer rational explanations for so many of the earth's major tectonic features that the basic assumption appears to be justified with only two minor modifications: *most* large-scale deformation occurs in narrow zones between plates that are *nearly* rigid.

It may appear at first glance that the hypothesis of plate tectonics is no more general than the hypothesis of continental drift. It describes the particular behavior of individual objects. But plate tectonics cannot, like continental drift, be explicated fully in a detailed historical account. There is in plate tectonics a crucial theoretical, and therefore general, dimension which is not reducible to a description of events. That theoretical dimension is not provided by a geologic hypothesis formulated within the last decade. It comes from the familiar and inviolable "superparadigm." By virtue of something being a member of the class "plates," we can attribute certain theoretical properties to that thing; properties such as rigidity and mass. Providing plates with properties of significance within physical theory permits geologists to invoke that theory with all its immense inferential efficacy in their discussions of plates and continental drift.

Revolution

Up to this point I have discussed continental drift as a ruling hypothesis in geology. I have said nothing about how it came to be accepted so widely. It is too soon to attempt to write a history of geology and geophysics during the past 50 years. But in pursuing my purpose I should like to consider briefly the radical shift in opinion about continental drift in the light of Kuhn's theory of scientific growth. According to Kuhn (1962, p. 52-53),

> Discovery commences with the awareness of anomaly, *i.e.*, with the recognition that nature has somehow violated the paradigm-induced expectations that govern normal science. It then continues with a more or less extended exploration of the area of anomaly. And it closes only when the paradigm theory has been adjusted so that the anomalous has become expected.

The recognition of anomaly does not, by itself, entail revolution. Anomalies may be regarded as puzzles rather than as counterinstances for a paradigm. Gravitational theory is not abandoned because a heavy object rises upon being released. An attempt is made instead to show that a trick has been performed which has no force as a falsifying event. The conventional account of what Wilson (1968) has suggested we call the "Wegenerian revolution" holds that the evidence for continental drift has become so overwhelming that no reasonable man any longer can ignore it. Geomagnetism has been singled out as providing an especially compelling body of evidence. It is a source of counterinstances to the hypothesis of the fixity of continents. But as Kuhn pointed out in the following passage (1962, p. 79), counterinstances to every theory may be recognized.

> Excepting those that are exclusively instrumental, every problem that normal science sees as a puzzle can be seen, from another viewpoint, as a counterinstance and thus as a source of crisis. Copernicus saw as counterinstances what most of Ptolemy's other successors had seen as puzzles in the match between observation and theory. Lavoisier saw as a counterinstance what Priestley had seen as a successfully solved puzzle in the articulation of the phlogiston theory. And Einstein saw as counterinstances what Lorentz, Fitzgerald, and others had seen as puzzles in the articulation of Newton's and Maxwell's theories. Furthermore, even the existence of crisis does not by itself transform a puzzle into a counterinstance. There is no such sharp dividing line. Instead, by proliferating versions of the paradigm, crisis loosens the rules of normal puzzle-solving in ways that ultimately permit a new paradigm to emerge. There are, I think, only two alternatives: either no scientific theory ever confronts a counterinstance, or all such theories confront counterinstances at all times.

Events do not identify themselves intrinsically as being either puzzles or counterinstances. The distinction can be made only from some theoretical vantage point. During the first half of the 20th century the driftists confronted geology with states and events which they considered to be counterinstances to the hypotheses of the fixity of continents. The opponents of drift regarded these events as puzzles to be accounted for within the context of the fixity of continents. The configuration of the coast of the South Atlantic, which figures so importantly in the arguments of the driftists, was not a counterinstance that entailed the abandonment of the fixity of continents. It wasn't even much of a puzzle. It was an accident.

By the middle of the 1950s American and western European geologists began to accept continental drift in increasing numbers. What had been considered puzzles for the hypothesis of the fixity of continents became counterinstances. Perhaps it was at this time that the crisis, which Kuhn considered to be a necessary prerequisite to revolution, occurred. I do not wish to dwell upon this for the reason given previously. We geologists cannot at this time stand off and dispassionately examine a movement in which we are so intimately and enthusiastically involved. By the middle of the 1960s, in any case, the western geologic community had accepted a new conceptual scheme to "define the legitimate problems and methods of a research field for succeeding generation of practitioners" (Kuhn, 1962). For Anderson (1972), to cite just one example, the crisis is over. Anomalously low heat flow on the flanks of slow spreading midoce-

an ridges is no counterinstance to the sea-floor-spreading hypothesis. It has nothing whatever to do with testing the hypothesis. It is a puzzle to be solved within the context of that hypothesis. In solving the puzzle, or in resolving the paradox, Anderson discovered something of geologic interest.

CONCLUSION

History is written within a conceptual framework which imposes a limit upon what we may suppose to have occurred in the past. To accept such a limit is simply to operate within the realm of rational discourse. In any historical discipline questions arise as to how much restriction should be imposed upon past events. The great debates of the 19th century concerning the uniformitarian principle were directed to this question. By the middle of the 20th century the question had been tacitly answered. Geologists would be guided in their historical deliberations by the most comprehensive principles expressed in the chemical and physical theory of their day. The acceptance of this superparadigm, while guarding against chaos in historical discourse, provided a means of escaping the limitation to "what can actually be observed" that so many geologists sought to impose. Theory permits the geologist to decide what is possible and what is not. But history goes beyond a consideration of what is possible to a consideration of "what actually happened." Physical theory exerts its pervasive influence because geologists have agreed that, no matter what the events of history appear to be, an effort must be made to bring them into accord with what physical theory dictates as possible. This task is accomplished by altering the specific conditions and events of the past to secure such accord. An account of the conditions and events of the past is a historical chronicle. The attempt to resolve paradoxes is not some dramatic but incidental activity. It occupies a central role in geologic methodology.

Changes in physical theory accomplished outside of geology may effect the character of historical inference in a dramatic way. But geologists have a way of altering the history of the earth without tampering with the theoretical apparatus which they presuppose. When the boundary conditions introduced during geologic inference are temporally and spatially widespread, and when they are comprehensive in the sense that they are invoked in many explanations, they take on the characteristics of a paradigm. It is just the sort of paradigm that we would expect to be developed within a discipline so radically historical as geology.

It has not been my intention to provide yet another consideration of the adequacy of Kuhn's views as a general theory of the history of science, but only to consider their applicability to the recent events in geology. In my judgment Kuhn's theory illuminates the last decade in the history of geology. It should be pointed out, however, that the acceptance of continental drift and plate tectonics, although having many of the aspects of a revolutionary change, might as well be regarded as the latest step in an evolutionary movement away from a cyclic view of earth history which began at least as long ago as the middle of the 19th century. This latest step permits geologists to give an account of earth history as unidirectional, irreversible change within the confines of an immutable natural order which finds its most explicit expression in physical theory.

REFERENCES CITED

Anderson, R. N., 1972, Petrologic significance of low heat flow on the flanks of slow-spreading midocean ridges: Geol. Soc. America Bull., v. 83, no. 10, p. 2947-2956.

Chamberlin, R. T., 1928, Some of the objections to Wegener's theory, *in* Theory of continental drift: Am. Assoc. Petroleum Geologists, p. 83-87.

Cox, A., 1973, Plate tectonics and geomagnetic reversals: San Francisco, W. H. Freeman, 702 p.

Hallam, A., 1973, A revolution in the earth sciences: Oxford, Clarendon Press, 127 p.

Hanson, N. R., 1958, Patterns of discovery: Cambridge, Cambridge Univ. Press, 240 p.

Kitts, D. B., 1974, Physical theory and geological knowledge: Jour. Geology, v. 82, no. 1, p. 1-23.

Kuhn, T. S., 1962, The structure of scientific revolutions: Chicago, Univ. Chicago Press, 172 p.

Longwell, C. R., 1928, Some physical tests of the displacement hypothesis, *in* Theory of continental drift: Am. Assoc. Petroleum Geologists, p. 145-157.

Munyan, A. C., 1963, Introduction to polar wandering and continental drift, *in* Polar wandering and continental drift: Soc. Econ. Paleontologists and Mineralogists Spec. Pub. 10, p. 1-3.

Popper, K. R., 1957, The poverty of historicism: Boston, Beacon Press, 166 p.

Wegener, A., 1924, The origin of continents and oceans: London, Methuen, 212 p. (Translated by J. G. A. Skerl from 3d German edition; published in 1922.)

Wilson, J. T., 1968, Static or mobile earth—the current scientific revolution: Am. Philos. Soc. Proc., v. 112, no. 5, p. 309-320.